PHYSIOLOGICAL BIOCHEMISTRY

PHYSIOLOGICAL BIOCHEMISTRY

By

Dr. Lata Bhattacharya

Professor
School of Studies in Zoology & Biotechnology
Vikram University
Ujjain (M.P.)
(India)

DISCOVERY PUBLISHING HOUSE PVT. LTD.
NEW DELHI-110 002

First Published-2009

ISBN: 978-81-8356-490-8

Published by:
DISCOVERY PUBLISHING HOUSE PVT. LTD.
4831/24, Ansari Road, Prahlad Street
Darya Ganj, New Delhi-110002 (India)
Phone: 23279245 • Fax: 91-11-23253475
E-mail: dphbooks@rediffmail.com • dphtemp@indiatimes.com
web: www.discoverypublishinggroup.com

Printed at:
Sachin Printers
Delhi

Preface

Biochemistry today has made spectacular progress in unraveling the mysteries of animate nature. This progress has allowed us to gain deeper insight into the principles of vital activity and has to a very significant extent stimulated the development of applied disciplines, especially medicine. The present title *"Physiological Biochemistry"* is intended for those who wish to understand living organisms, especially man. Biochemistry is essential for this purpose, but it would be almost impossible for a student to survey on his own the massive body of existing knowledge, constantly augmented by a remarkable torrent of brilliant discoveries. The purpose of the book, then is to organize our knowledge into something that can be comprehended in a relatively short time and still convey a reasonable complete picture of the chemical structure and function of man. Readability without sacrifice of coverage has been a prime goal. An important device in gaining that goal is to keep attention constantly focused on function, with repeated use of rationalization to show that the chemical facts are not isolated, but part of a whole.

Biochemistry has two major goals as a fundamental science, it treats the vital functions from the standpoint of physical chemistry, and as an applied discipline, it points out practical applications for the wealth of scientific knowledge it has acquired. The dual purpose has been, as far as possible, take into account in the writing of this book. We have also tried to summarize our pedagogical experience in teaching biochemistry to students specializing in medicine and pharmacy. The material of this book has been organized according to the principle of functionally in order to trace the close relationship between the functions of the living organism and the structures of its constituents molecules as well as the chemical and physico-chemical process in which they are involved.

The aim of this book is to present a core of biochemical knowledge that is desirable for undergraduate and postgraduate students and also those involved in the field of medical, microbiology, biotechnology and pharmaceutical. Every attempt has been made to keep abrest of the advances in the subject and at the same time to include the fundamentals.

To make the work more comprehensive and informative, the author has consulted many authoritative books, research journals, abstracts, monographs etc. He is grateful to all those great scholars whose work are cited or substantially reproduced.

There can be no claim to originality except in the manner of treatment and much of the information has been obtained from the books and scientific journals available in the different libraries.

The author expresses his thanks to his friends and colleagues whose continue inspirations have initiated him to bring out this book.

The author expresses his gratitude to Mr. Wasan and Staff of M/s Discovery Publishing House Pvt. Ltd., for their whole hearted co-operation in the publication of this book.

In the mean time, the author will remain sincerely responsible for any shortcomings of the book and be grateful to the readers for their suggestions and constructive criticism for the continuous betterment of the book. He takes this opportunity to appeal to the readers to send their suggestions straightway to his publisher.

Author

CONTENTS

CHAPTER 1

Introduction

The evidence for the role of endotoxemia and increased intestinal permeability to endotoxins in alcoholic liver disease (ALD) has been reviewed previously. The level of plasma endotoxin is increased in patients with ALD, and a correlation between endotoxemia and the severity of ALD has been established. This correlation is further confirmed by endotoxemia in experimental alcoholic liver damage in rats and mice. Administration of antibiotics to deplete the colonic luminal source of endotoxins attenuates alcohol-induced endotoxemia and liver damage suggesting that the intestinal luminal microflora as a source of endotoxemia.

The three possible mechanisms involved in alcoholic endotoxemia are bacterial overgrowth, alcohol-induced Kupffer cell dysfunction, and an alcohol-induced increase in intestinal permeability to endotoxins; increased intestinal permeability appears as a major contributor to endotoxemia. In the normal gut, the epithelial barrier function prevents the diffusion of toxins, pathogens and allergens from the gut lumen into intestinal tissue and liver. Alcohol administration increases gastrointestinal permeability to macromolecules both in normal subjects as well as in patients with ALD.

In some patients with ALD, the increased intestinal permeability persists even after 1-2 week of sobriety, suggesting that a more sustained barrier dysfunction is present in these patients. Demonstration of increased intestinal permeability in alcoholics showing the symptoms of liver disease and not in those alcoholics without liver disease indicates a close correlation between intestinal permeability and ALD. An increased intestinal permeability to macromolecules was also demonstrated in experimental models of alcoholic liver damage. Amelioration of alcoholic liver damage in rats by zinc or oat bran supplementation also associated with attenuation of alcohol-induced intestinal permeability to macromolecules. Acetaldehyde, the first product of ethanol metabolism, is highly toxic, and it appears to play a key role in cell injury in multiple organs, such as liver, heart, skeletal muscle, lung, and brain.

Our recent studies indicated that acetaldehyde disrupts intestinal epithelial barrier function and increases intestinal permeability to endotoxins and other macromolecules. Acetaldehyde disrupts intestinal epithelial tight junctions (TJ) and adherens junctions (AJ) by a tyrosine kinase-dependent mechanism. Acetaldehyde-induced increase in permeability was mediated by the inhibition of protein tyrosine phosphatases such as PTP1B and increase in tyrosine phosphorylation of TJ and AJ-proteins such as occludin, ZO-1, E-cadherin, and better catenin. Gastrointestinal mucosal protective factors such as epidermal growth factor (EGF) and L-glutamine effectively prevents acetaldehyde-induced disruption of TJ and AJ and increase in permeability to endotoxins and macromolecules.

Most these studies were conducted using Caco-2 cell monolayers, an extensively used model of the intestinal epithelium. However, the effects of acetaldehyde on the TJ and AJ and its prevention by EGF and L-glutamine were confirmed in human colonic mucosa. Therefore, valuable information can be derived from studies that use Caco-2 cell monolayers as an experimental model of the intestinal epithelium.

MATERIALS

Cell Culture

1. Dulbecco's Modified Eagle's Medium (DMEM; Gibco/BRL, Bethesda, MD) supplemented with 10% fetal bovine serum (FBS, HyClone, Ogden, UT).
2. L-Glutamine, 200 m*M*, l00x stock solution (Gibco/BRL).
3. Penicillin, 10,000 units/mL-streptomycin, 10,000μg/mL, 1000x (Gibco/BRL).
4. Gentamycin, 50 mg/mL (1000x, Gibco/BRL).
5. MEM nonessential amino acids solution, 10 m*M* (100x, Gibco/BRL).
6. Transwell inserts (Costar/Corning, Corning, NY).
7. Trypsin-EDTA: Trypsin solution (0.25%) and ethelenediamine tetraacetic acid (EDTA; 1 m*M)* from Gibco/BRL.

Acetaldehyde Exposure

1. Acetaldehyde: 99.5% (A.C.S reagent, Sigma-Aldrich, St. Louis, MO).
2. Sealing tape: 3 M comply indicator tape (Costar).
3. Phosphate-buffered saline (PBS-1): 10 g/L sodium chloride, 0.25 g/L potassium chloride, 1.435 g/L disodium hydrogen phosphate, 0.25 g/L potassium dihydrogen phosphate, pH 7.4 containing 0.134 g/L calcium chloride, 0.1 g/L magnesium chloride, 1.8 g/L glucose, and 6 g/L bovine serum albumin. Diluted from 10x PBS (Sigma-Aldrich, St. Louis, MO.

Measuring Transepithelial Electrical Resistance TER)

1. Millicell-ERS resistance monitoring system and electrodes (Millipore, Billercia, MA).
2. 70% ethanol-prepared by diluting absolute ethanol in order (Mallinckrodt, Phillipsburg, NJ).
3. PBS, pH 7.4 (Gibco/BRL).

Measuring Paracellular Permeability

1. Inulin-FITC (Sigma, St. Louis, MO), 10 mg/mL in PBS (stock solution).
2. FITC-lipopolysaccharide (LPS), isomer 1 ~ 98% (HPLC) (Sigma-Aldrich, St. Louis, MO).
3. Fluorescence plate reader: FLx 800 microplate fluorescence reader (Bio-TEK Instruments Inc., Winoonski, VT).
4. 96-well fluorescence plates.

Immunofluorescence Staining and Confocal Microscopy

1. PBS-2: 18.5 g/L sodium chloride, 5.7 g/L disodium hydrogen phosphate, 1.4 g/L sodium dihydrogen phosphate, pH 7.4. Prepared by diluting 10x PBS (Sigma-Aldrich, St. Louis, MO).

2. 3% paraformaldehyde: dissolve 3 g of paraformaldehyde (Sigma-Aldrich, St. Louis, MO) in 100 mL of PBS-2 with a drop of sodium hydroxide heated at 60°C. The pH is re-adjusted to 7.4.
3. Acetone:Methanol (1:1): mix equal volumes of acetone and methanol (both from Fisher-Scientific, Fairlawn, NJ).
4. Triton X100 (0.2%): prepared by dissolving Triton X100 (Sigma-Aldrich, St. Louis, MO) in PBS.
5. Tris-buffered saline containing Tween 20 (TBST): dilute 10X TBST (20 mM Tris-HCl, 0.15 *M* sodium chloride, pH 8.0, and dissolving 5 ml/L of Tween-20.
6. 4% nonfat milk: dissolve 4 g of blotting grade nonfat dry milk (BIORAD, Hercules, CA) in TBST.
7. Primary antibodies: mouse monoclonal anti-occludin and rabbit polyclonal anti-ZO-1 antibodies (Zymed Laboratories, S. San Francisco, CA).
8. Secondary antibodies: AlexaFlour-488- conjugated goat anti-mouse IgG anti-body (2 mg/ml) from INVITROGEN-Molecular Probes (Eugene, OR), and Cy3-conjugated anti-rabbit IgG from Sigma-Aldrich (St. Louis, MO).
9. Mounting fluid: dissolve 25 g DABCO (l,4-Diazabicyclo-[2.2.2]octane; Sigma-Aldrich, cat. no. D-2522) in 25 mL of 10x PBS-2 and mixing 225 mL of glycerol.
10. Nail polish: Nail Enamel/Vernis-clear (REVLON, Mississauga, ON).
11. Glass slides: precleaned Gold Seal Rite-ON microslides (Gold Seal, Portsmouth, NH).
12. Cover slips (Microscope cover glass, Fischer brand, Fisher Scientific, Pittsburgh, PA).
13. Humidifying chamber: you can use a plastic box with wet paper towels on the bottom and two plastic pipets to rest the plate with inserts.
14. Confocal microscope: LSM 5 PASCAL high-performance confocal laser scanning microscope (CARL ZEISS Micro imaging Inc, Thornwood, NY). Others such as BIORAD and Nikon confocal microscopes can be used.
15. Software: LSM 5 PASCAL software is used in both Zeiss LSM 5 PASCAL and BIORAD confocal microscopes. Image J software and LSM reader or Biorad reader plugins can be freely downloaded from NIH web site. Adobe Photoshop can be purchased from Adobe Systems Inc. (San Jose, CA).

Lactate Dehydrogenase (LDH) Release Assay

1. Cytotoxicity detection kit (Roche Applied Science, cat. no. 11 644 793 001). This kit has 2 reagents, "Catalyst" and "Dye solution." You have to mix them to use. This mixture (assay medium) is stable for several weeks at 4°C.
2. 96-well microplate.

WST Assay

1. Cell proliferation reagent WST-1 (Roche Applied Science, cat. no. 11644807001). For assay reagent, mix WST-1 reagent and PBS-1 (1:9) just before use. You need 150 μL of assay reagent for cells on 6.5-mm transwells and 400 μL for cells on 12-mm transwells.
2. PBS-1.
3. 96-well microplates.

METHODS

Caco-2 cell line was derived from colon adenocarcinoma. When grown in culture, Caco-2 cells form a monolayer of epithelial cells, which differentiate into columnar cells and polarize both morphologically and functionally. The phenotypes resemble that of normal ileal epithelial enterocytes. These cells organize TJ and AJ and develop barrier to the diffusion of macromolecules. These cells have been extensively used as a model to study the properties of differentiated enterocytes, the barrier function, and study of oral bioavailability of therapeutic drugs.

Studies in our laboratory found this to be a useful model to study acetaldehyde-induced endotoxin absorption, as most of the key observations made using this monolayer were reproduced in human colonic mucosa. The barrier function of the epithelium can be evaluated by measuring TER, which evaluates the barrier to ion flux, and by unidirectional flux of extracellular fluid markers such as mannitol and inulin. Extracellular markers are transported through the paracellular space, but not through transcellular route, unless the holes in the epithelium are formed due to dislodging of individual cells. TER can be modulated by changes in both paracellular permeability and transcellular ion permeability through ion channels.

Therefore, it is crucial to measure both parameters simultaneously. The integrity of TJ can be evaluated by visualizing the junctional distribution of TJ-specific proteins such as occludin and ZO-1 by immunofluorescence staining and confocal microscopy, followed by densitometric analysis. Disrupted junctional organization of occludin and ZO-1 indicates the disruption of TJ as the cause of increased paracellular permeability. Finally, it is important to confirm that increased permeability is not caused by loss of cell viability. Cell viability can be evaluated by measuring by LDH release into extracellular medium and WST assay to evaluate the mitochondrial activity.

Caco-2 Cell Monolayers on Transwell Inserts

1. Caco-2 cells are first grown in T75 flasks to achieve 90-95% confluence.
2. Cells are fed with medium every 48 h.
3. Cells from the flasks are lifted up in trypsin/EDTA solution. Cells are washed in DMEM and resuspended in medium and seeded onto Transwell inserts of different sizes (6.5 mm, 12 mm, or 24 mm diameters). Medium is replaced every 48 hrs. TER is monitored in representative wells for the development of barrier function. At 10-14 d after seeding, cells are used for permeability experiments.
4. Before all experiments cell monolayers are washed two times with PBS-1, and preincubated for 60 min with PBS-1, 0.2 mL and 0.8 mL on upper and lower wells, respectively.

Acetaldehyde Treatment

1. Prepare acetaldehyde solutions only just few minutes before the use. PBS-1 of volumes required for preparing stock solution of acetaldehyde (10%) and solution for use (0.5%), and chill them at −20°C for 10 min before mixing with acetaldehyde. When you are ready for acetaldehyde treatment, mix 2 mL of acetaldehyde with 18 mL of prechilled PBS− 1 and mix quickly to prepare 0.5% acetaldehyde.

2. In a 24-well cluster place transwell inserts with cell monolayers in the middle two rows as shown in Fig. 1.1. Place 0.5% acetaldehyde solution (1 ml/well) to the wells on the outer rows.
3. Close the lid and seal by taping on all sides and place the plate in an incubator.

Measuring TER

1. Sterilize chopstick electrodes by dipping in 70% ethanol for 10 min.
2. Remove, air-dry and soak in PBS-1. Keep the electrodes soaked in PBS-1 through the end of experiment.
3. Calibrating electrodes: To test the electrodes, dip it in PBS-1, turn mode switch to "V" and turn power on. Adjust the voltage to 0.0 with a screwdriver by pressing the "Test" button. Turn the mode to "R" and adjust resistance to 0.0.

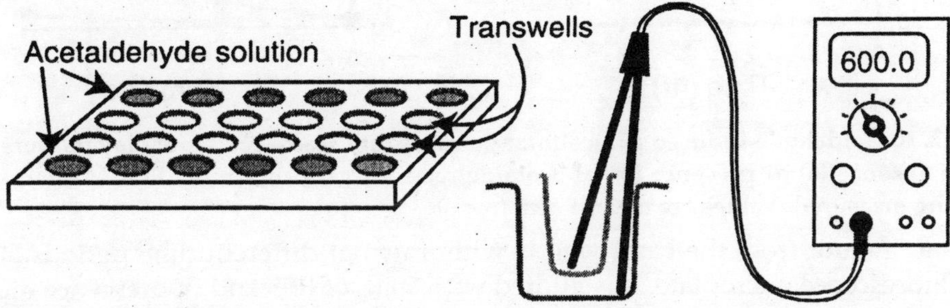

Fig. 1.1. Diagrammatic representation of 24-well cluster with acetaldehyde solution in outer wells and transwell inserts in inner wells. Diagram also shows the use of Millecel-ERS resistance monitoring meter.

4. Insert electrodes through one of the three holes on the wall of the inserts so that the longer electrode sits at the bottom of the lower well and the shorter electrode immersed in buffer in upper well Fig. l. Set the mode button to "R" and press "Measure" button to record the TER values.
5. Normally, TER value can be measured in the same monolayer at different time intervals. However, time course of TER changes during the acetaldehyde treatment cannot be studied in the same monolayer because the plate is sealed during the acetaldehyde treatment. Therefore, TER can be measured only at three time points: before preincubation, after preincubation, and at the end of experiment. A separate monolayer should be used for measuring TER values at different time periods.
6. The polycarbonate membranes at the bottom of transwell inserts pose some level of electrical resistance, which is usually 30 Ohm.cm^2. This can be measured in empty wells soaked in PBS-1 for 30 min. This background values should be subtracted from all values.
7. TER values can be presented as Ohms.cm^2 by multiplying the recorded values by the surface area of the monolayer, which is 0.33 cm^2 for 6.5-mm transwell inserts. TER can also be presented as percent of baseline values to compare the TER of individual monolayers.
8. Acetaldehyde treatment induces gradual decrease in TER, whereas TER in control cell monolayers remains unaffected.

Measuring Paracellular Permeability

1. Stock solution of FITC-inulin is diluted with PBS-1 to obtain a concentration of 0.5 mg/mL and placed in the upper well at the beginning of preincubation (0.2 mL for 6.5-mm transwells), whereas 0.8 mL of PBS-1 without inulin is placed in the lower well.

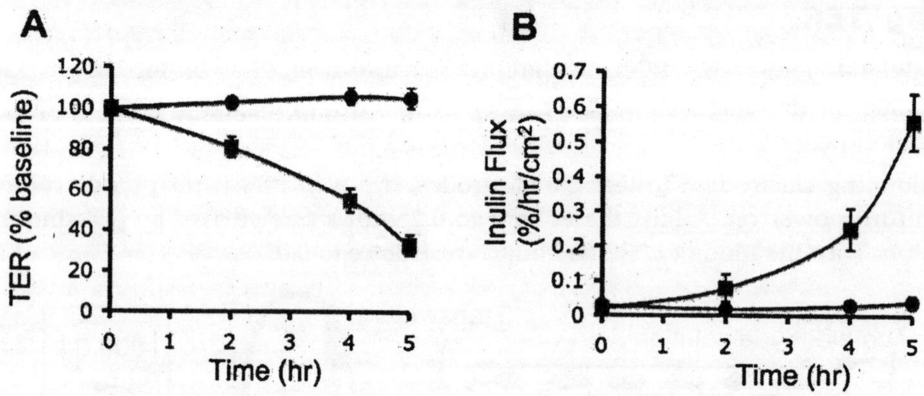

Fig. 1.2. Acetaldehyde-induced paracellular permeability. Caco-2 cell monolayers were incubated in the absence (•) or presence (■) of acetaldehyde for varying times. TER (*A*) and inulin flux (*B*) were measured. Values are mean ± sem (n = 6).

2. A 50-μL aliquot from the basal well is withdrawn at different time periods and placed in a fluorescence microplate. It is diluted with 50μL of PBS and fluorescence measured in a 96-well fluorescence plate reader. At the end of experiment, a 10-μL aliquot of buffer from upper well is withdrawn to measure the original amount of inulin administered to upper well.
3. The blank fluorescence reading is measured by using 100 μLof PBS-1. The fluorescence value measured after preincubation is subtracted from values for corresponding monolayer.
4. Inulin flux values can be presented as the absolute flux (μg inulin/ml/hr/cm^2) of inulin from upper to lower well using the following formula: Inulin flux (μg inulin/ml/hr/cm^2) = $[B \times V_B]/[V \times h \times a]$,

   where B is inulin in aliquot of basal well, V_B is total volume of lower buffer, V is volume of aliquot used, h is time at which flux was measured and a is surface area of the cell monolayer.

   It can also be measured as percent of inulin loaded to the upper well as follows: Percent flux/hr/cm^2 = $[B \times V_B \times V2 \times h \times a \times 100]/[V1 \times A_A]$,

   where B is inulin in lower buffer aliquot, V_B is volume of lower buffer, V2 is aliquot used from lower buffer, h is hour of incubation, a is surface area of cell monolayer, V1 is aliquot used from upper buffer, A is inulin in upper well, and V_A is volume of upper buffer.
5. Endotoxin permeability can be measured as described previously by replacing FITC-inulin with FITC-conjugated LPS.
6. Acetaldehyde treatment increases inulin flux in a time-dependent manner, whereas the rate of flux remains unaffected in control cell monolayer Fig. 1.2B.

Immunofluorescence Staining and Confocal Microscopy

1. At varying times after acetaldehyde treatment cell monolayers are washed in PBS-2 and placed 1 mL of prechilled (−20°C) acetone:methanol (1:1 v/v) onto the cells and allowed to sit at 0°C for 5 min. Aspirate the solvent completely and air dry.
2. Dried samples can be stored at −20°C for several weeks before further process.
3. Rehydrate acetone:methanol-fixed cell monolayers by soaking it in PBS-2 for 15 min. Wash monolayers once with PBS-2.
4. Permeabilization: Place 1 mL of 0.2% Triton X I00 solution in PBS-2 (cold), and incubate at 4°C for 5 min. Wash 3X with PBS-2 (1 mL) for 10 min each.
5. Blocking: Cells are blocked to prevent nonspecific binding of antibodies to cell components. Place 1 mL of 4% milk in TBST and incubate at room temperature for 30 min
6. Incubation with primary antibodies: prepare primary antibodies, anti-occludin, and anti-ZO-1 antibodies, in 4% milk to obtain final concentration of 5 μg/mL and 3 μg/mL, respectively. Place 100 μL of antibody solution on top of the cell monolayer. Incubate in humidifying chamber in dark for 1 h.
7. Aspirate antibody solution and wash cell monolayers in 1% milk three times, 10 min each.
8. Incubation with secondary antibodies: dilute secondary antibodies, Alexafluor-488 conjugated goat polyclonal anti-mouse IgG and Cy3-conjugated goat polyclonal anti-rabbit IgG, in 4% milk (1:100 dilution). Place 100 μl of antibody solution on top of the cell monolayers. Incubate in humidifying chamber in dark for 1 h. Wash tissues with PBS-2 three times, 10 min each.
9. Mounting: cut the membrane at the bottom of transwell inserts with stained cell monolayer and place it on top of a glass slide. Place a drop of mounting fluid on top of membrane. Place the cover slip carefully on top of the mounting fluid, and allow it to spread to all sides. Seal the cover slip with nail polish. Store slides in cold dark area until visualization by microscope.
10. Mounted slides can be stored at 4°C in dark for several months without loosing fluorescence. Use 40x oil immersion objective in a confocal microscope and collect images with a maximum of 500 gain, at 10% laser at a 1-μm setting. The laser can be increased if the stain is weak, but keep in mind that higher laser can result in quenching of fluorescence. But, do not increase gain to obtain higher intensity, because it reduces the quality of the image. Store images as pic or Ism files. Stack the sections of images using Image J software and save as RGB tiff files. These files can be opened in Adobe Photoshop software, and processed to optimize the image quality. Using this software you can also superimpose the occludin (green) and ZO-1 (red) stains for co-localization.
11. In untreated control cell monolayers, the occludin and ZO-1 stains co-localize at the intercellular junctions. Acetaldehyde treatment results in redistribution of both occludin and ZO-1 from the intercellular junctions into the intracellular compartments. Fig. 1.3.

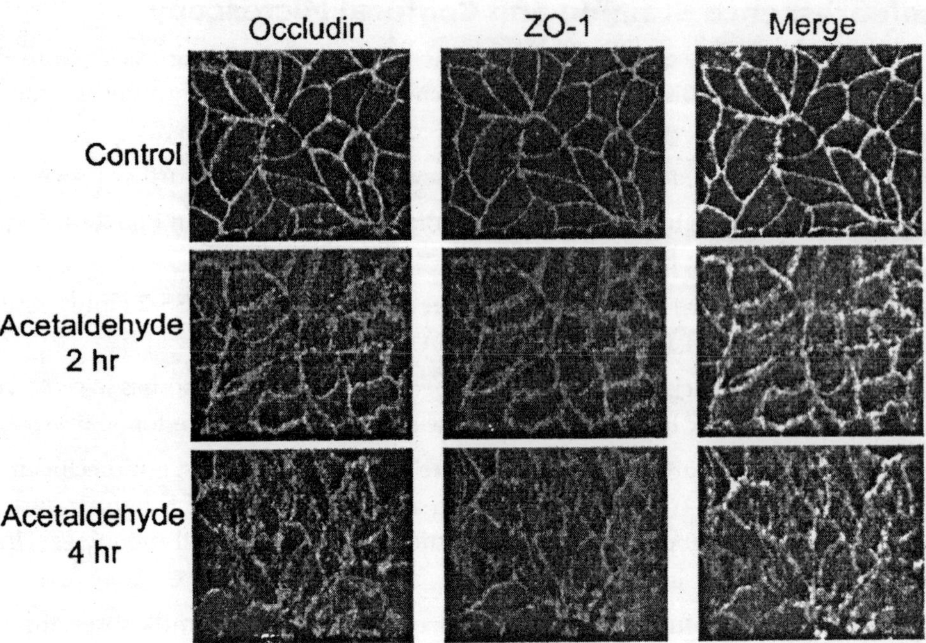

Fig. 1.3. Acetaldehyde disrupts tight junctions. Caco-2 cell monolayers were incubated with or without (control) acetaldehyde for varying times. Cell monolayers were fixed and double-labeled for occluding and ZO-1 by immunofluorescence method. Images were collected by confocal microscopy.

LDH Release Assay

1. Transfer buffers from both the upper and lower wells to microfuge tubes at the end of experiments or treatments. These samples can be stored at −20°C until the assay.
2. Place 100 μL of assay reagent in a 96-well microplate. Add 100 μL of sample (incubation buffer from the upper well).
3. Prepare blanks using 100 μL of assay reagent and 100 μL of PBS-1. For positive control, use buffer from cells treated with 2% Triton X I00 for 10 min.
4. Incubate the microplate at RT for 30 min and measure the absorbance of sample at 490 nm and at 600 nm (as reference to be subtracted from A490).
5. Quantitation and comparison: A490 values obtained can be directly compared between control cells and acetaldehyde-treated cells. Alternatively, percent toxicity can be measured by following formula: A490 for sample divided by A490 for positive control multiplied by 100.

WST Assay

1. Aspirate buffers from both upper and basal wells of transwell inserts.
2. Place 150 μL or 400 μL of assay medium in transwells of 6.5 mm or 12 mm diameters, respectively.
3. Incubate cells at 37°C for 15 min.

4. Withdraw buffer from upper well into microfuge tubes.
5. Mix solutions and transfer 100-μL solution to 96-well microplate.
6. Measure the absorbance of sample at 450 nm and 600 nm (as reference).
7. Measure blank reading using 100 μL of assay reagent. For positive control, use buffer from cells treated with 2% Triton X100 for 10 min.
8. Quantitation and comparison: A450 values obtained can be directly compared between control cells and acetaldehyde-treated cells. Alternatively, percent toxicity can be measured by following formula: A450 for sample divided by A450 for positive control multiplied by 100.

NOTES

1. All solutions should be prepared in Nanopure water (< 18.2 Ohms resistivitity).
2. Paraformaldehyde is a toxic chemical when inhaled. Therefore, preparing 3% paraformaldehyde should be prepared in a fume hood. It dissolves better at alkaline pH and warmer temperature. Therefore, drops of 4 N sodium hydroxide solution should be added and heated at 60°C on a heating magnetic stirrer. Once the solution appears clear, readjust the pH back to 7.4 by using 1 N hydrochloric acid.
3. Acetone:methanol mixture can be stored at −20°C; however, it is important that is kept in an explosion-proof freezer for safety. Failure to prechill this reagent before use can result in loss of antigenicity of proteins of interest.
4. Caco-2 cells used within 10 passages after thawing. Original Caco-2 cell line and its subclones such as Caco-2_{bbe} can be used for acetaldehyde treatment. However, the sensitivity to certain regulatory factors such as EGF may vary between strains.
5. Caco-2 monolayers develop maximal TER by 7 days after seeding, which is followed during the next week by differentiation into polarized columnar epithelial cells. Cellular responses to many factors depend on the differentiation status of the cells. Therefore, it is important to conduct experiments in a window of time after seeding.
6. EGF, a gastrointestinal mucosal protective factor, prevents disruption of TJ by various injurious factors such as acetaldehyde and hydrogen peroxide. Serum contains many such mucosal protective factors. Therefore, preincubation for one hour in PBS-1 allows clearance of any serum effects.
7. For 12-mm transwells, 0.8 mL and 1.5 mL, and for 24-mm transwells, 1.5 mL and 2.5 mL of PBS-1 are added to upper and lower wells, respectively. These volumes were chosen to match the level of buffers in upper and lower wells, so that hydrostatic pressure on the cell monolayers is minimized.
8. Small-size transwells of 6.5 mm in diameter are good enough to study the permeability. However, if proteins for biochemical studies are needed one can grow cells in large transwells (12 mm or 24 mm diameters). For 12-mm transwells, 12-well clusters are used. Place transwells in the middle row and acetaldehyde solution in the outer rows. In 24-mm transwells, place transwells in the middle two wells and acetaldehyde in the outer four wells.

9. Acetaldehyde is volatile (boiling point is 18°C) at room temperature and evaporates from solution very rapidly at 37°C. Therefore, it is important to expose cells to vapor-phase acetaldehyde, in a sealed chamber. Acetaldehyde concentration of 0.5% in the reservoir wells results in 400 μ*M* acetaldehyde in the buffer incubating the cell monolayers because of absorption of acetaldehyde vapors.
10. After certain period of incubation, the cells should be equilibrated to room temperature by placing at RT for 5 min. TER tends to increase as the temperature of cell monolayers decrease.
11. The samples double labeled for red and green fluorescence can be scoped simultaneously. However, depending on the filters used, there can be bleeding of fluorescence from one channel to the other channel. In such case, it is safer to take the image with settings for one channel first, and then the other. Always take the image of one with the weaker stain first to minimize the loss of other fluorescence due to quenching.

CHAPTER

2

Photo Physiology

Light plays pervasive role in human life. The earth is bathed in light from the sun, and from this light comes not only warmth but also energy for all living organisms. Of the 3×10^4 kJ m^{-2} of light energy falling on the earth each day,[1,2] ~30 kJ m^{-2} are captured by photosynthesis.[3] Light penetrating the atmosphere allows us to see and provides color to our environment. It controls the flowering of plants, the germination of seeds and spores, the greening of seedlings, and the daily cycles of many organisms. High in the stratosphere ultraviolet light reacts with oxygen to create a protective blanket of ozone. The ultraviolet light that is not screened out by the ozone layer kills bacteria, tans our skin, and often mutates our DNA, inducing many cancers. Organisms, from bacteria to higher plants, display *phototaxis,* the ability to move toward a source of light or to orient themselves with respect to a source of light. In plants the chloroplasts assume an orientation that maximizes efficiency of light absorption. Plants grow toward light *(phototropism),* and some organisms avoid light. Many organisms emit light.

Many of our most important experimental techniques involve the use of light or of other forms of electromagnetic radiation of a wide range of energies. X-rays, ultraviolet light, infrared light, and micro-waves all serve in the study of biomolecules.

A. PROPERTIES OF LIGHT

Light is a form of electromagnetic radiation and possesses characteristics of both waves and particles (*photons*). The energy of a photon is usually measured by the frequency (or by the wavelength in a vacuum to which it is inversely related. At the high-energy end (off the scale of the figure to the right) are cosmic rays and gamma rays, while at the low-energy end radio waves extend to wavelengths of many kilometers. The narrow range of wavelengths from about 100 nm to a few micrometers, which is the subject of this chapter, includes the ultraviolet, visible, and near infrared ranges. The second line of Fig. 2.1 shows this region expanded. Note that the range light reaching the earth's surface is narrow, largely being confined to wavelengths of 320—1100 nm. The human eye responds to an even more limited range of 380—760 nm, in which all of the colors of the rainbow can be found. The aromatic rings of proteins and nucleic acids absorb maximally at 280 and 260 nm, respectively. Even though these wavelengths are largely screened out by the ozone layer of the stratosphere, enough light penetrates to cause mutations and to damage the skin of the unwary sunbather.

Table 2.1. Some Properties of Light

Velocity of light in a vacuum	$c = 2.998 \times 10^8$ m s^{-1}
Velocity of light in a medium	$c' = c/n$ where n = refractive index
Wave number (in nm)	$\overline{v} = 1/\lambda$; $\overline{v}$ (in cm^{-1}) $= 10^7/\lambda$
Frequency	$v = c/\lambda = c\overline{v}$
	v (in hertz) $= 2.998 \times 10^{10}\ \overline{v}$ (cm^{-1}) in a vacuum
Energy of quantum	$E = hv = hc\overline{v}$
	E (joules) $= 1.986 \times 10^{-23}\ \overline{v}$ (cm^{-1})
	E (eV) $= 1.240 \times 10^{-4}\ \overline{v}$ (cm^{-1})
Energy of einstein	$E = Nhv = Nhc\overline{v}$
	$= 6.023 \times 10^{23} hc\ \overline{v}$
	E (joules) $= 11.961\ \overline{v}$ (cm^{-1})
	E (kcal) $= 2.859 \times 10^{-3}\ \overline{v}$ (cm^{-1})

The energy of a quantum of light is proportional to the *wave number* or *frequency.* The wave number $\overline{v}$ is the reciprocal of wavelength and is customarily given in units of cm^{-1} *(reciprocal centimeters).* Most of the absorption spectra in this book are plotted against wave number in cm^{-1}. The frequency v in *hertz* equal to *c'v,* where *c'* is the velocity of light in a medium. (The velocity of light in a vacuum is designated *c* and is equal to 3.00×10^8 m s^{-1}.) The energy of a quantum of light *E* is equal to *hv,* where *h* is Planck's constant, 6.626×10^{-34} J s^{-1}. From a chemical view-point, we are more interested in the energy of one *einstein, i.e.,* one "mole" of light (6.023×10^{23} quanta). The energy in kJ per einstein is 11,960 $\overline{v}$ (in cm^{-1} vacuum). The lower three scales of Fig. 2.1 also show the relationships of $\overline{v}$, *v,* and *E* to wavelength.

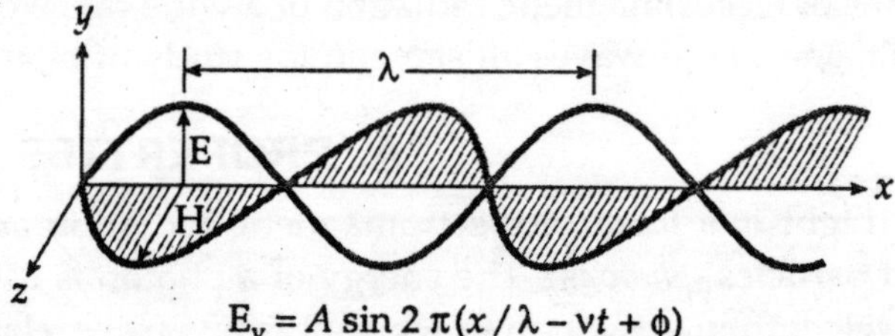

The light wave is characterized by oscillating electrical and magnetic fields. For propagation of light in the *x* direction the electric field Vector E, which is customarily plotted in the *y* direction, is a function of the wavelength λ, and the time.

$$E_y = A \sin 2\pi (x/\lambda - vt + \phi) \quad ...(2.1)$$

The magnetic vector H is right angles to the electric vector and is given by Eq. (2.2)

$$H_z = (\varepsilon/\mu) - A \sin 2\pi (x/\lambda - vt + \phi)$$

The velocity of propagation of light in a medium, *c'* depends upon both ε, the *dielectric constant* of the medium, and the *magnetic permeability.*

Velocity in a medium: $c' = c/(\varepsilon/\mu) - = c/n$ ...(2.3)

The *refractive index* of a medium relative to a vacuum is given the symbol *n.* It is the factor by which the velocity of light in a vacuum is diminished in a medium. It is a function of wavelength. For the 589 nm sodium line *n* is 1.00029 for air and 1.33 for water at 25°C.

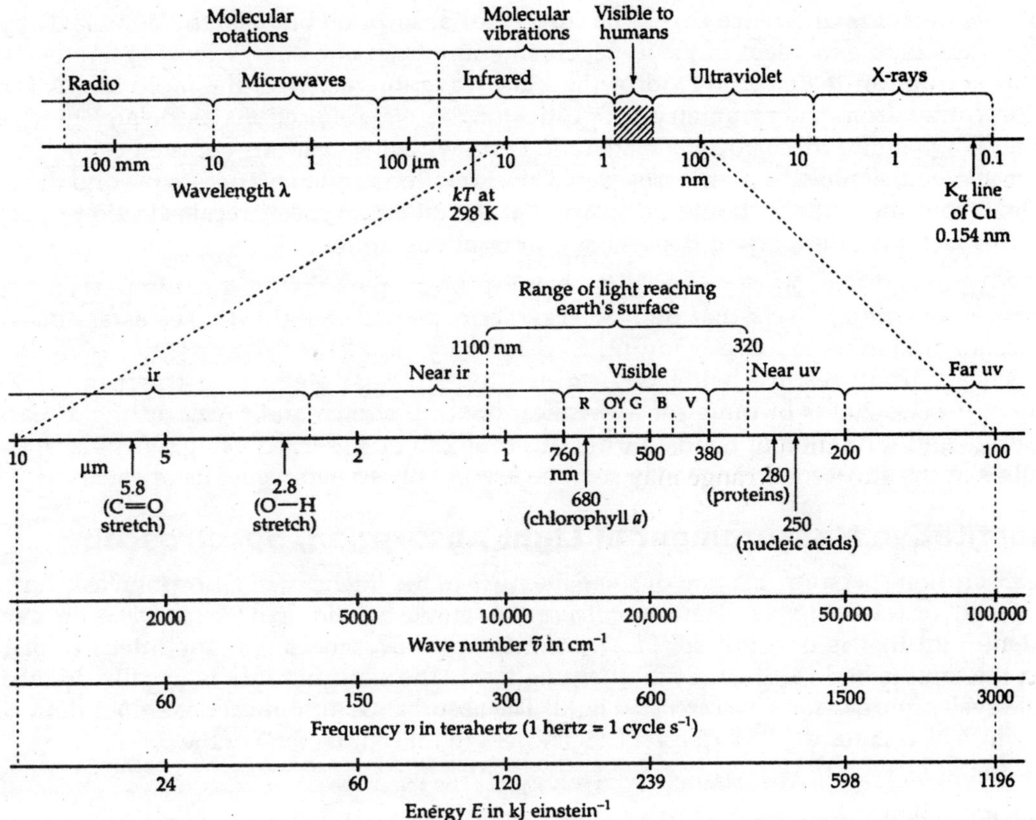

Fig. 2.1. A part of the electromagnetic spectrum. The letters V, B, G, Y, O, R over the visible part of the spectrum refer to the colors of the light. The position marked "K_a line of Cu" is the wavelength of X-rays and most widely employed in X-ray diffraction studies of proteins and other organic materials.

The term in Eq. 2.1 and 2.2 is a *phase factor.* Most light is called *incoherent* because ϕ varies for the many photons making up the beam. *Coherent light* produced by *lasers* contains photons all with the same phase relationship. If the electric vectors of all the photons in a beam of light are in the same plane (as will be the case for light emerging after passage through certain kinds of crystals), the light is called *plane polarized.* The direction of polarization is that of the electric vector E. *Circularly polarized light,* in which the electric vector rotates and traces out either a left-handed or right-handed helix, can also be generated. A beam of left-handed circularly polarized light, together with a comparable beam polarized in the right-handed direction, is equivalent to a beam of plane polarized light. Conversely, plane polarized light can be resolved mathematically into right- and left-handed circularly polarized components.

B. ABSORPTION OF LIGHT

Absorption of light is fundamental to all aspects of photochemistry and provides the basis for absorption spectroscopy. Light absorption is always *quantized.* It can take place only when the energy *hν* of a quantum is equal to the difference in energy between two energy levels of the absorbing molecule.

$$E_2 - E_1 = h\nu$$

Not only must the difference $E_2 - E_1$ be correct for absorption but also there must always be a change in the dipole moment of the molecule in going from one energy level to another. Only when this is true can the electric field of the light wave interact with the molecule. A further limitation comes from the symmetry properties of the wave functions associated with each energy level. Quantum mechanical considerations show that transitions between certain energy levels are allowed, while others are forbidden. Consideration of such matters is beyond the scope of this book, but the student should be aware that the quantum mechanical selection rules that express this fact are an important determinant of light absorption.

Many types of light source are used in chemical measurements. Of great importance is the recent development of lasers that deliver very short pulses of light. Pulses as short as five femtoseconds and even less are being utilized for very rapid spectroscopy and excitation of fluorescence. Structures are being determined by ultrafast electron diffraction or X-ray diffraction. It takes 200 fs or more for a chemical bond to stretch and break during a reaction. The cleavage and formation of bonds during this time can be observed using 5-fs pulses. Lasers with pulses in the attosecond range may soon be used to observe movements of electrons.

1. Quantitative Measurement of Light Absorption, Spectroscopy

An absorption spectrum is a plot of some measure of the intensity of absorption as a function of wave-length or wave number. The *transmittance* of a sample held in a *cell* (or *cuvette*) is the fraction of incident light that is transmitted, *i.e.*, transmittance = I/I_0 where I_0 is the intensity of light entering the sample and I is that of the emerging light. The transmittance is usually defined for a single wavelength, *i.e.*, for *monochromatic* light. The absorbance (or optical density) is defined by Eq. 2.5, which also states the *Beer-Lambert law.* The length (in centimeters) of the

$$\text{Absorbance} = A = \log_{10}(I_0/I) = \varepsilon cl \qquad \text{...(2.5)}$$

light path through the sample is l, c is the concentration in moles per liter, and is the *molar extinction coefficient* (molar absorptivity or molar absorption coefficient), whose units are liter mol^{-1} cm^{-1} (or M^{-1} cm^{-1}). The reader can derive Eq. 2.5 by assuming that in a thin layer of thickness dx the number of light quanta absorbed is proportional to the number of absorbing molecules in the layer. Integration from $x = 0$ to l gives the Beer-Lambert law. Equation 2.5 generally holds very well for solutions containing single ionic or molecular forms. However, it is usually valid only for monochromatic light. Furthermore, making precise measurements of absorbance is not easy. At $A = 1$ only 10% of the incident light is transmitted, and the utmost care is required to obtain a value of A good to within ±0.05. At $A = 2$ only 1% of incident light is transmitted, and the value of A will be much less reliable. Very low absorbances are also difficult to measure. In view of the importance that spectrometry has played in biochemistry, it may seem surprising that the first reliable commercial laboratory ultraviolet-visible spectrophotometers became available in 1940 and the first commercial infrared spectrometer in 1942.

2. The Energy Levels of Molecules

The energy of molecules consists of *kinetic* (translational), *rotational, vibrational,* and *electronic* components. The corresponding rotational, vibrational, and electronic energy levels are always quantized. Light quanta of wavelengths 0.2-20 mm (50-0.5 cm^{-1}; frequencies of 1.5×10^{12} to 1.5×10^{10} s^{-1}) with energies of 0.6 – 0.006 kJ/einstein are sufficient to excite molecules from a given rotational energy level to a higher one. Spectra in this "far infrared" or "microwave" region often consist of a closely spaced series of lines. For example, the rotational spectrum of gaseous

HCl is a series of lines at 20.7 cm^{-1} intervals beginning at that wave number and reaching a maximum at about 186 cm^{-1} (54 μm). The energies involved in absorption of such light are far lower than energies of activation for common chemical reactions and lower than the average translational energy of molecules in solution at ordinary temperatures (3/2 k_bT or 3.7 kJ/mol at 25°C). However, they are still much higher than energies involved in the nuclear transitions of NMR spectra). Compare 500 Hz for a proton resonating at 1 ppm in a 500 MHz NMR spectrometer with the $10^{10} - 10^{12}$ s^{-1} frequencies of microwave spectra.

Vibrational energies range from about 6 to as much as 100 kJ mol^{-1} with corresponding wave numbers of ~500 – 8000 cm^{-1}. The resulting absorption spectra are in the infrared region. Excited electronic energy levels are ~120 – 200 kJ mol^{-1}, and the spectral transitions are at 10,000 – 100,000 cm^{-1} (1000 – 100 nm wavelengths) in the visible and ultraviolet region.

3. Infrared (IR) and Raman Spectra

Absorption in the near infrared region is dominated by changes in vibrational energy levels. A typical wave number is that of the "amide A" band at 3300 cm^{-1} (3.0 μm wavelength), approximately 10^{14} s^{-1}. First let us consider the stretching vibrations of a diatomic molecule. The two nuclei of the molecule can be thought of as connected with a spring. The energy of oscillation is approximately that of a harmonic oscillator. Application of quantum theory shows that the discrete energy levels that can be assumed by the oscillator are equally spaced. The difference between each pair of successive energy levels is $h\nu$, where ν is the frequency of light that must be absorbed to raise the energy from one level to the next. In the ground state (unexcited state) the molecule still possesses a *zero-point energy*, $E = 1/2\, h\nu_0$, equal to half the energy needed to induce a transition.

While the harmonic oscillator is a good approximation to the behavior of a molecule in the lower vibrational energy states, marked deviations occur at higher energies. At the lower energy levels the change in the distance between the atomic centers during the course of the vibration amounts to only ±10% or less, but as the energy becomes greater the bond stretches more and the motion becomes *anharmonic*. The energy states of molecules are often in the form of *Morse curves* in which energy is plotted against internuclear distance. As the internuclear distance becomes very short, the energy rises steeply. As the bond is stretched, there comes a point at which addition of more energy ruptures the bond. A diatomic molecule will dissociate into atoms and more complex molecules into fragments. Vibrational energy levels can be portrayed as horizontal lines at appropriate heights on the Morse curve.

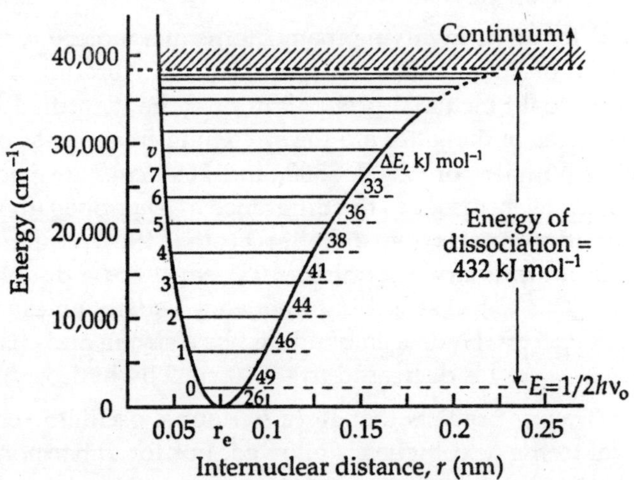

Fig. 2.2 The potential energy of the hydrogen molecule as a function of internuclear distance, and the position of its vibrational energy levels. ΔE values are energy differences between successive levels; v designates vibrational quantum numbers.

Because there are many rotational energy levels corresponding to each vibrational level, IR spectra contain absorption bands resulting from simultaneous changes in both the vibrational and rotational energy levels of molecules. Instead of single peaks corresponding to single transitions in vibrational energy, progressions of sharp bands at closely spaced intervals are observed. An example is provided by the band corresponding to the stretching frequency of the H—Cl bond in gaseous HCl at 2886 cm^{-1} (3.46 μm). There is actually no band at this wave number but a series of almost equally spaced bands on either side of the fundamental frequency from ~2600 to ~3100 cm^{-1} at intervals of ~21 cm^{-1} *i.e.,* the wave number of the rotational frequency seen in the microwave spectrum. The effect is to broaden the band as seen in a low-resolution spectrum. This is only one cause of the broadening of IR bands in solution. Another cause is interaction with solvent to provide a heterogeneity in the environments of the absorbing molecules.

The IR spectra of diatomic molecules are relatively easy to interpret, but for more complex substances the infrared absorption bands often cannot be associated with individual chemical bonds. Instead, they correspond to the *fundamental vibrations* (normal vibrations) of the *molecule.* Fundamental vibrations are those in which the center of gravity does not change. For a molecule containing *n* atoms, there are $3n - 6$ such vibrations. They are sometimes dominated by a vibration of a single bond, but often involve synchronous motion of many atoms. The fundamental vibrations of a molecule are described by such words as *stretching, bending* (in-plane and out-of-plane), *twisting,* and *deformation.* Rarely are all $3n - 6$ bands seen in an infrared spectrum. Some of the vibrations, *e.g.,* the symmetric stretching of the linear CO_2 molecule, are not accompanied by any change in dipole moment, while other bands may simply be too weak to be observed clearly.

Vibrations involving many atoms in a molecule, *i.e., skeletal vibrations,* are often found in the region of 700 – 1400 cm^{-1} (14 – 7 μm). Vibrational frequencies that are dominantly those of individual functional groups can often be identified in the range 1000 – 5000 cm^{-1} (10 – 2 μm). Examples of the latter are the stretching frequencies of C–H, N–H, and O–H bonds, which have wave numbers of ~2900, 3300, and 3600 cm^{-1}, respectively. The energy (and frequency) of the vibrations increases as the difference in electronegativity between the two atoms increases. When a bond connects two heavier atoms, the frequency is lower, *e.g.,* the wave number for C–O in a primary alcohol is ~1053 cm^{-1}. For a double bond it increases; for C = O it is ~ 1700 cm^{-1}. This C = O stretching frequency usually gives rise to one of the strongest bands observed in IR spectra. Hydrogen bonding has a strong and characteristic effect. Thus, the O–H frequency at ~3600 cm^{-1} is decreased to –3500 cm^{-1} by hydrogen bonding.

Theory predicts that for a harmonic oscillator only a change from one vibrational energy level to the next higher is allowed, but for anharmonic oscillators weaker transitions to higher vibrational energy levels can occur. The resulting "overtones" are found at approximate multiples of the frequency of the fundamental. Combination frequencies representing sums and differences of frequencies of individual IR bands may also be seen. The intensities of these bands are low, but their presence at relatively high energies in the near IR region (4000 – 12,500 cm^{-1}) means that they may be easier to observe than the fundamental frequencies in the more crowded IR region. Development of the very sensitive *Fourier-transform infrared spectroscopy* (FTIR) has made it possible to record the complex IR spectra of macromolecules in dilute aqueous solutions rapidly. The water spectrum is substracted digitally. FTIR has been utilized to study amide groups in peptides, carboxyl groups in proteins, conformations of sugar rings in DNA, and the ionization state of phosphate groups. New computational methods involving use of two-dimensional representations provide simplifications in interpretation of IR spectra. Another variant is *total reflection FTIR,* a technique that records spectra of thin films and has permitted the recording of transient changes in protein spectra with microsecond time resolution.

Vibrational frequencies of amide groups. The IR absorption bands of amide groups, which are present in both proteins and in the purine and pyrimidine bases, have attracted a great deal of attention. The *amide I* band at ~1680 cm^{-1} is associated with an in-plane normal mode of vibration that involves primarily C = O stretching. (The band is designated I′ if the N—H has been exchanged to form N–^{2}H). The *amide II* band at ~ 1500^{-1} and the *amide III* band at ~ 1250 cm^{-1} both arise from inplane modes that involve N—H bending, while the higher frequency *amide A* band at ~ 3450 cm^{-1} involves N—H stretching. It is shifted to ~ 3300 cm^{-1} when the N—H is hydrogen bonded. Examples of IR spectra of proteins, including the amide bands A, I, and II. The band shapes are complex. Those of the amide bands I and III depend upon the conformation of the peptide chain. For example, amide groups in α helices have an amide I band about 20 cm^{-1} higher than do those in β structures.

If peptide chains can be oriented in a regular fashion, it may be useful to measure *infrared linear dichroism.* Absorption spectra are recorded by passing plane polarized light through the protein in two mutually perpendicular directions, with the electric vector either parallel to the peptide chains or perpendicular to the chains. Such a pair of spectra is shown in Fig. 2.3A for oriented fibrils of insulin. In this instance, the insulin molecules are thought to assume a β conformation and to be stacked in such a way that they extend transverse to the fibril axis (a cross-β structure). When the electric vector is parallel to the fibril axis, it is perpendicular to the peptide chains. Since the amide I band is dominated by a carbonyl stretching motion that is perpendicular to the peptide chains in the β structure, this band is enhanced when the electric vector is also perpendicular to the peptide chains and

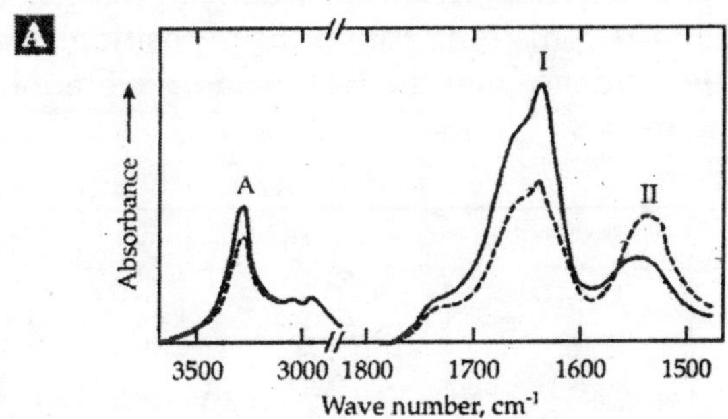

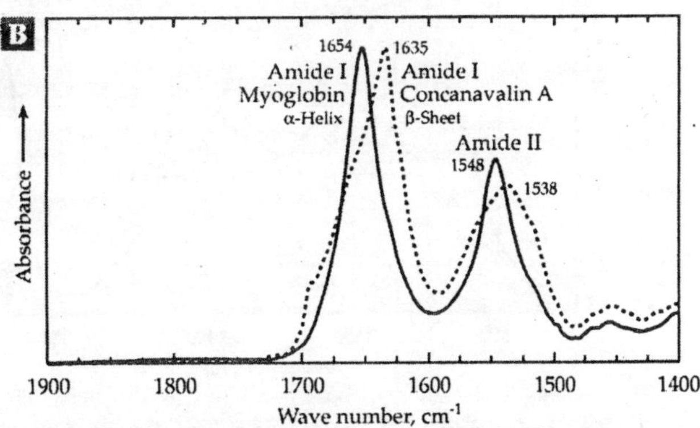

Fig. 2.3 Infrared absorbance spectra of the amide regions of proteins. (A) Spectra of insulin fibrils illustrating dichroism. Solid line, electric vector parallel to fibril axis; broken line, electric vector perpendicular to fibril axis. From Burke and Rougvie. Courtesy of Malcolm Rougvie. (B) Fourier transform infrared (FTIR) spectra of two soluble proteins in aqueous solution obtained after subtraction of the background H_2O absorption. The spectrum of myoglobin, a predominantly α-helical protein, is shown as a continuous line. That of concanavalin A, a predominantly β-sheet containing protein, is shown as a broken line.

is diminished when the electric vector is parallel to the peptide chains (perpendicular to the fibril axis. The same is true of the amide A band which is dominated by an N—H stretch. On the other hand, the dichroism of the amide II band is the opposite because it tends to be dominated by an N—H bending, which is in the plane of the peptide group but is longitudinal in direction. In *isotope-edited FTIR,* heavy atoms such as ^{13}C are introduced to shift IR bands and assist in their identification. The method can be combined with measurement of linear dichroism of oriented peptides.

The loss of the amide II band in D_2O is one of the major tools for studying protein dynamics. In some cases the four main secondary structures, α helix, β sheet, β turn, and random coil, can be distinguished. The amide bands of pyrimidines can also be observed in IR spectra of pyrimidines. Figures 2-4A show the spectrum of 1-methyluracil in H_2O and also in D_2O. Notice that the amide II band is totally lacking in D_2O. The same figure also shows the IR spectrum of 1-methyluracil containing ^{18}O in the 4 position. The shift of 9 cm^{-1} in the amide II position is part of the evidence that the NH bending vibration is extensively coupled to C = O and C = C stretching modes.

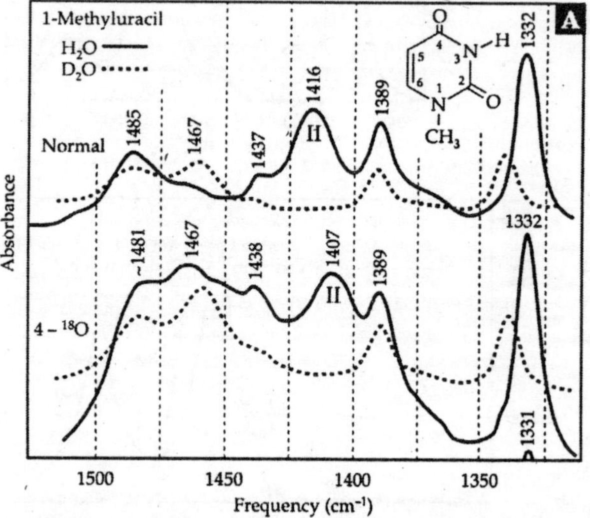

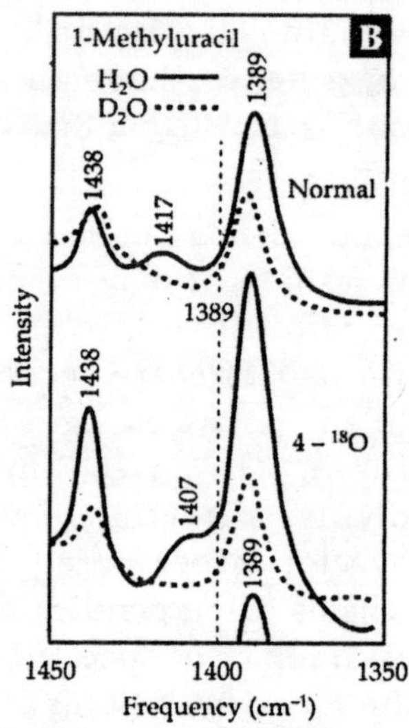

Fig. 2.4 (A) Infra-red and (B) Raman spectra of 1-methyluracil in H_2O and D_2O. Spectra for normal 1-methy-luracil and for the specific isotopic derivatives with ^{18}O in the 4 position are shown.

Raman spectra. In a collision between a photon and a molecule, the photon may undergo *elastic collision* in which the photon loses no energy but changes its direction of travel. Such scattering is known as *Rayleigh scattering* and forms the basis for a method of molecular mass determination. Sometimes *inelastic* collisions occur in which both the molecule and the photon undergo changes in energy. Since such changes must be quantized and involve vibrational and rotational levels of the molecule, the spectrum of the scattered light (*Raman spectrum*) contains much of the same information as an ordinary IR spectrum. However, the selection rules are not the same. Some transitions are *"infrared active"* and others are *"Raman active."* IR-active transitions can occur only when the dipole moment varies with time as a molecule vibrates, whereas Raman-active transitions require that the polarizability change with time. For this reason, it, is useful to measure both IR and Raman spectra on the same sample. Until recently, Raman spectroscopy was not used much in biochemistry because of the low intensity of the scattered light. However, with laser excitation the technique is practical and can be applied to aqueous or non-aqueous or non-aqueous solutions, and to solid or dispersed forms of macromolecules.

Both amide I and amide III bands are seen in Raman spectra of proteins. Lippert *et al.* devised the following method for estimating the fractions of α-helix, β sheet, and random coil conformations in proteins. The amide I Raman bands are recorded at 1632 and 1660 cm^{-1} in D_2O (amide I′). The amide III band, which is weak in D_2O, is measured at 1240 cm^{-1} in H_2O. The intensities of the three bands relative to the intensity of an internal standard (the 1448 cm^{-1} CH_2 deformation) are related to those of standard poly-L-lysine in known conformations. See also Craig and Gaber. The Raman spectrum of 2-methyluracil is shown in Fig. 2-4B. Note the low intensity of the amide II band relative to that of the amide I band, a characteristic of Raman spectra. Linear dichroism observed by polarized Raman microspectrophotometry has provided information about orientation of indole rings of tryptophan in filamentous virus particles.

In *resonance Raman spectroscopy* a laser beam of a wavelength that is absorbed in an electronic transition is used. The scattered light is often strongly enhanced at frequencies differing from that of the laser by Raman frequencies of groups within the chromophore or of groups in another molecule adjacent to the chromophore. The resonance effect not only increases the sensitivity of Raman spectroscopy but also allows a person to study specifically the vibrational spectrum of a selected aromatic group or other structure within a macromolecule. Problems associated with the technique are fluorescence, which may be 10^6 times as strong as the Raman emission, and photo-chemical damage from the intense laser beam. Fluorescence is often quenched with KI.

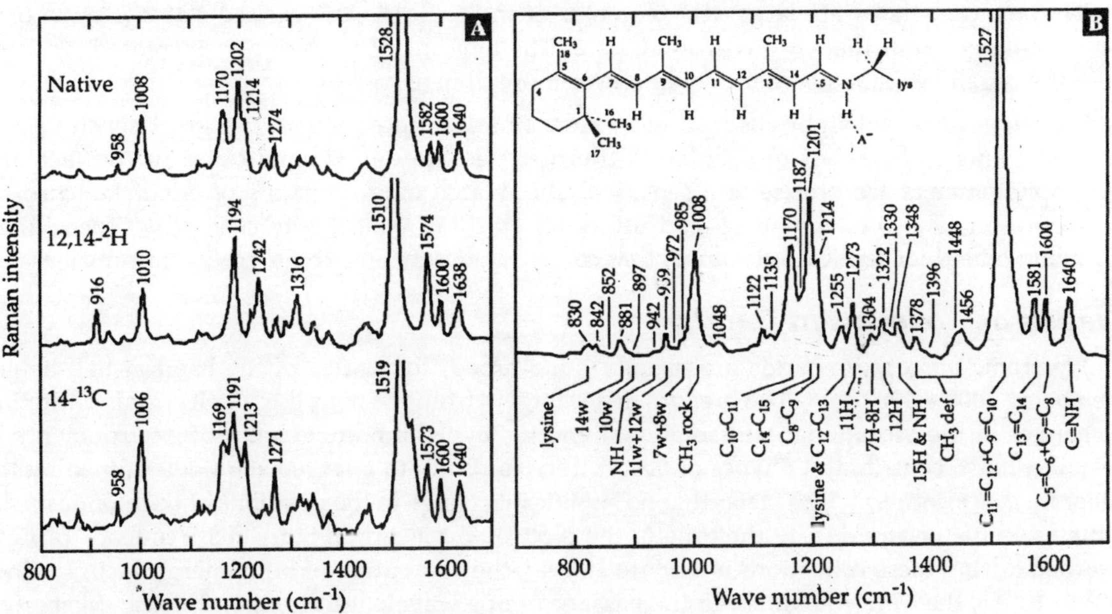

Fig. 2.5 Resonance Raman spectra. (A) Of the retinaldehyde-containing bacteriorhodopsin bR_{568} (see Fig. 2-45) and its 12,14-²H and 14-¹³C isotopic derivatives. (B) Of bR_{568} labeled with the dominant internal coordinates that contribute to the normal modes.

If the exciting laser has a frequency ν_0 and the frequency of a vibrationally excitation in a molecule is ν_1 the Raman spectrum will contain a pair of bands, the stronger one or "stokes" band of energy h ($\nu_0 - \nu_1$) and a weaker one or "anti-stokes" band of h ($\nu_0 + \nu_1$). Special techniques such as *coherent anti-stokes* Raman scattering provide a means of getting around the fluorescence problem. Raman spectroscopy is also possible with excitation of ultraviolet absorption bands. It can be applied to peptides, tryptophan or tyrosine residues of proteins, nucleic acid bases, heme proteins, other metalloproteins, flavin coenzymes, pyridoxal phosphate, flavoproteins, carotenoid-containing proteins, and to substrates undergoing cleavage in the active site of papain. Resonance Raman spectroscopy is very useful in the study of adducts of CO, NO, and O_2 with heme proteins because coupled vibrational modes of both the porphyrin rings and axial ligands can be observed. Resonance Raman spectra are strongly amplified for samples adsorbed to specially prepared colloidal silver particles.

4. Electronic Spectra

Biochemists make extensive use of spectroscopy in the ultraviolet (UV) and visible range. Visible light begins at the red end at ~12,000 cm^{-1} (800 nm) and extends to 25,000 cm^{-1} (400 nm). The ultraviolet range begins at this point and extends upward, the upper limit accessible to laboratory spectrophotometers being ~55,000 cm^{-1} (180 nm). The energies covered in the visible-UV range are from ~ 140 to ~ 660 kJ/mol. The latter is greater than the bond energy of all but the strongest double and triple bonds. It is understandable that UV light is effective in inducing photochemical reactions. Even the lower energy red light, which is used by plants in photosynthesis, contains enough energy per einstein to make it feasible to generate ATP, to reduce $NADP^+$, and to carry out other photochemical processes. Although the energies of light absorbed in electronic transitions are large, the geometry of molecules in the excited states is often only slightly altered from that in the ground state. The amount of vibration is increased, and the molecule usually expands moderately in one or more dimensions.

The significance of light absorption in biochemical studies lies in the great sensitivity of electronic energy levels of molecules to their immediate environment and to the fact that spectrophotometers are precise and sensitive. The related measurements of circular dichroism and fluorescence also have widespread utility for study of proteins, nucleic acids, coenzymes, and many other biochemical substances that contain intensely absorbing groups or *chromophores.*

Shapes of Absorption Bands

Electronic absorption bands are usually quite broad, the width of the band at half-height often being 3000 – 4000 cm^{-1}. The breadth arises largely from the coupling of electronic excitation to changes in the vibrational and rotational energy levels. Inhomogeneity of environments in the solvent also contributes. Shapes of absorption bands are to a large extent determined by the *Franck-Condon principle,* which states that no significant change in the positions of the atomic nuclei of the molecule occurs during the time of the electronic transition. Since the frequency of light absorbed during these transitions is ~10^{15} to 10^{16} s^{-1}, the absorption of light energy occurs within 10^{-15} to 10^{-16} s, the time equivalent to the passage of one wavelength of light. During this period the vibrational motions of the nuclei are almost insignificant because of the much lower frequencies of vibration. Two types of potential energy curves for excited states of molecules are shown in Fig. 2.6. In the first the geometry of the molecule is little changed between ground state and excited state. At room temperature most molecules are in the lowest energy states of at least the most energetic of the various vibrational modes of the molecule ($3/2\ k_BT \sim 300\ cm^{-1}$). Therefore, the most probable transitions occur from the lowest vibrational states of the ground electronic states. The most probable inter-nuclear distance for a molecule in the ground state is the equilibrium distance r_e. Since that distance is the same in all of the vibrational levels of the electronically excited state, transitions to any of these states may occur. The transition to the first vibrational level of the excited state is most likely. The result is an absorption spectrum in which the sharp band representing the "0 – 0 transition" is most intense and in which there are progressively weaker bands corresponding to the 0 – 1, 0 – 2, 0 – 3, etc., transitions. Many organic dyes with long series of conjugated double bonds have spectra of this type.

A second type of spectrum is illustrated in Fig. 2.6. In this instance, the molecule has expanded in the excited state, and r_e is greater than in the ground state. The Franck–Condon principle suggests that a transition is likely only to those vibrational levels of the excited state in which the internuclear distance is compressed for a significant fraction of the time, approximately to that of r_e in the ground state. Examination of Fig. 2.6 explains why the resulting absorption spectra tend to have weak 0 – 0 bands and stronger bands corresponding to transitions to higher levels.

For real spectra of polyatomic molecules the situation is more complex. Some molecules in the ground state do occupy higher vibrational levels of the less energetic modes. Therefore, there will be weaker lines, some of which lie on the low-energy side of the 0 – 0 transition. Since in polyatomic molecules there are several normal modes of vibration, there will be other progressions of absorption bands paralleling those shown in Fig 2.6 and filling in the valleys between them. All of the band are broadened by rotational coupling and by interactions with solvent.

An example of a molecule giving a spectrum of the type shown in Fig 2.6 is toluene. The vapor phase spectrum contains a large number of sharp lines, some of which can be seen in the low-resolution spectrum of Fig. 2.7. Several progressions can be identified. One begins with the intense 0 – 0 line at $37.48 \times 10^3\ cm^{-1}$ and in which spacing of $\sim 930\ cm^{-1}$ between lines corresponds to a vibration causing symmetric expansion of the ring (ring breathing frequency), a frequency that can also be observed in the infrared spectrum. Other progressions beginning at the 0 – 0 line involve additional modes of vibration with frequencies (in the excited electronic state) of 460, 520, and $1190\ cm^{-1}$. Additional weaker bands are *"buried"* in the valleys in Fig. 2.7. When the spectrum of toluene is measured in solution, the sharp lines are broadened, but there are still indications of vibrational structure.

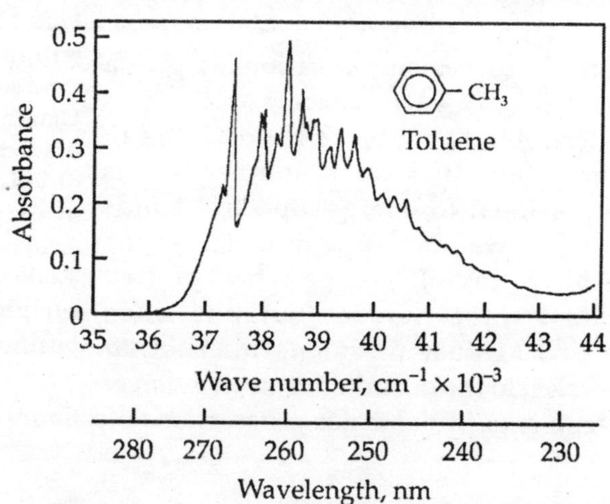

Fig. 2.6. The spectrum of the first electronic transition of toluene vapor at low resolution.

Gaussian curves (normal distribution functions) can sometimes be used to describe the shape of the overall envelope of the many vibrationally induced subbands that make up one electronic absorption band, *e.g.*, for the absorption spectrum of the copper-containing blue protein of *Pseudomonas*. Gaussian bands are appropriate. They permit resolution of the spectrum into components representing individual electronic transitions. Each transition is described by a *peak position, height* (molar extinction coefficient), and *width* (as measured at the half-height, in cm^{-1}). However, most absorption bands of organic compounds are not symmetric but are skewed toward the high-energy side. It is best to fit such bands with a skewed function such as the *log normal* distribution curve. In addition to position, height, and width, a fourth parameter provides a measure of *skewness*. Computer-assisted fitting with log normal curves gives precise values for the positions, widths, and intensities. In general, the peak position is somewhat to the high-energy side of the 0 – 0 transition.

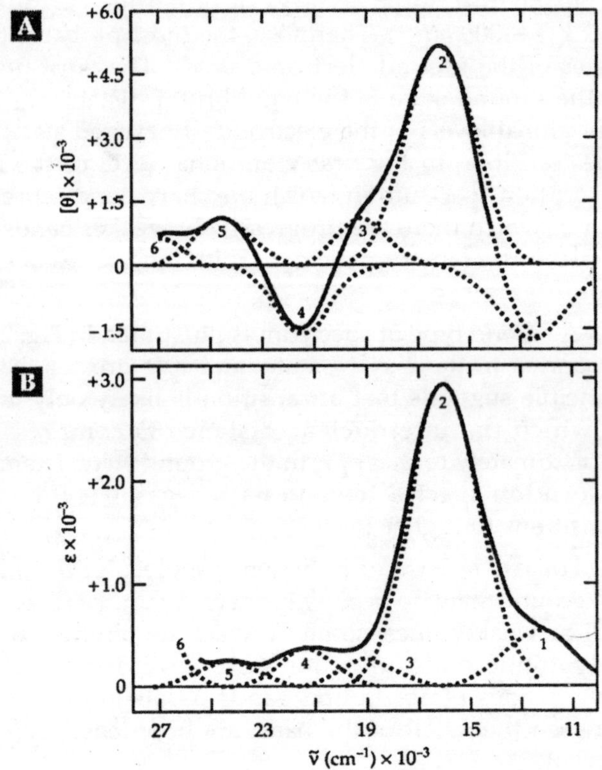

Fig. 2.7. Resolution of the visible circular dichroism (ellipticity) spectrum (A) and absorption spectrum (B) of the *Pseudomorias* blue protein into series of overlapping Gaussian bands (—). The numbers 1 to 6 refer to bands of identical position and width in both spectra. Absorption envelopes resulting from the sum of the set of overlapping Gaussian bands (—) correspond within the error of the measurement to the experimental spectra. The *dashed part* of the CD envelope above 700 nm was completed by a curve fitter with the use of a band in the position of *band 1* of the absorption spectrum.

Absorption spectra plotted as a linear function of wavelength are sometimes fitted with Gaussian curves. However, Gaussian curves only occasionally give a good fit for such spectra, and it is undesirable to measure bandwidths in nanometers. It is wave number that is proportional to energy. Spectral bands tend to have similar widths across the visible-ultraviolet range when plotted against wave number but not when plotted against wavelength. Another approach to the quantitative description of spectra is to fit the major progressions of vibrational subbands with series of narrow Gaussian curves.

Classification of Electronic Transitions

The intense 600-nm absorption band of the copper blue protein in Fig. 2.8 is attributed to a $d - d$ transition of an electron in the metal ion from one d orbital to another of higher energy. The intensity is thought to arise from transfer of an electron from a cysteine thiolate to the copper. The electronic transitions in most organic molecules are of a different type. Transitions lying at energies <55,000 cm^{-1} are classified as either $\pi - \pi^*$ or $n - \pi^*$. In the $n - \pi^*$ transitions an electron is moved from a bonding π molecular orbital to an antibonding (π^*) orbital. Such a transition is present in ethylene at 61,540 cm^{-1} (162.5 nm) with a maximum molar extinction coefficient ε_{max} of ~15,000 M^{-1} cm^{-1}. An $n - \pi^*$ transition results from the raising of an electron in an unshared pair of an oxygen or nitrogen atom into a π^* antibonding orbital. These transitions are invariably weak. For example, acetone in H_2O shows an $n - \pi^*$ transition at 37,740 cm^{-1} (265 nm). The value of ε_{max} is ~240 and the width is about 6400 cm^{-1}. A characteristic of $n - \pi^*$ transitions is a strong shift to lower energies as the compound is moved from water into less polar solvents. For example, the peak of the acetone band lies at 36,920 cm^{-1} in methanol and at 35,970 cm^{-1} (278 nm) in hexane. Such a solvent shift is often taken as diagnostic of an $n - \pi^*$ transition, and it is often stated that the $\pi - \pi^*$ bands shift in the opposite direction upon change of solvent character. However, the latter is not true for many of the polar chromophores found in biochemical substances. Thus, the $\pi - \pi^*$ bands of tyrosine also shift to lower energies when the molecule is moved from water into hexane. However, the magnitude of the shift is much less than for the $\pi - \pi^*$ band of acetone.

A molecule can have several different excited states of increasing energies. In benzene and its derivatives there are three easily detectable $\pi - \pi^*$ transitions. The first is a weak band centered at ~260 nm in toluene and ~275 nm in tyrosine with $\varepsilon = 10^2$ to 10^3. The second is a band at a higher frequency (at 1.35 0.10 times the frequency of the first band) with ε_{max} often as high as 10^4. The third band is found at still higher energies with ε_{max} reaching 5×10^4. The excited-state energy levels represented by these transitions were labeled ^1L_b, ^1L_a, and ^lB_a by Platt. Other authors described the levels in terms of the symmetries of the molecular orbitals, the ground state being $^1A_{1g}$, and the three excited states $^1B_{2u}$, $1B_{1u}$, and $^1E_{1u}$. In these symbols the superscript 1 indicates that the excited states are *singlet* in nature; that is, the electrons remain paired in the excited states. Absorption of visible and ultraviolet light almost always leads to singlet excited states initially. For more complex ring systems the number of possible electronic transitions increases, but attempts are often made to relate these transitions back to those of benzene.

The intensities of electronic transitions vary greatly. The area ($\mathcal{A}$) under the absorption band, when ε is plotted against wave number, is directly proportional to a dimensionless quantity called the *oscillator strength f*.

$$f = \frac{2.303\ m_e c^2}{\pi\ Ne^2} F\mathcal{A} = 4.32 \times 10^{-9}\ F\mathcal{A} \quad ...(2.6)$$

In this equation m_e and e are the mass and charge of the electron, c is the velocity of light, N is Avogadro's number, and $\mathcal{A}$ is the area in a plot of ε vs ν in cm^{-1}; F is a dimensionless correction factor that is related to the refractive index of the medium and is near unity for aqueous solutions. If the area is approximated as that of a triangle of height ε_{max} and width (at half-height) W, we find that for a typical absorption band of $\varepsilon_{max} = 10^4$ and $W = 3000$ cm^{-1}, $f = 0.13$.

The oscillator strength is related to the probability of a transition and can become approximately 1 only for the strongest electronic transitions. However, it is rarely this high. For example, the oscillator strength is ~10^{-4} for Cu^{2+} and ~2×10^{-3} for the toluene absorption band shown in Fig. 2.7. The low intensity of absorption bands of benzene derivatives is related to the fact that these transitions are quantum mechanically forbidden for a completely symmetric molecule. It is only because of coupling with asymmetric vibrations of the ring that the ^1L_b transition of benzene becomes weakly allowed. In the benzene spectrum the 0 – 0 transition is completely absent, and only those progressions involving uptake of an additional 520 cm^{-1} of a nonsymmetric vibrational energy are observed. In the case of toluene and phenylalanine, the asymmetry of the ring introduced by the substituents permits the 0 – 0 transition to occur and leads to a higher oscillator strength than that observed with benzene. The ^1L_a transition of benzenoid derivatives is also partially forbidden by selection rules, and only the third band begins to approach an oscillator strength of one.

Use of Plane Polarized Light

The intensity of a spectral transition is directly related to the *transition dipole moment* (or simply the transition moment), a vector quantity that depends upon the dipole moments of the ground and excited states. For aromatic ring systems, the transition dipole moments of the $\pi - \pi^*$ transitions lie in the plane of the ring. However, both the directions and intensities for different $\pi - \pi^*$ transitions within a molecule vary.

The transition moment has a dimension of length (usually given in angstroms) and can be thought of as a measure of the extent of the charge migration during the transition. Light is absorbed best when the directions of polarization (*i.e.,* of the electric vector of the light) and of the transition moment coincide. This fact can easily be verified by light absorption measurements on crystals. As with infrared spectra of oriented peptide chains the electronic spectra of crystals display a distinct dichroism. Crystals of coenzyme-containing proteins are very appropriate for spectroscopy with polarized light because the chromophores are spaced far enough apart to avoid electronic interaction and have absorbances low enough to record with crystals of the order of 0.1 mm thickness.

In contrast to $\pi - \pi^*$ transitions, the $n - \pi^*$ transitions of heterocyclic compounds and carbonyl-containing rings are often polarized in a direction perpendicular to the plane of the ring. Linear dichroism of cytosine, adenine, and other nucleic acid bases has been measured on single crystals and in partially oriented polymer films. Magnetically induced linear dichroism provides 'a new tool for study of metalloproteins.

Relationship of Absorption Positions and Intensity to Structures

While quantum mechanical calculations permit prediction of the correct number and approximate positions of absorption bands, they are imprecise. For this reason, electronic spectroscopy also relies upon a combination of empirical rules and atlases of spectra that can be used for comparison purposes. The following may help to orient the student. The position of an absorption band shifts *bathochromically* (to longer wavelength, lower energy) when the number of conjugated double bonds increases. Thus, *butadiene* absorbs at 46,100 cm^{-1} (217 nm) vs the 61,500 cm^{-1} of ethylene. As the number of double bonds increases further, the bathochromic shifts become progressively smaller (but remain more nearly constant in terms of wavelength than wave number). For *lycopene* with 11 conjugated double bonds the absorption bands is located at 21,300 cm^{-1} and displays distinct vibrational structure. Certain ring molecules such as the porphyrins and chlorophylls have spectra that can be related back to those of the linear polyenes. Note that the porphyrin α and β bands represent vibrational structure of a single electronic transition, whereas the intense Soret band results from a different transition.

Substituted benzenes almost invariably absorb at lower energies than the parent hydrocarbon. The stronger the electron withdrawing or donating ability of the substituent, the larger the bathochromic shift. The magnitude of the shift has been correlated with the Hammett 6 constants. Thus, the first absorption band of tyrosine in water is shifted 2600 cm^{-1} toward the

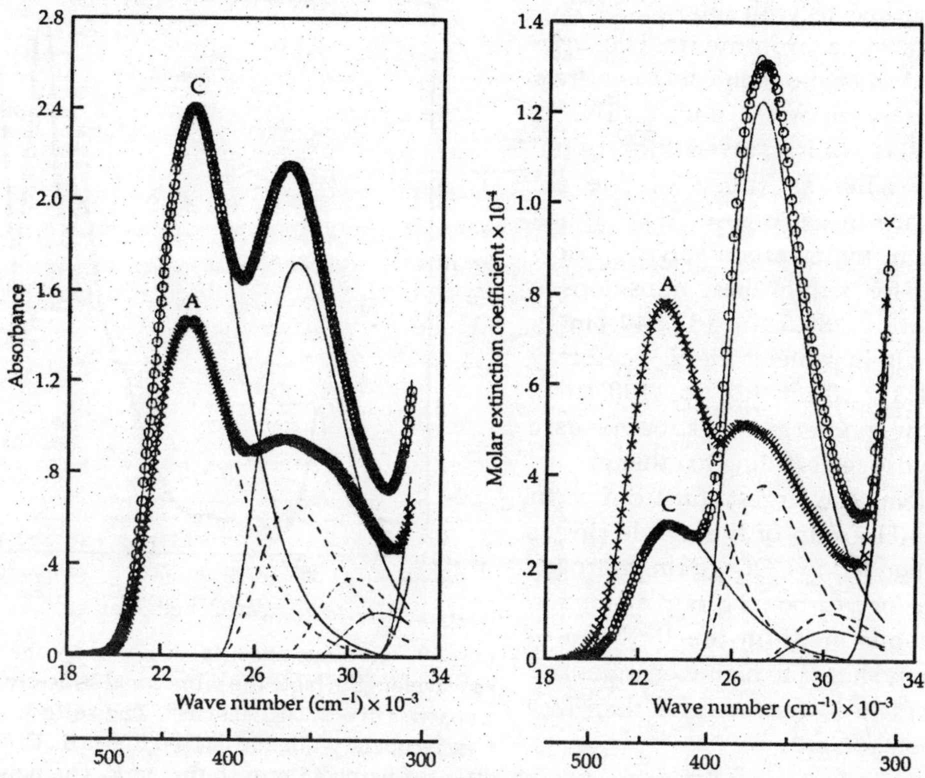

Fig. 2.8. Polarized absorption-spectra of orthorhombic crystals of cytosolic aspartate aminotransferase. The light beam passed through the crystals along the *b* axis with the plane of polarization parallel to the *a* axis (A) or the *c* axis (C). Left, native enzyme at pH 5.4; right, enzyme soaked with 300 mM 2-methylaspartate at pH 5.9. The band at ~ 430 nm represents the low pH protonated Schiff base form of the enzyme. Upon soaking with 2-methylaspartate the coenzyme rotates ~30° to form a Schiff base with this quasisubstrate. The result is a large change in the *c/a* polarization ratio. The ~ 364 nm band in the complex represents the free enzyme active site in the second subunit of the dimeric enzyme.

red from that of benzene, while that of the dissociated tyrosine anion is shifted 4700 cm^{-1}, very roughly in proportion to the 6_p values of Box 6-C. Especially large shifts are observed when functional groups of opposite types (that is, an electron donating group vs an electron accepting group) are both present in the same ring. The effects of ortho and meta substituent pairs are closely similar (in contrast to the differing electronic effects of ortho and meta pairs in chemical reactivity). Substituent pairs in para positions yield somewhat different spectral shifts. When there are more than two substituents, the two strongest groups often dominate in determining the character of the spectrum. Useful empirical rules have been developed.

Spectra of Proteins and Nucleic Acids

Most proteins have a strong light absorption band at 280 nm (35,700 cm^{-1}) which arises from the aromatic amino acid tryptophan, tyrosine, and phenylalanine. The spectrum of phenylalanine resembles that of toluene whose 0 – 0 band comes at 37.32 × 10^3 cm^{-1}. The vibrational structure of phenylalanine can be seen readily in the spectra of many proteins. The spectrum of tyrosine is also similar but the 0 – 0 peak is shifted to a lower energy of ~35,500 cm^{-1} (in water). Progressions with spacings of 1200 and 800 cm^{-1} are prominent. The low-energy band of tryptophan consists of two overlapping transitions ^1L_a and ^1L_b. The ^1L_b transition has well-resolved vibrational subbands, whereas those of the ^1L_a transition are more diffuse. Tryptophan derivatives in hydrocarbon solvents show 0 – 0 bands for both of these transitions at approximately 289.5 nm (34,540 cm^{-1}). However, within proteins the ^1L_a band may be shifted 3 – 10 nm (up to 1100 cm^{-1}) toward lower energies, probably as a result of hydrogen bonding to other groups in the protein. The largest shifts can occur when the NH group of the indole ring is hydrogen bonded to COO^-, a ring nitrogen of histidine, or a carbonyl group of amides. In an aqueous medium the ^1L_a band of tryptophan is shifted to higher energies and the ^1L_a band to lower energies than in a hydrocarbon solvent.

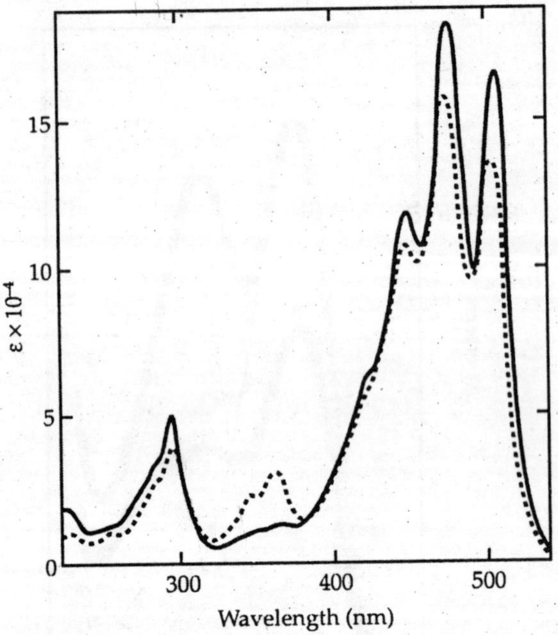

Fig. 2.9. The absorption spectrum of lycopene (plotted vs wavelength). Note the vibrational structure, which has a spacing of ~ 1200-1500 cm^{-1}. The solid line is for all translycopene while the dashed line is that of the sample after refluxing 45 min in the dark. The new peak at ~ 360 nm arises from isomers containing some cis double bonds.

In addition to the three aromatic amino acids, disulfide bonds absorb in the near ultraviolet region as indicated. Since the absorption characteristics depend upon the dihedral angles in the disulfide bridges, it is difficult to accurately evaluate the contribution of this chromophore to the 280-nm band.

Tyrosine, tryptophan, and phenylalanine all have additional transitions in the high-energy UV region of the spectrum. Even more intense are the absorption bands of the amide groups, which become significant above 45,000 cm^{-1}. These include a weak $n - \pi^*$ transition at ~45,500 cm^{-1} (210 nm) overlapped by a strong $\pi - \pi^*$ transition at ~52,000 cm^{-1} (192 nm). Histidine also has absorption bands in this region.

As with polypeptides, the light absorption properties of polynucleotides reflect those of the individual components. The spectra of the purine and pyrimidine bases as ribonucleosides are shown in diagram. The number of individual electronic transition their origins are not immediately obvious, but many measurements in solutions and in crystals, as well as theoretical computations have been made. Cytosine has $\pi - \pi^*$ transitions at ~275, 230, 200, and 185 nm, the two highest energy bands being overlapped. Adenine derivatives have seven $\pi - \pi^*$ transitions. Spectra of flavins contain at least four intense transitions.

Whereas proteins have their low energy absorption band at ~280 nm, polynucleotides typically have maxima at ~260 nm (38,500 cm^{-1}). A phenomenon of particular importance in the study of nucleic acids is the *hypochromic effect*. In a denatured polynucleotide the absorption is approximately the sum of that of the individual components. However, when a double helical structure is formed and the bases are stacked together, there is as much as a 34% depression in the absorbance at 260 nm. This provides the basis for optical measurement of DNA melting curves. The physical basis for the hypochromic effect is found in dipole-dipole interactions between the closely stacked base pairs.

Difference Spectra and Derivative Spectra

Changes in light-absorbing properties of proteins and nucleic acids are often measured as a function of some quantity such as pH, temperature, ionic environment, or the presence or absence of another interacting molecule. The induced changes in the spectrum are small but can be seen if the *difference* between the two spectra, one {*"unperturbed"* and the other in the presence of some "perturbant," is recorded. The perturbant might be an additional reagent, an altered solvent (*e.g.*, with added glycerol, D_2O), a change in pH, or temperature. The difference spectrum shown in Fig. 2.11B arises from the binding of an inhibitor succinate together with a substrate carbamoyl phosphate to the catalytic subunit of aspartate carbamoyl transferase. The difference spectrum appears as a pair of peaks and a valley in the aromatic amino acid region. With proper interpretation (caution!) difference spectra can be used to infer something about the change in environment of aromatic amino acids in a protein.

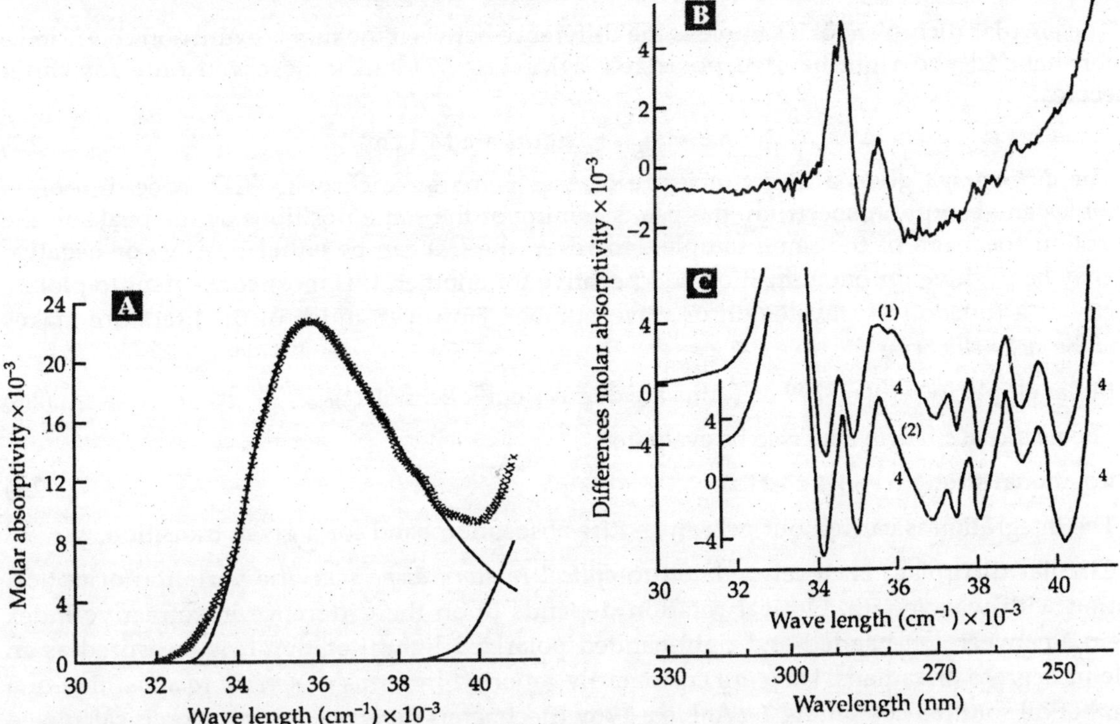

Fig. 2.10. The difference spectrum for the catalytic subunit of asparta carbamoyltransferase in the presence and absence of succinate and carbamoyl phosphate. (A) The spectrum of the unperturbed enzyme (points) fitted with two log normal curves (solid line). (B) The difference between the spectrum of enzyme plus 0.09 M succinate and 4.3 mM carbamoyl phosphate and that shown in (A) (compare with published difference spectrum for intact aspartate carbamoyltransferase). (C) Curve 1, "fine structure plot" obtained by subtracting the spectrum in A from the smooth curve obtained by summing the two log normal curves. Curve 2, a similar plot for enzyme plus succinate and carbamoyl phosphate.

Difference spectra are usually recorded by placing the unperturbed spectrum in the *reference* light beam of a spectrophotometer and the perturbed solution in the *sample* beam in carefully matched cuvettes. However, the spectrum shown in Fig. 2.11B was obtained by recording the two spectra independently and subtracting them with the aid of a computer. The same data have been treated in another way by fitting two log normal curves to the absorption bands and plotting the differences between the mathematically smooth fitted curve and the experimental points taken at close intervals as shown in Fig. 2.11C. The two *"fine structure plots"* obtained in this way are an alternative way of representing the same data that gave rise to the difference spectrum. The method has the advantage that information about the overall band shape is obtained from the computer-assisted curve fitting process. Thus, the binding of succinate and carbamoyl phosphate caused an almost insignificant shift (of 20 cm^{-1}) in the overall band position and a very slight broadening. The principal effect is an enhancement in the vibrational structure at 34,600 cm^{-1} in the 0 – 0 band of the two tryptophan residues present in the subunit. The cause of this change is not entirely obvious, a weakness of difference spectroscopy. Another way of examining a spectrum such as that in Fig. 2.11A is to plot a mathematical derivatives of the curve. Both second and fourth derivatives yield curves similar to the difference plots of Fig. 2.11C.

5. Circular Dichroism and Optical Rotatory Dispersion

The circular dichroism of a sample is the difference between the molar extinction coefficients for left-handed and right-handed polarized light (Eq. 2.7) and is observed only for chiral molecules.

$$\Delta\varepsilon = \varepsilon_L - \varepsilon_R \text{ (units are } M^{-1}\ cm^{-1}) \quad ...(2.7)$$

The *dichrograph* gives a direct measure of $\Delta\varepsilon$. A circular dichroism (CD) spectrum often resembles an absorption spectrum, the peaks coming at the same positions as the peaks in the absorption spectrum of the same sample. However, the CD can be either positive or negative and may be positive for one transition and negative for another. It is most convenient to plot $\Delta\varepsilon$ directly as a function of wavelength or wave number. However, much of the literature makes use of the *molar ellipticity*

$$\text{Molar ellipticity} = [\theta] = 3299\ \Delta\varepsilon \text{ (units are degrees } cm^2 \text{ decimole}^{-1}) \quad ...(2.8)$$

The *rotational strength* may also be evaluated.

$$\text{Rotational strength} = \int [(\Delta\epsilon)/\lambda]\ d\lambda \quad ...(2.9)$$

The integration is carried out over the entire absorption band for a given transition.

Circular dichroism is closely related to *optical rotatory dispersion,* the variation of optical rotation with wavelength. Optical rotation depends upon the difference in refractive index ($\eta_L - \eta_R$) between left-handed and right-handed polarized light. Rotation α is measured as an angle in degrees or radians. Data are customarily reported in terms of *specific rotation,* that of a hypothetical solution containing 1 g /ml in a 1 dm (decimeter) tube. Specific rotation is calculated (Eq. 2.10) from the observed rotation, the concentration c' in g ml^{-1}, and the length of the tube l' in decimeters.

$$\text{Specific rotation} = [\alpha] = \alpha_{obs}/\ c'\ l' \quad ...(2.10)$$

The *molecular rotation* is defined by Eq. 2.11 in which M_r is the molecular mass and c and l are in moles per liter and cm, respectively.

Molecular rotation = $[\phi] = 100\ \alpha_{obc}/cl = [\alpha]\ M_r/100$ ...(2.11)

It is often multiplied by a factor of $3/(\eta^2 + 2)$ to correct for a minor effect of the polarizability of the field acting on the molecules. The rotation in the radians per centimeter of light path can be related (Eq. 2.12) directly to the wavelength of the light and the refractive indices η_L and η_R.

α (radians/cm) = $[\alpha]\ c'/1800 = \pi/\lambda\ [\eta_L - \eta_R]$ ...(2.12)

In contrast to circular dichroism, optical rotary dispersion (ORD) extends far from absorption bands into spectral regions in which the compound is transparent. As an absorption band is approached, the optical rotation increases in either the positive or negative direction. Then, within the absorption band it drops abruptly through zero and assumes the opposite sign on the other side of the band (the Cotton effect). Although the occurrence of optical rotation in nonabsorbing regions of the spectrum provides an advantage to ORD measurements, the interpretation of ORD spectra is more complex than that of CD spectra. In principle, the two can be related mathematically and both are able to give the same kind of chemical information.

The CD in the *d* – *d* bands of the blue copper protein (Fig. 2.8) arises in part from the fact that within the protein the copper ion is in an asymmetric environment. For a similar reason, the aromatic amino acids of proteins often give rise to circular dichroism. In the case of tyrosine, the sign of the CD bands can be either positive or negative but is the same throughout a given transition. The CD bands are similar in shape to the absorption bands. The behavior of phenylalanine is more complex. The progression of vibrational subbands at 930 cm^{-1} intervals above the 0 – 0 band all have the same sign, and the intensities relative to that of the 0 – 0 band are similar to those in absorption. However, the vibrations of wave numbers equal to that of the 0–0 transition plus 180 and 520 cm^{-1} sometimes give rise to CD bands of the opposite sign, and the relative intensity relationships are variable.

The binding of a symmetric chromophore to a protein or nucleic acid often induces CD in that chromophore. For example, the bands of enzyme-bound and pyridoxamine phosphates positively dichroic in CD, but the band of the quinonoid intermediate at 20,400 cm^{-1} (490 nm) displays negative CD. When *"transimination"* occurs to form a substrate Schiff base, the CD is greatly diminished. While the coenzyme ring is known to change its orientation, it is not obvious how the change in environment is related to the change in CD.

A series of octant rules make it possible to predict the sign and magnitude of CD to be expected for $n - \pi^*$ transitions of simple carbonyl compounds. Theoretical approaches to the CD and ultraviolet absorption of proteins in the high-energy ultraviolet region have also been developed. In a regular β structure, in an α helix or in a crystalline array, the transitions of adjacent amide groups may be *coupled,* the excitation energy being delocalized. This *exciton* delocalization leads to a splitting (Davydov splitting) into two transitions of somewhat different energies and polarized in different directions. The amide absorption band at 52,600 cm^{-1} is split in an α helix into components at ~48,500 and 52,600 cm^{-1}. Furthermore, low-energy $\pi - \pi^*$ and $n - \pi^*$ states are close together in energy, a fact that allows mixing of the two states and appearance of rotational strength in the $\pi - \pi^*$ band with a sign opposite to that in the $n - \pi^*$ band.

Both the sign and intensity of the CD bands of peptides also depend upon conformation. Well-defined differences are observed among α helices, β structure, and random-coil conformations. Measurements may be extended into the "vacuum ultraviolet" region—up to 60,000 cm^{-1} in aqueous solutions. A useful empirical approach is to deduce spectra of helices, β

structures, and unordered peptide chains from measured spectra together with an examination of actual structures obtained by X-ray crystallography. Note that the CD curve for the α helix has a deep minimum at 222 nm, whereas the β form has a shallower minimum. The random structure has almost no CD at the same wavelength. The approximate helix content of a protein is often estimated from the depth of the trough at 222 nm in the CD spectrum. Better predictions can be made by using a computer-assisted comparison of an experimental CD spectrum with those of a series of proteins of known 3-D structure.

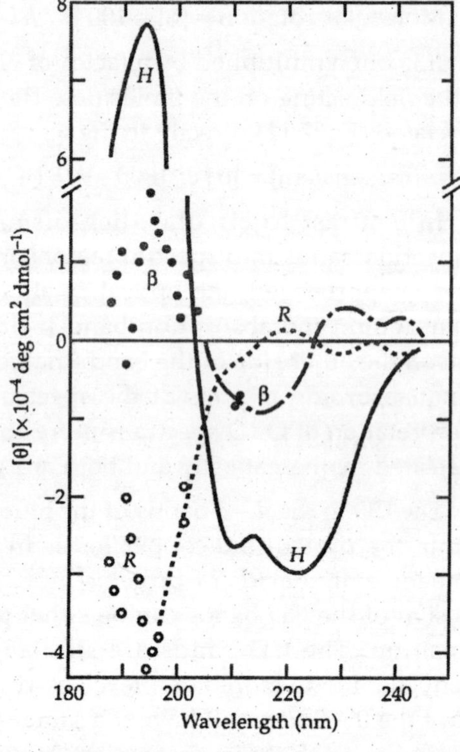

Fig. 2.11. Circular dichroism of the helix (H), β, and unordered (R) form computed from the CD of five proteins. Points are plotted when a smooth curve could not be drawn.

The circular dichroism of polynucleotides at 275 nm is a linear function of both the helix winding angle and the base pair twist. Measurement of CD spectra on large polynucleotides or large molecular aggregates is complicated by differential light scattering of right and left circularly polarized light. However, the phenomenon can also provide new structural information. Progress has been made in attempts to predict the optical rotation of molecules from quantitative values for the polarizabilities of individual atoms.

Vibrational circular dichroism involves IR absorption bands. The technique has been applied to sugars, oligosaccharides, proteins, and nucleic acids. The related *vibrational Raman optical activity* has also been applied to polyribonucleotides.

6. Photoacoustic Spectroscopy

Photoacoustic spectra are recordings of the energy emitted as heat after absorption of monochromatic light. The sample is placed in a closed photoacoustic cell. The light beam, which is chopped at an audiofrequency, induces a periodic heating and cooling of the gas in contact with the sample in the cell. This is sensed as sound by a sensitive microphone. The resulting electrical signals are sent to a computer for analysis. The output is an absorption spectrum resembling that measured optically. The samples do not have to be transparent. A related technique is photoacoustic calorimetry.

7. X-Ray Absorption and Mössbauer Spectroscopies

The importance of X-ray diffraction to biochemistry is obvious, but techniques related to absorption of X-rays and γ-rays have also come into widespread use. Abbreviations such as *XANES* and *EXAFS* are common in the metalloprotein literature. The names arise from the sharp increase in the absorption coefficient for X-rays as their energy is increased to what is called the K *absorption edge.* At slightly lower energies absorption of an X-ray by an atom leads to expulsion of an electron or the raising of an electron to an excited state. Absorption of X-rays will expel all except the inner 1s electrons. As the energy is increased further, the stepwise increase in absorption that constitutes the edge is observed. At higher energies the absorption decreases. However, with a high-resolution instrument distinct oscillations are observed on the high-energy side of the edge, extending for ~20 eV. This is X-*ray absorption near-edge structure* (XANES). When an X-ray absorbing atom in a molecule is surrounded by other atoms, a fine structure that depends upon the nature of these atoms and their distances from the absorbing atom is observed over a range of several hundred electron volts above the edge. This *is extended X-ray absorption fine structure* (EXAFS).

The EXAFS technique has been especially useful for metalloproteins. It has often provided the first clues as to the identity of atoms (O, N, S) surrounding a metal atom and either covalently bonded to it or coordinated with it. Interpretations are often difficult, and a common approach is to try to simulate the observed spectrum by calculation from a proposed structure.[118] Tautomerism in crystalline Schiff bases has been studied by near-edge X-ray absorption fine structure (NEXAFS) employing soft X-rays.

Mössbauer spectroscopy, also called recoil-free nuclear resonance absorption, depends upon resonant absorption of γ-rays emitted by a radioactive source by atomic nuclei. The phenomenon was initially difficult to observe, but the German physicist Mössbauer devised a way in which to record the absorption of a quantum of energy equal to the difference in two energy states of the atomic nucleus. The method depends upon a Doppler effect observed when the sample or source moves. Consequently, Mossbauer spectra, are plots of absorption versus velocity. Mössbauer spectroscopy has been applied to numerous metalloproteins, especially those containing iron centers. It is a major tool in investigation of Fe-S proteins. Since Fe is "silent" in Mössbauer spectroscopy, proteins are often enriched with Fe for observation.

C. FLUORESCENCE AND PHOSPHORESCENCE

An electronically excited molecule is able to lose its excitation energy and return to the ground state in several ways. One of these is to reemit a quantum of light as fluorescence. The intensity and spectral properties of fluorescent emission can be measured by illuminating a sample in a cuvette with four clear faces with the measuring photomultiplier set at right angles to the exciting light beam. In absorption spectrophotometry we measure a difference between the light intensity of the beam entering the sample and that emerging from the sample. In fluorescence spectroscopy we measure the absolute intensity of the light emitted. Although this intensity is small, the measurement can be made extremely sensitive, far more so than can light absorption. For this reason, fluorescence is widely used for detection and analysis, *e.g.,* in DNA sequencers. Enzyme kinetics can be studied with fluorescent substrates at very low concentrations. Fluorescent antibodies, DNA chips, and numerous bioassay and imaging methods are dependent upon measurement of fluorescence. Fluorescence can also yield a wealth of information about the chemical and physical properties of electronically excited states of molecules.

1. Excitation and Emission Spectra

Measurements of the intensity of fluorescence at any wavelength vs the wavelength of monochromatic light used to excite the fluorescence give a fluorescence *excitation spectrum.* The excitation spectrum is an example of an *action spectrum,* which is a measure of any response to absorbed light. At very low concentrations of pure substances, action spectra tend to be identical to absorption spectra. However, since the observed response (fluorescence in this case) is proportional to light absorbed, action spectra should be compared to plots of 1 – T (where T = transmittance, vs wavelength rather than to plots of ε vs λ. The two plots are proportional at low concentrations.

A fluorescence *emission spectrum* is a record of fluorescence intensity vs wavelength for a constant intensity of exciting light. Excitation and emission spectra for a flavin and for the indole ring of tryptophan are both given in Fig. 2.13. The heights of the emission spectra have been

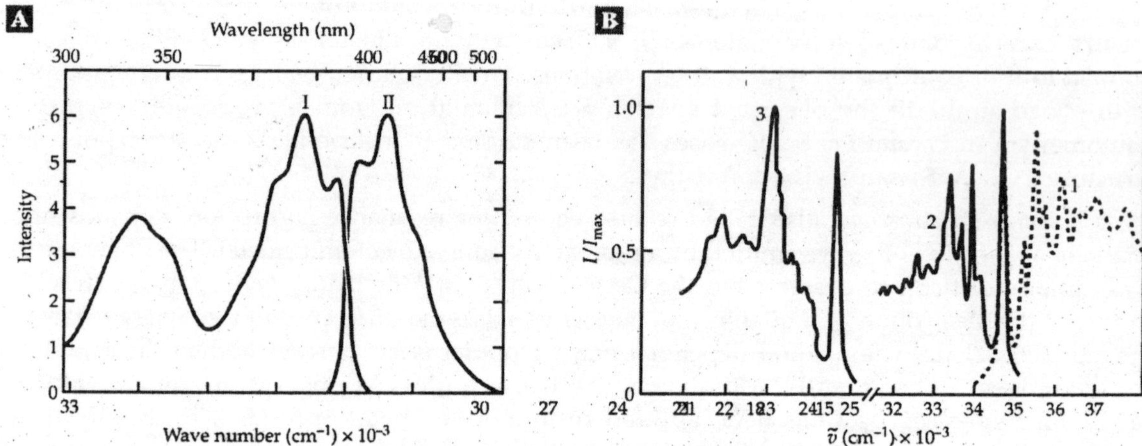

Fig. 2.12. (A) Corrected emission and excitation spectra of riboflavin tetrabutyrate in *n*-heptane. Concentration, about 0.4 mg l^{-1}. Curve 1: excitation spectrum; emission at 525 nm. Curve 2: emission spectrum; excitation at 345 nm. From Kotaki and Yagi. (B) Indole in cyclohexane, T = 196 K. 1, Fluorescence excitation spectrum; 2, fluorescence spectrum; and 3, phosphorescence spectrum.

adjusted to the same scale as that of absorption. The fluorescent emission is always at a lower energy than that of the absorbed light. The excitation and emission spectra overlap only slightly, and the emission spectrum is an approximate mirror image of excitation spectrum. Absorption usually leads to a higher vibrational energy state after light absorption than before. However, most of the excess vibrational energy is dissipated before much fluorescent emission occurs. The excited molecule finds itself in the lowest vibrational state of the upper electronic state, and it is from this state that the bulk of the fluorescent emission takes place. Furthermore, whereas absorption usually occurs from the lowest vibrational state of the ground electronic level, fluorescence can populate many excited vibrational states of the ground electronic state. Consequently, as indicated in the figure, the fluorescent emission spectrum consists of a series of subbands at lower energies than those observed in the absorption spectrum. The two spectra have only the 0 adjusted to the same scale as that 0 – 0 transition in common. As can be seen from Fig. 2.14 even the two 0 – 0 transitions do not coincide exactly. The peak of emission is shifted toward slightly lower energies than that of absorption because during or immediately following absorption of a photon there is some rearrangement of solvent molecules around the absorbing molecule to an energetically more stable arrangement. Just as excess vibrational energy is dissipated in the excited state, so relaxation of these solvent molecules around the excited chromophore leads to a small shift in energy. A similar relaxation occurs in the ground state of a molecule that has just emitted a photon as fluorescence. This also contributes to the shift in position of the 0 – 0 band in fluorescence.

Several molecular properties can be measured using emission and excitation spectra. These include fluorescence lifetime, efficiency, anisotropy of the emitted light, mobility of chromophores, rates of quenching, and energy transfer to other chromophores.

Fluorescence Lifetimes

Why are some molecules fluorescent, while others are not? The possibility for fluorescent emission is limited by the radiative lifetime τ_r, which is related by Eq. 2.13 to the first-order rate constant k_f for exponential decay of the excited state by fluorescence.

$$\tau_r = 1/k_f \qquad ...(2.13)$$

The radiative lifetime is a function of the wavelength of the light and of the oscillator strength of the transition. For molecules absorbing in the near UV, the approximation of Eq. 2.14 is often made.

$$1/\tau_r \sim 10^4\varepsilon_{max} \qquad ...(2.14)$$

Thus, if $\varepsilon = 10,000$, the radiative lifetime (the time in which the fluorescence decays to $1/e$, its initial value) is $\sim10^{-8}$ s (10 ns). If the absorption is more intense, the lifetime is shorter, and if it is less intense, it is longer. Other modes of deexcitation compete with fluorescence; therefore, the shorter the radiative lifetime the more likely that fluorescence will be observed.

The actual lifetime τ of an excited molecule is usually less than τ_r because of the competing nonradiative processes. The sum of their rate constants can be designated k_{nr}. The *fluorescence efficiency* (or *quantum yield*) ϕ_F is given by Eq. 2.15.

$$\phi_F = k_f/k_f + k_{nr}) = k_r\tau$$

where

$$\tau^{-1} = k_f + k_{nr} \qquad ...(2.15)$$

For a highly fluorescent molecules such as riboflavin, ϕ_f may be 0.25 or more. For tryptophan in water it is about 0.14, and in proteins it varies from near zero to 0.35.

Time-resolved Fluorescence Spectroscopy. The fluorescence lifetime τ can be measured with either of two different types of fluorometer. *Pulse fluorometers* use pulsed lasers that can deliver pulses of light lasting as little as one picosecond or less. This permits rapid excitation and permits the direct observation of emitted light, using photon countings, over the entire range of time from a few picoseconds to milliseconds required for decay of the fluorescence. The observed value of τ for riboflavin 5'-phosphate (ε_{max} = 12,200 at 450 nm) at 25°C is ~5 ns. That for tryptophan is 3 ns.

Phase fluorometers utilize continuous irradiation by a beam of light that is sinusoidally modulated. If the frequency of the modulation is set correctly, there will be a phase difference in the modulation of the fluorescent emission that will depend upon τ. Phase fluorometry can yield the same information as does pulse fluorometry. By using two or more modulation frequencies the decay rates and fluorescence lifetimes for various chromophores in a protein can be observed. For example, the protein *colicin A* contains three tryptophans W86, W130, and W140. Their fluorescence decays with lifetimes τ_1, τ_2, τ_3 of ~0.6 – 0.9 ns, 2.0 – 2.2 ns, and 4.2 – 4.9 ns at pH 7. While τ_3 originates mainly from W140, both of the other tryptophans contribute to τ_1 and τ_2. Changes in fluorescence intensity with pH reflect a pK_a value of 5.2. Tryptophan, which often occurs at only one or a few places in a protein, is a useful fluorescent probe for study of protein dynamics. The optical properties of 7-azatryptophan, 2-azotryptophan, and 5-hydroxytryptophan are even better because their absorption maxima occur at longer wavelengths. These amino acids can be biosynthetically introduced in place of tryptophan in proteins. The maximum fluorescence of tryptophan in one protein is at 350 nm, but for 7-azatryptophan in the same protein it was shifted to 380 nm.

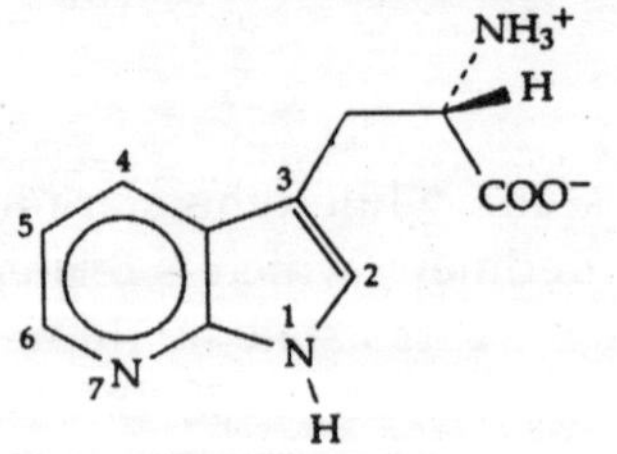

7-Azatryptophan

Triplet States, Phosphoresence, and Quenching

In addition to emitting fluorescent radiation, molecules can often pass from the excited singlet state to a lower energy *triplet state,* in which two electrons are now unpaired and the molecule assumes something of the character of a diradical. This process, known as *intersystcm crossing,* competes directly with fluorescence and shortens the fluores-cence lifetime. The triplet state is long-lived (*e.g.*, for tryptophan in water at 20°C it is 1.2 ms) and is responsible for much of the photochemical behavior of molecules. It also gives rise to the delayed light emission known as *phosphorescence.* Other processes that compete with fluorescence are *photochemical reactions* of the singlet excited state and *internal conversion.* The latter is the process by which a molecule moves from the lowest vibrational state of the upper electronic level to some high vibrational state of the unexcited electronic level. This is the principal means of depopulating the electronic state and competes directly with fluorescence.

The rate of relaxation by nonradiative pathways can be increased by addition of *quenchers.* Quenching of fluorescence occurs by several mechanisms, many of which involve collision of the excited chromophore with the quenching molecule. Some substances such as iodide ion are especially effective quenchers. The fluorescence efficiency of a substance in the absence of a quencher can be expressed in terms of the rate constants for fluorescence (k_f), for nonradiative decay (k_{nr}), and for phosphorescence (k_p) :

$$\phi_f = k_f/(k_f + k_{nr} + k_p) \qquad ...(2.16)$$

In the presence of a quencher, Q, there is an additional rate process for relaxation. The ratio of the fluorescence efficiency in the absence of (ϕ_f^0) and the presence of a quencher is given by the *Stern-Volmer equation.*

$$\phi_f^0/\phi_f = 1 + K[Q] = 1 + k_Q\tau_0[Q]$$

The constant K is known as the Stern-Volmer quenching constant; k_Q is the rate constant for the quenching reaction, and τ_0 the lifetime in the absence of quencher. Fluorescence quenching of tryptophan in proteins by acrylamide or O_2 has been used to determine whether tryptophan side chains are accessible to solvent or are *"buried"* in the protein. The long-lived phosphorescence of tryptophan can be studied in a similar way.

A recent application of fluorescence quenching is the development of *"molecular beacons"* for detection of viruses such as the AIDS viruses HIV-1 and HIV-2. A single-stranded oligonucleotide is synthesized with a 25- or 33-nucleotide sequence complementary to a sequence in the target viral RNA. At the ends of this sequence are two 6-nucleotide arms with complementary sequences

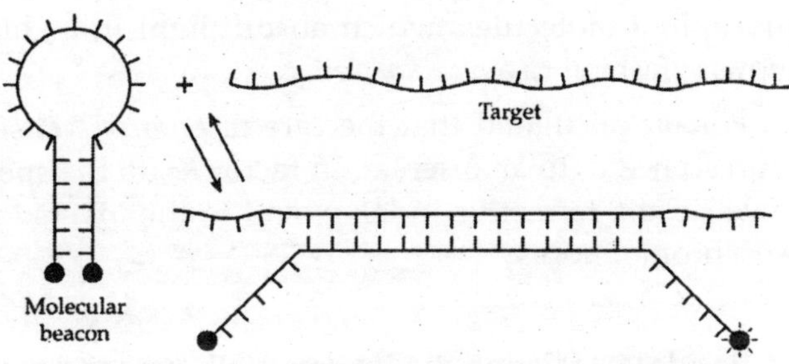

that will form a stable double-helical stem at the annealing temperature used for PCR amplification of the viral nucleic acid. The end of one arm carries a covalently bonded fluorescent dye, *e.g.*, a fluorescein or rhodamine derivative. The other arm carries a potent covalently linked fluorescence quencher such as 4-(4′-dimethylaminophenylazo) benzoic acid. When the arms form a duplex, the quencher will be next to the fluorophore and no fluorescence will be seen upon irradiation with light of a suitable exciting wavelength. However, if viral DNA is present it will hydridize with the central polynucleotide, keeping the fluorophore and quenchers far apart and allowing the beacon to signal the presence of a virus. As few as ten retroviral genomes could be detected. By using a series of molecular beacons with different colored fluorescence and specific for different viruses, it is possible to test for more than one virus simultaneously.

Anisotropy

Light emitted from excited molecules immediately after absorption is always partially polarized, whether or not the exciting beam consists of plane polarized light. When light polarized in a vertical plane is used for excitation, part of the emitted light (of intensity I_v) will have its electric vector parallel to that of the exciting light. The remainder of intensity I_h, will be polarized in a horizontal plane. The *polarization* P of the emitted radiation is defined by Eq. 2.19 and the *anisotropy* R by Eq. 2.20. After excitation by a laser pulse both the fluorescence and its anisotropy decay with time and can be measured. The decay of R (but not of P) can usually be described as the sum

$$P = (I_v - I_h)/(I_v + I_h) \quad ...(2.19)$$

$$R = (I_v - I_h)/(I_v + 2\, I_h) \quad ...(2.20)$$

of simple exponential curves, which are readily obtained by phase fluorometry. These can in turn be related to specific types of motion, such as rotation of the emitting molecule or group. Rotation of tryptophan rings, both free and restricted, has been studied in a variety of proteins. However, interpretation is difficult. The rotational rates obtained from anisotropy measurements are strongly affected by the viscosity of the medium.

2. Fluorescence Resonance Energy Transfer (FRET)

Electronic excitation of one chromophore sometimes elicits fluorescence from a different chromophore that is located nearby. For example, excitation of a monomolecular layer of dye can induce fluorescence in a layer of another dye spaced 5 nm away. Excitation of tyrosine residues in proteins can lead to fluorescence from tryptophan, and excitation of tryptophan can cause fluorescence in dyes attached to the surface of a protein or in an embedded coenzyme. Such fluorescence resonance energy transfer (*FRET*) is expected for molecules, when the fluorescence spectrum of one overlaps the absorption spectrum of the other. The mechanism is not one of fluorescence emission and absorption but of nonradiative resonant transfer of energy. Resonant transfer of energy is of major biological significance in photosynthesis. Most of the chlorophyll molecules, which absorb light in a chloroplast, transfer the absorbed energy in a stepwise fashion to a *reaction center.*

Förster calculated that the rate of energy transfer k_t, should be proportional to the rate of fluorescence k_f, to an orientation factor K^2, to the spectral overlap interval J, to the inverse fourth power of the refractive index n, and to the inverse sixth power of the distance r separating the two chromophores.

$$k_t \propto k_f K^2 J n^{-4} r^{-6} \qquad \text{...(2.21)}$$

Besides predicting the inverse sixth power dependence of energy transfer, Förster provided a formula for calculating R_0, the distance between chromophores at which 50% efficient singlet-singlet energy transfer takes place. R_0 is commonly of the order of 2.0 nm. Making use of these relationships, Stryer proposed a method of measuring distances between chromophores. He calibrated the method by constructing a series of molecules containing various lengths of the rigid threefold polyproline helix to which dansyl groups were attached at one end and naphthyl

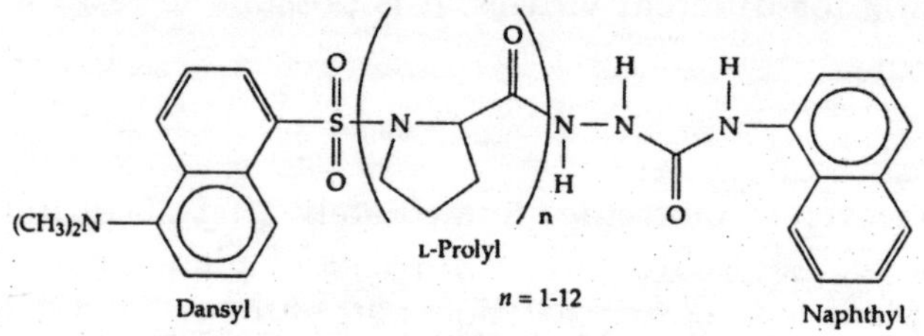

$(CH_3)_2N$ Dansyl L-Prolyl n = 1-12 Naphthyl

groups at the other. By exciting the naphthyl group, which has the higher energy absorption band and is strongly fluorescent, the characteristic lower energy emission of the dansyl group could be observed if energy transfer took place. Since the fluorescent emission band of the naphthyl group overlaps the absorption band of the dansyl group, efficient transfer was expected. The results of a plot of transfer efficiency against distance is shown in Fig. 2.15. The inverse sixth power dependence was followed quite accurately with a value of R_0 ~3.4 nm. Having calibrated his "spectroscopic ruler," Stryer turned his attention to biochemical macromolecules. Attaching the same kinds of fluorescent probe to the visual light receptor rhodopsin, Wu and Stryer were able to estimate distances between specific parts of the molecule and to draw some conclusions about the overall shape.

More recently the FRET technique has been widely applied to a broad range of biochemical problems. Sensitivity has been improved to the extent that fluorescence of single molecules can be detected. Use of *terbium* (Tb^{3+}) or *europium* (Eu^{3+}) ions, which can provide luminous labels for metal-binding sites, has provided another advance. These ions absorb light poorly and are therefore only weakly fluorescent. However, they can be excited by resonance energy transfer and become brilliantly luminous. This *luminescence resonance energy transfer* (*LRET*) is a variant of FRET, which allows distances up to ~10 nm, to be measured. Another advance is the ability to graft into specific proteins fluorescent tags such as the intact *green fluorescent protein* or an amino acid sequence such as CCXXCC in which the four –SH groups of the cysteines serve to trap an arsenic derivative of fluorescein. *Confocal laser scanning microscopy* is basic to many applications.

Specific applications of FRET and LRET include observation of myosin movement measurement of distances between binding sites on tubulin,[161] determining stoichiometry of subunit assembly in a γ-aminobutyrate receptor of brain, association of proteins in peroxisomes, study of hydridization of deoxyribonucleotides, verifying the handedness of various forms of DNA, and other studies of DNA and RNA.

Using the Förster equation the distance between the two calcium-binding sites in parvalbumin has been estimated by energy transfer from Eu(III) in one site to Tb(III) in the other to within 10 – 15% of the distance of 1.18 nm based on X-ray crystallography.

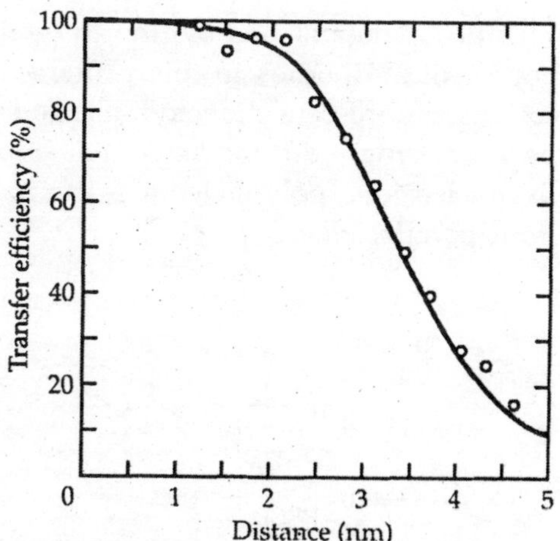

Fig. 2.13. Efficiency of energy transfer as a function of distance between α-naphthyl and dansyl groups at the ends of a polyprolyl "rod" (L-prolyl)$_n$. The observed efficiencies of transfer for $n = 1$ to 12 are shown as points. The solid line corresponds to an r^{-6} distance dependence.

3. Energy-Selective Spectroscopic Methods

At low enough temperatures vibrational fine structure of aromatic chromophores may be well resolved, especially if they are embedded in a suitable matrix such as argon or N_2, which is deposited on a transparent surface at 15 K. This *matrix isolation spectroscopy* may reveal differences in spectra of conformers or, of tautomers. In the latter example the IR spectra of the well-known amino-oxo and amino-hydroxy tautomers of cytosine can both be seen in the matrix isolation IR spectrum. Fig. 2.16 is an IR spectrum, but at low temperatures electronic absorption spectra may display sharp vibrational structure. For example, aromatic hydrocarbons dissolved in *n*-heptane or n-octane and frozen often have absorption spectra, and therefore fluorescence excitation spectra, which often consist of very narrow lines. A laser can be tuned to excite only one line in the absorption spectrum. For example, in the spectrum of the carcinogen 11-methylbenz(*a*)anthrene in frozen octane three major transitions arise because there are three different environments for the molecule. Excitation of these lines separately yields distinctly different emission spectra. Likewise, in complex mixtures of different hydrocarbons emission can be excited from each one at will and can be used for estimation of amounts. Other related methods of energy selective laser spectroscopy include *fluorescence line narrowing* and *spectral hole burning*.

4. Analytical Applications of Fluorescence

Because of the high sensitivity with which fluorescence can be detected, its measurement is important as an analytical tool. As a result of improved techniques *fluorescence microscopy* has become one of the most important of all tools in biological studies. New types of microscopes have increased resolution beyond what was thought possible. Studies such as those of lipid metabolism in the transparent zebrafish are possible using substrates that carry fluorescent labels. As mentioned in the preceding paragraph, many aromatic compounds can be detected by their fluorescence. The relatively weak fluorescence of proteins and nucleic acids can be greatly enhanced by the binding of a highly fluorescent dye to the macromolecules. Fluorescent antibodies are widely employed for this purpose. Fluorescent labels are rapidly replacing radioisotopes in analysis of nucleic acids. For example, biotin may be attached to a pyrimidine base of a nucleoside triphosphate by a long spacer arm. The modified base can then be incorporated enzymatically into polynucleotides, e.g., in the synthesis of probes used for hybridization. The attached biotin can be detected by binding to avidin or streptavidin and use of fluorescent antibodies to this protein. Fluorescent dyes can also be covalently attached to nucleotides. Fluorescent dideoxynucleoside triphosphates are used as chain terminators in DNA sequencing. Using a different dye that fluoresces at a different wavelength for each of the four dideoxynucleosides, polynucleotides can be sequenced automatically using a single column rather than four parallel lanes.

Biotin

Fluorescent "probes" such as 1-*ranilinonapthalene-8-sulfonate* or l, 6-diphenyl-l, 3, 5-hexatriene embedded in membranes, contractile fibers, etc., can reveal changes in mobility that accompany alterations in physiological conditions. For example, molecular changes occurring in membranes during nerve conduction and in mitochondria during electron transport can be observed. Another type of probe is examplified by 2'-(*N*, *N*-dimethylamino)-6-4-*trans*-cyclohex-amoic (DANCA). Its emission maximum shifts from 390 nm in cyclohexane to 520 nm in water, presumably as a consequence of increased polarization of the molecule in the excited state, as indicated by the green arrows in the accompanying structural formula. DANCA can be used to obtain some idea of the polarity of sites within macromolecules to which it binds.

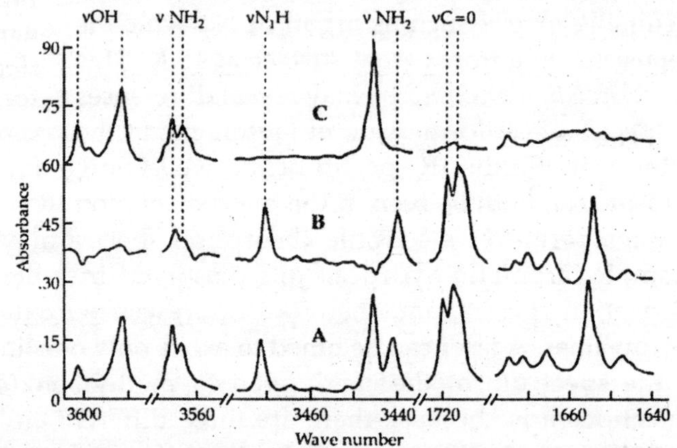

Fig. 2.14. Infrared spectrum of cytosine in the –OH, NH, and C = O stretching region in an argon matrix in 15K. (A) Initial spectrum of a sample containing both tautomers. (B) Spectrum of the amino-oxo tautomer calculated by substraction of the spectrum of the amino-hydroxy tautomer. (C) Spectrum of the aminohydroxy tautomer obtained by irradiating the initial sample in order to photochemically convert it into the aminohydroxy tautomer. Spectra (B) and (C) were also corrected for residual amounts of the second tautomer.

1-Anilinonaphthalene-8-sulfonate

DANCA

Study of calcium ions in living cells has been immensely aided by calcium fluorophores, which are often derivatives of EDTA. One of these, *calcein,* is also very specific toward Fe^{3+}. The natural calcium-dependent luminous protein, aequorins is also widely used.

Fe^{III} - calcein • H^+

Nontoxic but highly fluorescent dyes are used to study diffusion within and between cells. The so called Lucifer dyes have been employed to trace shapes and branching patterns in neurons.

Structural formula for Lucifer dyes
For Lucifer yellow
R = —NH—CO—NH—NH_2

The recently developed *fluorescence correlation spectroscopy* permits studies of molecular association in one femtoliter of solution using a confocal or two-photon microscope. Two lasers are used to excite two fluorophores of different colors, each one on a different type of molecule. Fluorescence of single molecules can be detected, and molecular associations can be detected by changes in the distribution of the fluctuations in fluorescence intensity caused by Brownian motion. A different type of advance is development of computer programs that analyze chromosomes stained with a mixture of dyes with overlapping spectra and display the result as if each chromosome were painted with a specific color. Yet another advance is development of *semiconductor nanocrystals* (or *"quantum dots"*) with narrow absorption bands and intense fluorescence. The wavelength of absorption and fluorescence depends upon the size of the crystals. For example, CdSe crystals of diameter 2–5 nm coated with silica or with a surfactant fluoresce across the visible range. They have a variety of uses in biological staining.

D. PHOTOCHEMISTRY

Because of their high energy, molecules in either the singlet or triplet photoexcited state undergo a greater variety of chemical reactions than do molecules in the ground state. Many of these photo-chemical reactions arise from the triplet state that is formed from the singlet by intersystem crossing. Selection rules forbid transitions between excited triplet state and ground state; therefore, the radiative lifetime of the triplet state is long. The diradical character of the triplet state also makes it unusually reactive. Despite its forbidden character, nonradiative deexcitation of the triplet state is possible, and phosphorescence is observed for most molecules at low temperatures if the solvent is immobilized as a glass. The intense light from lasers can also induce a variety of photochemical processes that arise from absorption of two or more photons.

One of the simplest of photochemical processes is the dissociation or uptake of a proton by an excited molecule as a result of a change in the pK_a value of a functional group. Various other bond cleavages may lead to dissociation into ions or radicals. Photoelimination and photoaddition reactions both occur. Molecules may be isomerized, a process of importance in visual receptors. Excited molecules may become strong oxidizing agents able to accept hydrogen atoms or electrons from other molecules. An example is the *photooxidation* of EDTA by riboflavin which undergoes photoreduction. A biologically more important example is in photosynthesis, during which excited chlorophyll molecules carry out *photoreduction* of another molecule and are themselves transiently oxidized. A frustrating aspect of investigation of photochemical reactions is that the variety of reactions possible often leads to a super-abundance of photochemical products, *e.g.*, see the thin layer chromatogram of cleavage products of riboflavin. However, biological photoprocesses are usually much more specific.

1. Chemical Equilibria in the Excited State

When pyridoxamine with a dipolar ionic ring structure and an absorption peak at 30,700 cm^{-1} (326 nm) is irradiated, fluorescence emission is observed at 25,000 cm^{-1} (400 nm). When basic pyridoxamine with an anionic ring structure and an absorption peak at 32,500 cm^{-1} (308 nm) is irradiated, fluorescence is observed at 27,000 cm^{-1} (370 nm), again shifted ~5500 cm^{-1} from the absorption peak. However, when the same molecule is irradiated in acidic solution, where the absorption peak is at 34,000 cm^{-1} (294 nm), the luminescent emission at 25,000 cm^{-1} is the same as from the neutral dipolar ionic form and abnormally far shifted (9000 cm^{-1}) from the 34,000 cm^{-1} absorption peak. The phenomenon, which is observed for most phenols, results from rapid dissociation of a proton from the phenolic group in the photoexcited state. A phenolic group is more acidic in the excited state than in the ground state, and the excited pyridoxamine cation in acid solution is rapidly converted to a dipolar ion.

The variation of fluorescence intensity with pH can provide direct information about the pK_a in the excited state. Förster suggested the following indirect procedure for estimating excited-state pK_a values for phenols. Let E_1 represent the energy of the 0 – 0 transition (preferably measured as the mean of the observed 0 – 0 transition energies in absorption and fluorescent emission spectra); let E_2 represent the energy of the 0 – 0 transition in the dissociated (anionic in the case of a phenol) form, while ΔH and ΔH^* represent the enthalpies of dissociation in the ground and excited states, respectively. It is evident from the diagram that Eq. 2.22 holds.

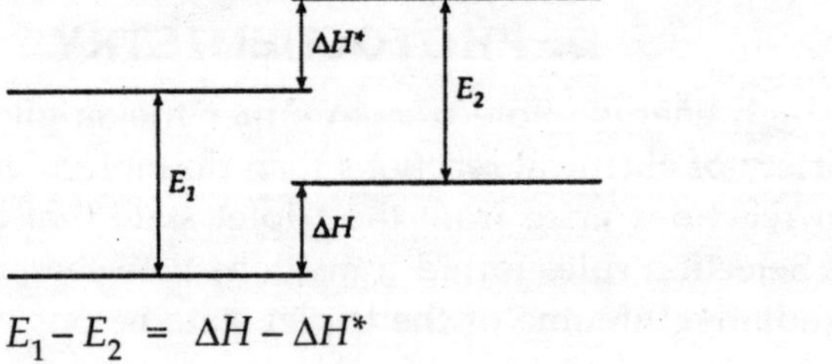

$$E_1 - E_2 = \Delta H - \Delta H^* \quad \text{...(2.22)}$$

If we assume that the changes in entropy for the reaction are the same in the ground and excited state,

$$\log_{10}(K^*/K) = Nh\,(\Delta\,\nabla\,)/(2.3RT)$$

or

$$pK^* = pK - (2.1 \times 10^3)\,\Delta\nu\,(\text{cm}^{-1}) \text{ at } 25^0\text{C} \qquad ...(2.23)$$

For example, a shift in the spectrum of the basic form by 1000 cm^{-1} to a lower wave number compared with the acid form corresponds to a decrease of 2.1 units in pK_a for dissociation of the acid form. Whereas it is desirable to use both absorption and fluorescent measurements to locate the approximate positions of 0 – 0 bands, absorption measurements alone are often used, and the positions of the band maxima are taken. Thus, for pyridoxamine the shift in absorption maximum from 34,000 cm^{-1} in the protonated form to 30,700 cm^{-1} in the dissociated form suggests that the pK_a of pyridoxamine of 3.4 in the ground state is shifted by 6.9 units to –3.5 in the excited state. Bridges *et. al.*, evaluated this same pK_a from the pH dependence of fluorescence as $pK^* \sim -4.1$.

While phenols and amines are usually more acidic in the singlet excited state than in the ground state, some substances, *e.g.*, aromatic ketones, may become more *basic* in the photoexcited state.

Observation of an abnormally large shift in the position of fluorescent emission of pyridoxal phosphate (PLP) in glycogen phosphorylase answered an interesting chemical question. A 330 nm (30,300 cm^{-1}) absorption band could be interpreted either as arising from an adduct of some enzyme functional group with the Schiff base of PLP and a lysine side chain (structure A) or as a nonionic tautomer of a Schiff base in a hydrophobic environment. For structure A, the fluorescent emission would be expected at a position similar to that of pyridoxamine. On the other hand, Schiff bases of the type indicated by structure B would be expected to undergo a photoinduced proton shift (phototautomerization) to form structure C with an absorption band at 430 nm (23,300 cm^{-1}) and fluorescent emission at a still lower energy. Since the observed fluorescence was at 530 nm, it was judged that the chromophore does have structure B.

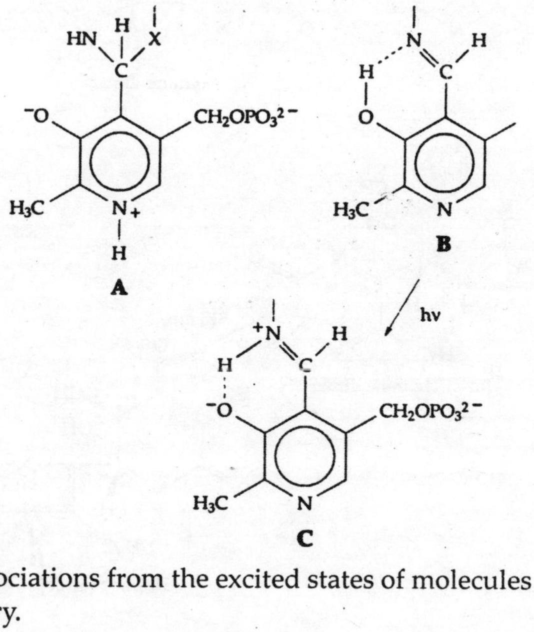

The rate of proton dissociations from the excited states of molecules can be measured directly by nano-second fluorimetry.

2. Photoreactions of Nucleic Acid Bases

Photochemical reactions of the purines and pyrimidines assume special significance because of the high molar extinction coefficients of the nucleic acids present in cells. Light is likely to be absorbed by nucleic acids and to induce photoreactions that lead to mutations. Both pyrimidines and purines undergo photochemical alterations, but purines are only about one-tenth as sensitive as pyrimidines. *Photohydration* of cytidine is observed readily. The reaction is the photochemical analog of the hydration of α, β-unsaturated carboxylic acids. Uracil derivatives also undergo photohydration.

A more important reaction is the photodimerization of thymine, a reaction also observed with uracil. A variety of stereoisomers of the resulting cyclobutane-linked structure are formed. The one shown in the equation predominates after irradiation of frozen thymine. Another important isomer is the 6 – 4 thymine dimer. Both of these types of *cyclobutane dimers* block DNA replication. This accounts for much of the lethal and rnutagenic effect of ultraviolet radiation on organisms. The matter is sufficiently important that a special "excision repair" process is used by cells to cut out the thymine dimmers. In addition, light-dependent *pholotyases,* act to reverse the dimerization reactions.

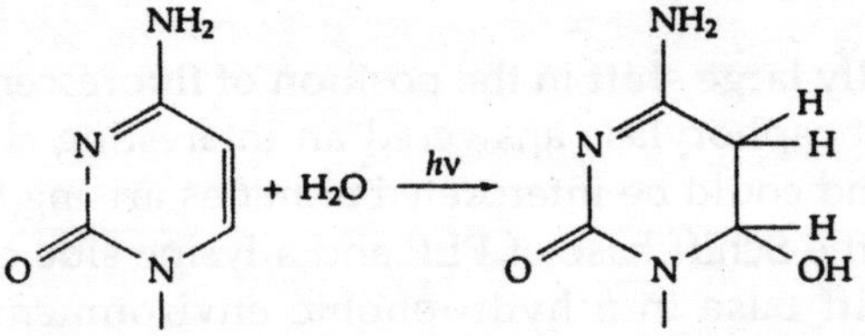

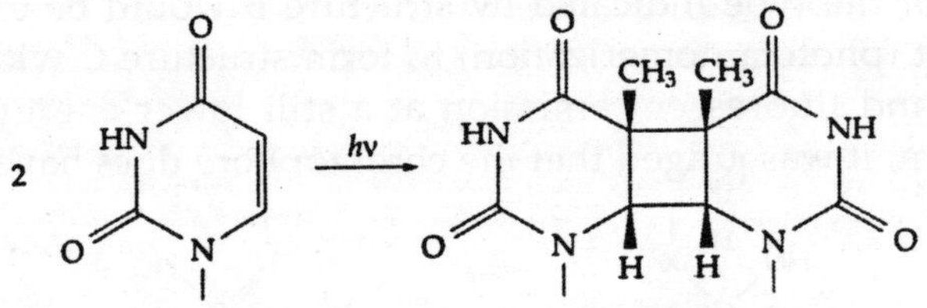

NH_2 N N O + H_2O $\xrightarrow{hv}$ NH_2 H H N H N OH O

2 HN O N O $\xrightarrow{hv}$ O O CH_3 CH_3 HN NH O N H H N O

Thymine dimer

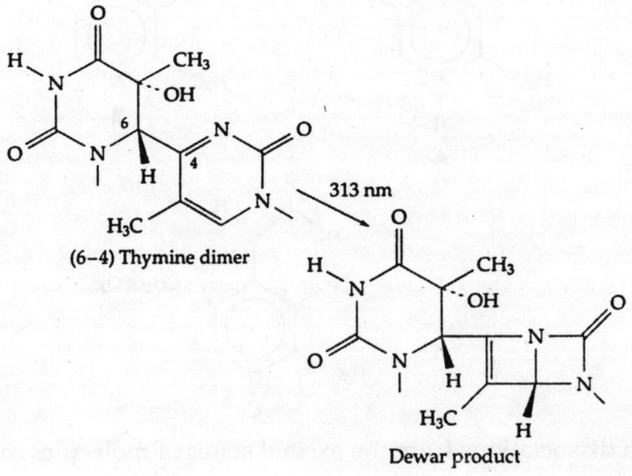

O H N CH_3 OH 6 N O O N H 4 N H_3C

(6–4) Thymine dimer

313 nm

O H N CH_3 OH N O O N H N H_3C H

Dewar product

Light can also cause addition and other crosslinking reactions between DNA and proteins or other cell constituents. One use of such reactions in the laboratory is DNA *"photo footprinting"* a technique which reveals contact regions between DNA and associated proteins. Another type of cellular damage is caused by photosensitization of DNA by a light-absorbing intercalating agent such as *8-methoxypsoralen.* DNA as well as adjacent proteins can be damaged.

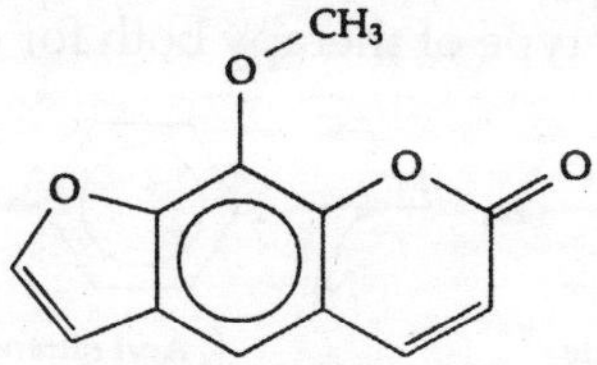

CH_3 O O O O

8-Methoxypsoralen

3. Sunburn, Cancer, and Phototherapy

Ultraviolet light is sometimes classified according to its energy and capacity for damaging cells as follows :

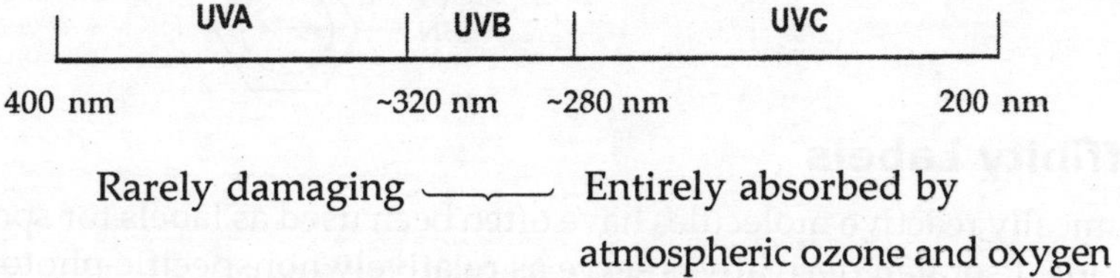

Most damaging causing sunburn and mutations.
Most is absorbed by the ozone layer.

The UVB radiation causes most damage to skin. UVA radiation is at least an order of magnitude less damaging and is usually harmless to human skin. The UVC solar radiation is all absorbed by ozone and dioxygen of the atmosphere. UVC radiation produced by ultraviolet lamps is usually all absorbed in the epidermis.[5]

Ultraviolet light damages proteins as well as DNA. Residues of Trp, Tyr, His, Cys, and Met are especially susceptible to photolysis, or photooxidation by O_2, or by singlet oxygen. Also damaged are unsaturated lipids, porphyrins, flavins, etc. Kynurenic acid and urocanic acid, an important ultraviolet filter in skin, are also decomposed by light.

Repeated sunburn ages skin and may induce cancer. However, light also has beneficial effects. It allows us to see, provides a source of vitamin D, induces enzymatic repair of some DNA damage, and provides all of our food energy, directly or indirectly, by photosynthesis.

Light is used in *phototherapy.* This happens most frequently in the irradiation of newborn babies with white light to isomerize bilirubin from the 4Z, 15Z form to more readily excreted forms such as the 4*E*, 15Z isomer. About half of newborns have some *jaundice* (elevated bilirubin), and if it is severe it must be treated promptly to avoid neurological damage. There are sometimes complications, but the treatment is usually effective. *Psoriasis* frequently treated by irradiation with UVB light, which is thought to inhibit growth of the abnormal skin cells. Ten to 35 treatments are usually required, and the condition may return after some years. An alternative treatment is irradiation with UVA light after ingestion of 8-methoxypsoralen or other psoralen derivative. Another skin condition that may respond to phototherapy is *vitiligo.* Treatment with UVA and psoralen derivatives may stimulate repigmentation. If depigmentation is extreme, the remaining pigmentation may be reduced by bleaching with the monobenzyl ester of hydro-quinone.

Photodynamic therapy is a cancer treatment that involves intravenous injection of a light-absorbing molecule such as a porphyrin, which may be taken up preferentially by cancer cells. Laser irradiation by deeply penetrating red light (650 – 800 nm wavelength) causes oxygen-dependent photosensitization. Improvements in lasers, in fiber optics, and in photosensitizers may lead to widespread use of this type of therapy both for cancer and for some other conditions.

Arylazide —N_3 —hν, −N_2→ Aryl nitrene —N̈:

R′—R(—C—R″)—H → R′—R(—C—R″)—HN̈—C$_6$H$_5$

4. Photoaffinity Labels

Photochemically reactive molecules have often been used as labels for specific sites in proteins and nucleic acids. Psoralen derivatives serve as relatively nonspecific photochemically activated crosslinking agents for DNA and double-stranded RNA. *Aryl azides* are converted by light to aryl nitrenes, which react in a variety of ways including insertion into C–H bonds. In some cases UV irradiation can be used to join natural substrates to enzymes or hormones to receptors. For example, progesterone, testosterone, and other steroids have been used for direct photoaffinity labeling of their receptors. Synthetic *benzophenones* have also been used widely as photoactivated probes.

5. Microphotolysis and Ultrafast Light-induced Reactions

Fluorescence microphotolysis, or photobleaching, has been widely used to study translational mobility of lipids and proteins in membranes. An attenuated laser beam may be focused down to the diameter of a cell or less. Then the intensity can be suddenly increased by several orders of magnitude, bleaching any fluorescent material present. The return of fluorescent material by free diffusion from a neighboring region (*fluorescence recovery after photobleaching*) or by diffusion through a membrane into a cell can then be observed. Diffusion coefficients of labeled biopolymers or of components of cells can be evaluated, and translation and metabolism of lipids and other components can be followed.

Laser-based techniques are being used for ultrafast observation of the results of a photochemical process, *e.g.*, the light-induced dissociation of CO from the hemoglobin • CO complex. A dissociating laser pulse can be as short as 100 fs (0.1 ps) or less. This is shorter than the time of vibrational motion of nuclei in an electronically excited state (~0.3 ps). Using IR spectroscopy, events that follow can be observed at intervals as short as 0.1 ns. X-ray diffraction measurements using 150 ps pulses have allowed direct observation of the CO dissociated from hemoglobin or myoglobin and its recombination with the same protein. Femtosecond dynamics of electron transfer along a DNA helix is also being studied.

6. Optical Tweezers, Light-directed Synthesis, and Imaging

The radiation pressure exerted by light is very weak. A bright laser beam of several milliwatts of power can exert only a few piconewtons (pN) of force. However, a force of 10 pN is enough to pull a cell of *E. coli* through water ten times faster than it can swim. In about 1986, it was found that a laser beam focused down to a spot of ~ one λ (~1 μm for an infrared laser) can trap and hold in its focus a refractile bead of ~1 μn diameter. This *"optical tweezers"* has become an important experimental tool with many uses. For example, not only are optical tweezers of utility in studying biological motors but also mechanical properties of all sorts of macromolecules can be examined. For example, DNA can be stretched and its extensibility measured. Actin filaments have even been tied into knots!

Light-directed solid state synthesis of peptides and oligonucleotides is another new tool. Development of this technology may provide new advances in preparation of DNA chips with a higher density of components than are now available.

X-rays and more recently NMR (MRI)-imaging have become well known to us. It might seem impossible to use visible light for a similar type of imaging. However, a laser beam can pass through a person's head. Is it possible, using computer-based techniques, to create an image from the emergent light? Efforts are being made to do exactly this. *Optical coherence tomography* using backscattered infrared light and related fast techniques have higher resolution than MRI, computerized tomography (CT), or ultrasound. Ultrabright synchrotron radiation is also being used in *infrared microspectrophotometry*.

E. PHOTOSYNTHESIS

The photochemical reduction of CO_2 to organic materials is the basic source of energy for the biosphere. Nevertheless, the process is limited to a few genera of photosynthetic bacteria eukaryotic algae, and higher green plants. Photosynthetic bacteria include the distinctly different purple, green, and bluegreen (cyanobacteria) groups, each of which has a different array of photosynthetic pigments. However, the basic mechanism of transduction of solar energy into chemical energy is the same in all of the bacteria and in green plants.

Photosynthesis involves the incorporation of CO_2 into organic compounds by reduction with NADPH with coupled hydrolysis of ATP. This is most often via the Calvin-Benson cycle. In a few organisms a reductive tricarboxylic acid cycle is employed. The idea that the chloroplasts of plants or the pigmented granules of photosynthetic bacteria generate NADPH or reduced ferredoxin plus ATP is now thoroughly accepted. However, it was not always obvious. Consider the overall equation for formation of glucose by photosynthesis in higher plants :

$$6\,CO_2 + 6H_2O \rightarrow 6\,O_2 + C_6H_{12}O_6 \qquad ...(2.28)$$

$$6\,CO_2 + {}_{12}H_2O^* \rightarrow 6O_2^* + C_6H_{12}O_6 + 6H_2O \qquad ...(2.29)$$

The stoichiometry of the reaction suggests that all 12 of the oxygen atoms of the evolved O_2 might come from CO_2 or that some might come from CO_2 and some from H_2O. In fact, water supplies both of the oxygen atoms needed for formation of O_2, as is indicated by the asterisks in Eq. 2.29. This possibility was suggested by van Niel in 1931. He pointed out that in bacterial photosynthesis no O_2 is produced, and that bacteria must have access to a reducing agent to provide hydrogen for the reduction of CO_2.

$$H_2A \xrightarrow{hv} A + 2[H] \qquad ...(2.30)$$

In this equation, H_2A might be H_2S (in the purple sulfur bacteria), elemental H_2, isopropanol, etc. From a consideration of these various reactions, van Niei concluded that in the O_2-producing cyanobacteria and eukaryotioplants water serves as the oxidizable substrate in Eq. 2.30 and is cleaved to form O_2 and to provide hydrogen atoms for reduction. This photochemical cleavage is the only known biological oxidation reaction of H_2O. No oxidizing agents present in living things are powerful enough to dehydrogenate water except for the photochemical *reaction centers* of photosynthetic organisms.

1. Two Photosystems, the Z Scheme, and Reaction Centers

It had long been known that for green plants light of wavelength 650 run was much more efficient than that of 680 nm. However, Emerson and associates in 1956 showed that a combination

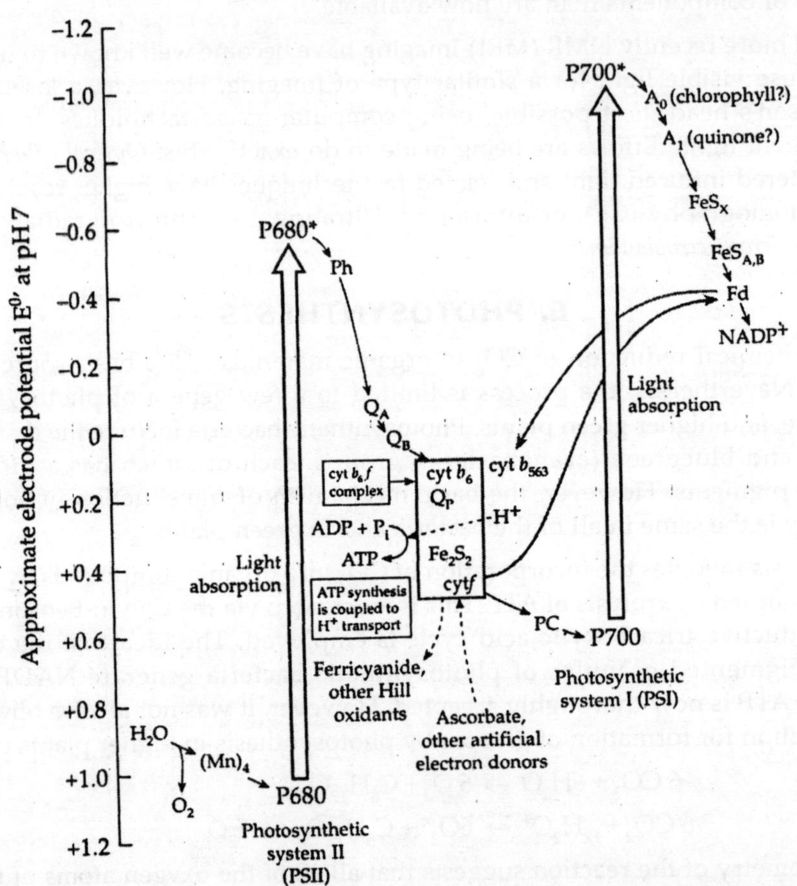

Fig. 2.15. The zigzag scheme (Z scheme) for a two-quantum per electron photoreduction system of chloroplasts. Abbreviations are P680 and P700, reaction center chlorophylls; Ph, pheophytin acceptor of electrons from PSII; Q_A, Q_B, quinones bound to reaction center proteins; PQ, plastoquinone (mobile pool); Cyt, cytochromes; PC, plastocyanin; A_0 and A_1 early electron acceptors for PSI, possibly chlorophyll and quinone, respectively; Fx, Fe_2S_2 center bound to reaction center proteins; F_A, F_B, Fe_4S_4 centers; Fd, soluble ferredoxin; and DCMU, dichlorophenyldimethylurea. Note that the position of P682, P700, Ph, Q_A, Q_B, A_0, and A_1 on the E_0' scale are uncertain. The E_0' values for P682 and P700 should be for the (chlorophyll/chlorophyll cation radical) pair in the reaction center environment. These may be lower than are shown.

of light of 650 nm *plus* that of 680 nm gave a higher rate of photosynthesis than either kind of light alone. This result suggested that there might be two separate photosystems. What is now known as *photosystem I* (PSI) is excited by far red light (~700 nm), while *ghotosystem II* (PSII) depends upon the higher energy red light of 650 nm. Additional evidence supported the idea. Hill had shown many years before that mild oxidizing agents such as ferricyanide and benzoquinone can serve as substrates for photoproduction of O_2, while Gaffron found that some green algae could be adapted to photooxidize H_2 to protons Fig. 2.30 and to use the electrons to reduce NADP. Thus, photosystem I could be disconnected from photosystem II. The powerful herbicide *dichlorophenyldimethylurea* (DCMU) was found to block electron transport between the two photosystems. In the presence of DCMU electrons from such artificial donors as ascorbic acid or an indophenol dye could be passed through photosystem I.

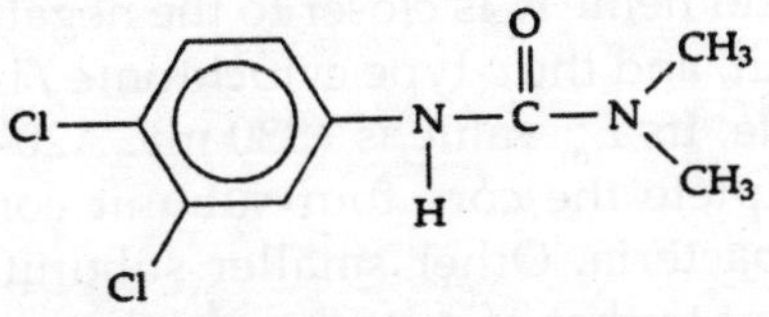

3-(3,4-Dichlorophenyl)-1,1-dimethylurea

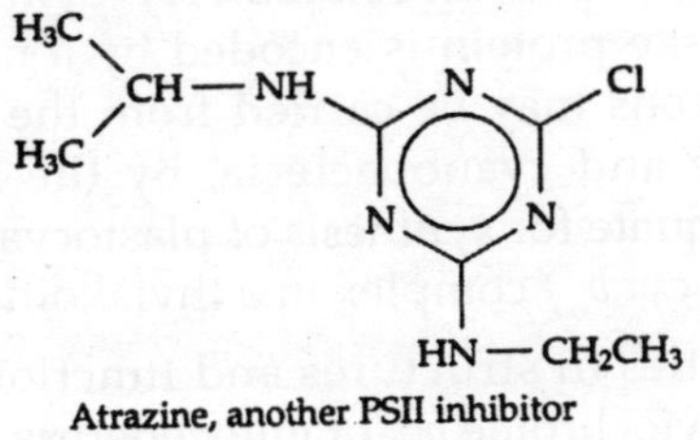

Atrazine, another PSII inhibitor

The Z Scheme

The result of these and other experiments was the development of the series formulation or zigzag scheme of photosynthesis which is shown in Fig. 2.17. Passage of an electron through the system requires two quanta of light. Thus, four quanta are required for each NADPH formed and eight quanta for each CO_2 incorporated into carbohydrate.

An important experiment of Emerson and Arnold employed very short flashes of light and measurement of the quantum efficiency of photosynthesis during those flashes. A striking fact was observed. At most, during a single turnover of the photosynthetic apparatus of the leaf, one molecule of O_2 would be released for each 3000 chlorophyll molecules. However, it could be calculated that for each O_2 released only about eight quanta of light had been absorbed. It followed that about 400 chlorophyll molecules were involved in the uptake of one quantum of light. This finding suggested that a large number of chlorophyll molecules act as a single light receiving unit (usually called a light-harvesting or antenna complex) able to feed energy to one *reaction center.* The concept is now fully accepted.

Electron Transport and Photophosphorylation

Two molecules of NADPH are required to reduce one molecule of CO_2 via the Calvin-Benson cycle, and three molecules of ATP are also needed. How are these formed? The Z scheme provides part of the answer. There is enough drop in potential between the upper end of PS1 and the lower end of PSII to permit synthesis of ATP by electron transport. It is likely that only one molecule of ATP is formed for each pair of electrons passing through this chain. Since, one and a half molecules of ATP are needed per NADPH, some other mechanism must exist for the synthesis of additional ATP. Furthermore, many other processes in chloroplasts depend upon ATP so that the actual need for photogenerated ATP may be larger than this.

Arnor demonstrated that additional ATP can be formed in chloroplasts by means of *cyclic photo-phosphorylation:* Electrons from the top of PSI can be recycled according to the dashed lines in Fig. 2.17. An electron transport system, probably that of the Z scheme, is used to synthesize ATP. As isolation of proteins and cloning of their genes progressed, it became clear that a complex of proteins known as *cytochrome* $b_6 f$ is closely related to the cytochrome bc_1 of mitochondria. As in that complex, cytochrome b_6 carries two hemes, designated b_h and b_1 with E_m of –84 and –158 mv, respectively. E_m values are for *Chlamydamonas.* Heme b_1 is closer to the positively charged membrane surface (*lumen side*) and heme b_h is closer to the negative surface (*stroma side*). In maize cytochrome b_6 is a 23-kDa subunit, and the *c*-type cytochrome *f* is a larger ~34-kDa subunit whose heme is close to the lumenal side. Its E_m value is +330 mv. A20-kDa Rieske Fe–S protein and an additional 17-kDa subunit complete the core four-subunit complex, which is found in green plants, green algae, and cyanobacteria. Other smaller subunits are also present. It is usually assumed that a *Q* cycle equivalent to that of mitochondrial complex III operates in the pumping of protons across the thylakoid membrane. However, at high rates of photosynthesis the electron transfer may bypass cytochrome *f*. The associated ATP synthase is also subject to complex regulatory mechanisms. The Rieske protein is encoded by a nuclear gene, but genes for other subunits are chloroplastic. Electrons may be carried from the cytochrome *f* subunit to PSI by plastocyanin or, in many algae and cyanobacteria, by the small *cytochrome* c_6. It is often synthesized when copper is indequate for synthesis of plastocyanin. Fig. 2.18 is a schematic view of PSI, PSII, and the intermediate cyt $b_6 f$ complex in a thylakoid membrane.

In spite of the close similarities in structures and function, there are distinct differences between cytochrome $b_6 f$ and the cytochrome bc_1 of mitochondria. Among these are the presence of stoichiometrically bound chlorophyll α and β-carotene in the photosynthetic complex. The function of the chlorophyll is uncertain, but the carotene is probably there to quench the chlorophyll triplet state, which would probably cause photodamage via formation of singlet O_2.

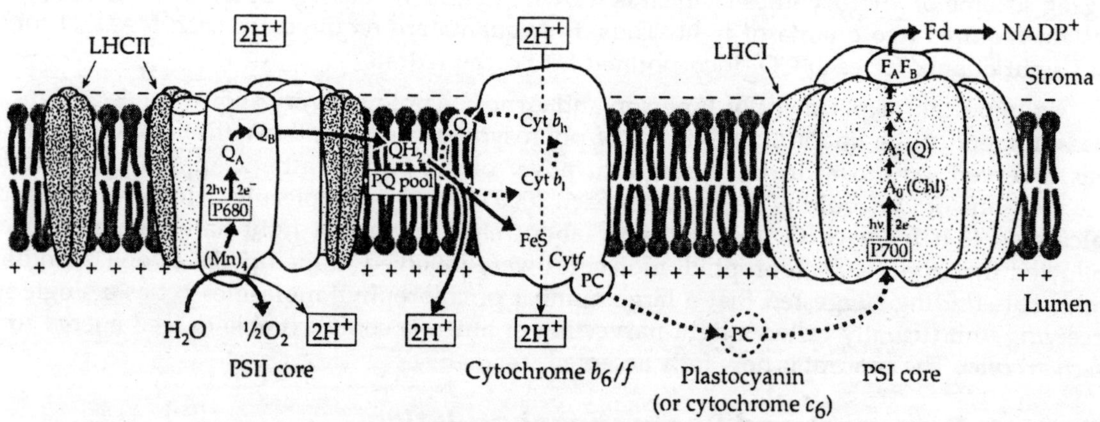

Fig. 2.16. Schematic view of photosynthetic reaction centers and the cytochrome b_6 *f* complex embedded in a thylakoid membrane. Plastocyanin (or cytochrome c_6 in some algae and cyanobacteria) carries electrons to the PSI core.

Bacterial Photosynthesis

What is the relationship of the Z scheme of Fig. 2.17 to bacterial photosyntheses? In photoheterotrophis, such as the purple *Rhodospirillum,* organic compounds *e.g.,* succinate, serve as electron donors in Eq. 2.30. Because they can utilize organic compounds for growth, these bacteria have a relatively low requirement for NADPH or other photochemically generated reductants and a larger need for ATP. Their photosynthetic reaction centers receive electrons via cytochrome *c* from succinate ($E^{o\prime}$ = + 0.03 V). The centers resemble PSII of chloroplasts and have a high midpoint electrode potential $E^{o\prime}$ of 0.46 V. The initial electron acceptor is the Mg^{2+}-free bacteriopheophytin whose midpoint potential is –0.7 V. Electrons flow from reduced bacteriopheophytin to menaquinone or ubiquinone or both via a cytochrome bc_1 complex, similar to that of mitochondria, then back to the reaction center P870. This is primarily a cyclic process coupled to ATP synthesis. Needed reducing equivalents can be formed by ATP-driven reverse electron transport involving electrons removed from succinate. Similarly, the purple sulfur bacteria can use electrons from H_2S.

In contrast, the reaction centers of green sulfur bacteria resemble PSI of chloroplasts. Their reaction centers also receive electrons from a reduced quinone via a cytochrome *bc* complex. However, the reduced form of the reaction center bacteriochlorophyll donates electrons to iron-sulfur proteins as in PSI. The latter can reduce a quinone to provide cyclic photophosphorylation. Cyanobacteria have a photosynthetic apparatus very similar to that of green algae and higher plants.

2. Chloroplast Structure

Chloroplasts come in various sizes and shapes, but all contain a small number of DNA molecules ranging in size from 120 – 160 kb. Complete sequences are known for DNA from chloroplasts of a liverwort (121,025 bp), tobacco (155,844 bp), maize (*Zea mays)* and other plants. The 140,387 bp DNA from maize chloroplasts is a circular molecule containing the genes for 23S, 16S, 5S, and 4.5S RNA, for 30 species of tRNA, and for 70 different proteins. Among them are subunits of RNA polymerase, NADH dehydrogenase, subunits of both PSI and PSII, rubisco (large subunit), cytochromes *b* and *f,* six subunits of ATP synthase, and others. As with mitochondria, some subunits of the enzyme complexes that provide the cell with energy (*e.g.,* ATP synthase) are encoded in the nucleus.

A characteristic of chloroplast genomes is the presence of a pair of large (in maize 22,348 bp) inverted repeat sequences. Since they can form a large hairpin structure with a very large loop, they may stabilize the gene sequence. The mutation rate within the repeat sequence is lower than in the single-stranded regions. The same genes are found in corresponding positions in both maize and rice. Both genomes have a number of sites of departure from the standard genetic code. These *"editing sites"* give rise to C → U transitions in the RNA transcripts.

Most land plants have similar chloroplast DNA sequences, but considerable divergence is observed among algae. For example, the red alga *Porphyra purpura* has 70 genes not found in chloroplasts of land plants. Each gene of the chloroplasts of the dinoflagellate *Heterocapsa triquetra* is carried on its own DNA minicircle. However, ~2000 chloroplast proteins are encoded by nuclear DNA. The corresponding proteins are synthesized on cytoplasmic ribosomes and are transported into the chloroplasts. Some of these proteins must pass through both the double membrane of the envelope and the thylakoid membrane. As in mitochondria an array of different transport proteins are required. They are distinctly different from the mitochondrial transport proteins and involve their own unique targeting mechanisms.

Chloroplast Membrane

Like the other energy-producing organelles, the mitochondria chloroplasts are surrounded by an outer double membrane or *envelope* and also contain an internal membrane system. Within the colorless stroma are stacks of flattened discs known as *grana.* The discs themselves (the *thylakoids)* consist of pairs of closely spaced membranes 9 ηm thick, each pair being separated by a thin internal space or loculus. At least 75 different proteins are present in the isolated membranes. There is also a high content of *galactosyl diacylglyccrol. digalactnsyl diacylglycerol,* and *sulfolipid.* Lipids account for half of the mass of thylakoid membranes.

Through the use of freeze-fracture and freeze-etching techniques of electron microscopy, it is possible to see, embeddded in the thylakoid membranes, particles which may represent *individual photosynthetic units* (also called *quantosomes*). They are about 20 nm in diameter, and at least many of them presumably contain a reaction center surrounded by light-collecting chlorophyll—protein complexes. Others may represent the cytochrome $b_6 f$ complex and ATP synthase, whose knobs also protrude into the stroma. A photosynthetic unit can also be defined chemically by the number of various types of molecules present in a chloroplast membrane for each four manganese atoms. Separate units contain PSI and PSII. These reaction centers appear to have a different distribution within the thylakoids, the PSI units being located principally in the unstacked membranes and the PSII units in the grana stacks.

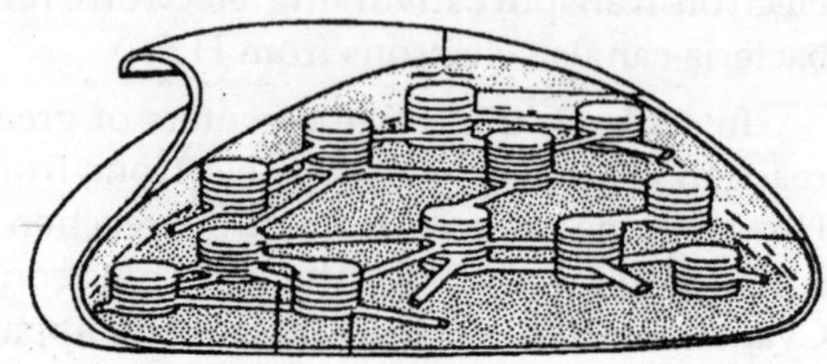

Fig. 2.17. Schematic drawing of a chloroplast.

Photosynthetic Pigments and their Environments

The chlorophylls are related in structure to the hemes but ring IV (D) is not fully dehydrogenated as in the porphyrins. The *chlorine* ring system is further modified in chlorophyll by the addition of a fifth ring (V) containing an oxo group and a methyl ester. Ring V has been formed by crosslinking between the propionic acid side chain of ring III and a methine bridge carbon to give the parent compound *pheoporphyrin.* Chlorophylls contain constituents around the periphery that indicate a common origin with the porphyrins. However, one of the carboxyethyl groups is esterified with the long-chain phytyl group in most of the chlorophylls. Chlorophyll *a* is the major pigment of chloroplasts and is a centrally important chromophore for photosynthesis in green plants. Most of the other chlorophylls, as well as the carotenoids and certain other pigments, are referred to as *accessory pigments.* Many of them have a light-receiving antenna function. Carotenoids are also photoprotectants. The relative numbers of pigment molecules in the photosynthetic units (average of PSI and PSII) of spinach chloroplasts are given in Table 2.2.

While the structure of chlorophyll *a* shown in Fig. 2.20 is the predominant one, other forms exist, *e.g.,* with

$$-CH_2CH_3 \text{ or } -\underset{\displaystyle OH}{\underset{|}{CH}}-CH_3$$

replacing the vinyl group on ring I or with vinyl or hydroxyethyl replacing the ethyl group on ring II. The same kind of variation occurs for chlorophyll *b*. In 80% acetone chlorophyll *a* has a sharp absorption band at 663 nm (15,100 cm^{-1}), but within chloroplasts the absorption maximum is shifted toward the red, the majority of the chlorophyll absorbing at 678 nm. Chlorophyll *b* is also nearly always present in green leaves. The absorption peak in acetone is at 635 nm (15,800 cm^{-1}). Chlorophyll *c* found in diatoms, brown algae (Phaeophyta), and dinoflagellates lacks the phytyl group. Chlorophyll *d* contains a formyl group on ring I.

Photosynthetic bacteria contain *bacteriochlorophylls* in which ring II is reduced. The absorption band is shifted to the red from that of chlorophyll *a* to ~770 nm. The most abundant chlorophylls of green sulfur bacteria, *bacteriochlorophylls c, d* and *e* (or *Chlorobium* chlorophylls), contain a hydroxyethyl group on ring I; ethyl, *n*-propyl, or isobutyl groups on ring II; often an ethyl group instead of methyl on ring III; and a methyl group on the methine carbon linking rings I and IV. A variety of polyprenyl side chains can replace the phytyl group of the chlorophylls of higher plants. The *pheophytins*, which are identical to the chlorophylls but lack the central magnesium

Table 2.2. Approximate Composition of Photosynthetic Units in a Spinach Chloroplast[a]

Component	*Number of molecules[b]*
Chlorophyll *a*	160
Chlorophyll *b*	70
Carotenoids	48
Plastoquinone A	16
Plastoquinone B	8
Plastoquinone C	4
α-Tocopherol	10
α-Tocopherylquinone	4
Vitamin K_2	4
Phospholipids	116
Sulfolipids	48
Galactosylglycerides	490
Iron	12 atoms
Ferredoxin	5
Cytochrome b_{563}	1
Cytochrome b_{559}	
Cytochrome *f*	1
Copper	6 atoms
Plastocyanin	1
Manganese	2 atoms
Protein	928 kDa

[a] Averaged for PSI and PSII. After Gregory, R. P. E (1971) *Biochemistry of Photosynthesis*, Wiley, New York [data of Luchtenthaler, H. K., and Park, R. B. (1963) Nature *(London)* 198,1070] and White, A., Handler, P., and Smith, E. L. (1973) *Principles of Biochemistry*, 5th ed., p. 528, McGraw-Hill, New York.

[b] Numbers of molecules assuming 2 Mn^{2+} ions per unit (4 for PSII and 0 for PSI).

ion, play an essential role in photosynthetic reaction centers. They can be formed in the laboratory by splitting the Mg^{2+} out from chlorophyll with a weak acid. Other derivatives are the *chlorophyllides* formed by hydrolysis of the methyl ester group and *chlorophyllins* formed by removal of both the methyl and phytyl groups.

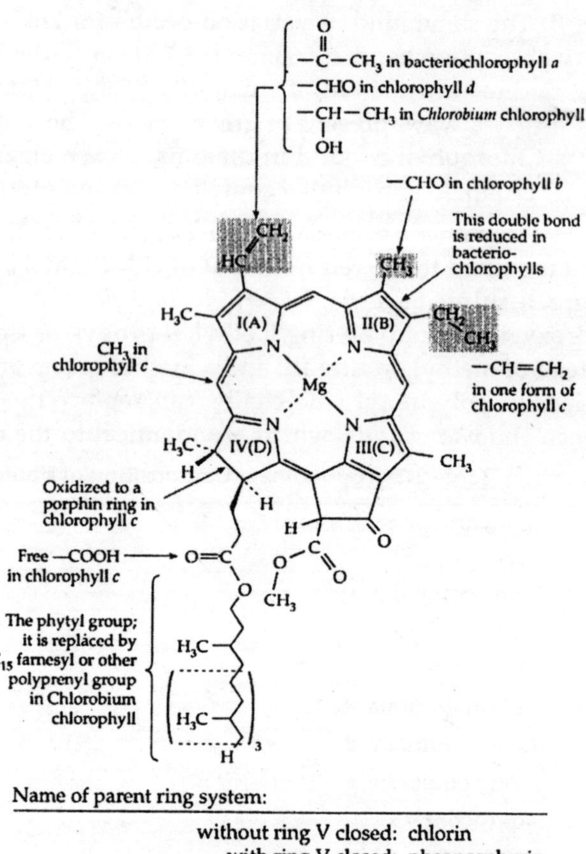

Name of parent ring system:

without ring V closed:	chlorin
with ring V closed:	pheoporphyrin
with free — COOH on ring V:	chlorophyllides
with free — COOH on both rings II and V:	chlorophyllins
with Mg_{2+} replaced by 2 H^+:	pheophytins

Fig. 2.18. Structures of the chlorophylls.

Since chlorophyll can be removed readily from chloroplasts by mild solvent extraction, it might appear that it is simply dissolved in the lipid portion of the membranes. However, from measurements of dichroism it was concluded that the chlorophyll molecules within the membranes have a definite orientation with respect to the planes of the thylakoids and are probably bound to fixed structures. The absorption spectrum of chlorophyll in leaves has bands that are shifted to the red by up to 900 cm^{-1} from the position of chlorophyll *a* in acetone. Most green plants contain at least four major chlorophyll bands at ~ 662, 670, 677, and 683 nm as well as other minor bands. This fact suggested that the chlorophyll exists in a number of different environments. As a result, the absorption is spread over a broader region leading to more efficient capture of light. Only a small fraction of the total chlorophyll is in the reaction centers; that for PSI absorbs at ~ 700 nm and that for PSII at ~682 nm.

Bacteriochlorophyll in *Chroniatium* has three absorption bands with peak positions at 800, 850, and 890 nm. The last includes the reaction center bacteriochlorophyll and is the only form that fluoresces. Recent studies have established that most if not all chlorophyll is bound to specific proteins, a fact that can account for the various overlapping absorption bands.

The *carotenes* and *carotenoids* are very important accessory pigments. The major component in most green plants is β-carotene. Green sulfur bacteria contain γ-carotene in which one end of the molecule has not undergone cyclization and resembles lycopene. Chloroplasts also contain a large variety of oxygenated carotenoids (xanthophylls). Of these, neoxanthin, violaxanthin. and *lutein* predominate in higher plants and green algae. *Euglena* and related microorganisms contain much antheraxanthin. A light-collecting protein from marine dinoflagellates contains both chlorophyll *a* and several molecules of the carotenoid peridinin. Brown algae and diatoms contain mostly *fueoxanthin* and *zeaxanthin* while the bacterium *Rhodospirillum rubrum* synthesizes spirilloxanthin.

It is a striking fact that there are no naturally occurring green plants that lack carotenoid pigments. Carotenoidless mutants are used in photosynthesis research, but they apparently cannot survive under natural conditions. Carotenoids not only participate as members of the light-receiving complex but also confer protection to chlorophyll against light-induced destruction by singlet oxygen. This accounts for the fact that carotenoids are usually intimately associated with chlorophyll in the pigment complexes.

A third class of accessory pigment of more limited distribution are *the open tetrapyrroles,* sometimes called *"plant bile pigments"* because of their relationship to the pigments of animal bile. Among these are the *phycocyanins,* which provide the characteristic color to cyanobacteria. They are conjugated proteins (biliproteins) containing covalently bound phycocyanobilin. The red *phycoerythrins* of the Rhodophyta contain bound phycoerythrobilin, an isomer of phyco-erythrobilin. There are four common isomeric *bilins,* each having a different number of conjugated double bonds. Together, they provide for a broad range of colors from blue to red. The bilins are attached to proteins via addition of a cysteine –SH group to the vinyl group of ring A of the bilin. A double attachment may be formed by addition of –SH groups to both vinyl groups. Isolated tetrapyrrole pigments tend to have a helical structure and to absorb light at lower wavelengths than do the protein-bound pigments which assume elongated conformations.

There are three major classes of conjugated phycobiliproteins, all of which are αβ heterodimers often

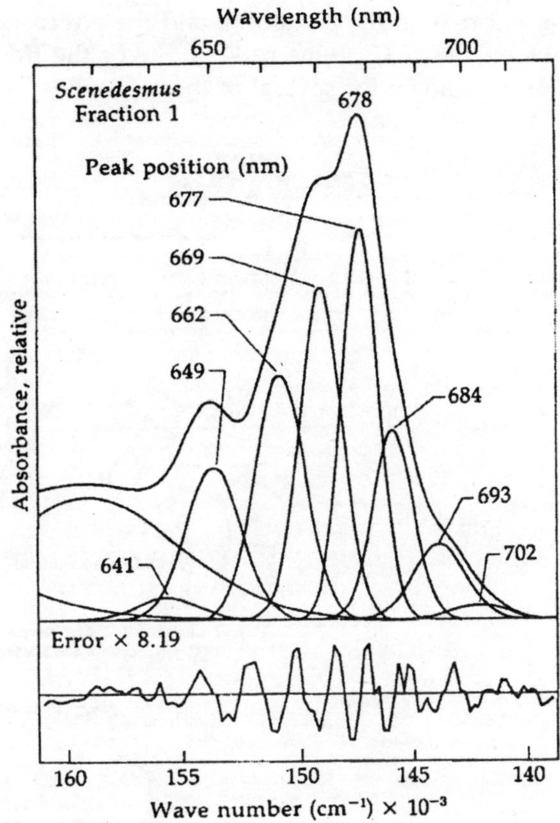

Fig. 2.19. Absorption spectrum of chlorophyll in a suspensim of chloroplast fragments from the green alga *Scenedesmus* showing the multicomponent nature of the chlorophyll environments.

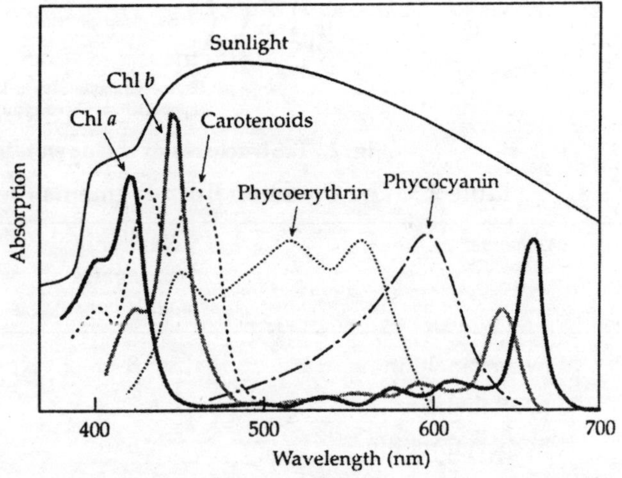

Fig. 2.20. Absorption spectra of chlorophylls and accessory pigments compared.

associated as $(\alpha\beta)_6$. The *allophycocyanins* carry one bilin per subunit, the phycocyamns carry one on the α and two on the β subunit, and the phycoerythrins carry two or three on the α subunit and three on the β. Cysteine α-84 is one of the frequent attachment sites. Three-dimensional structures are known for several of these proteins

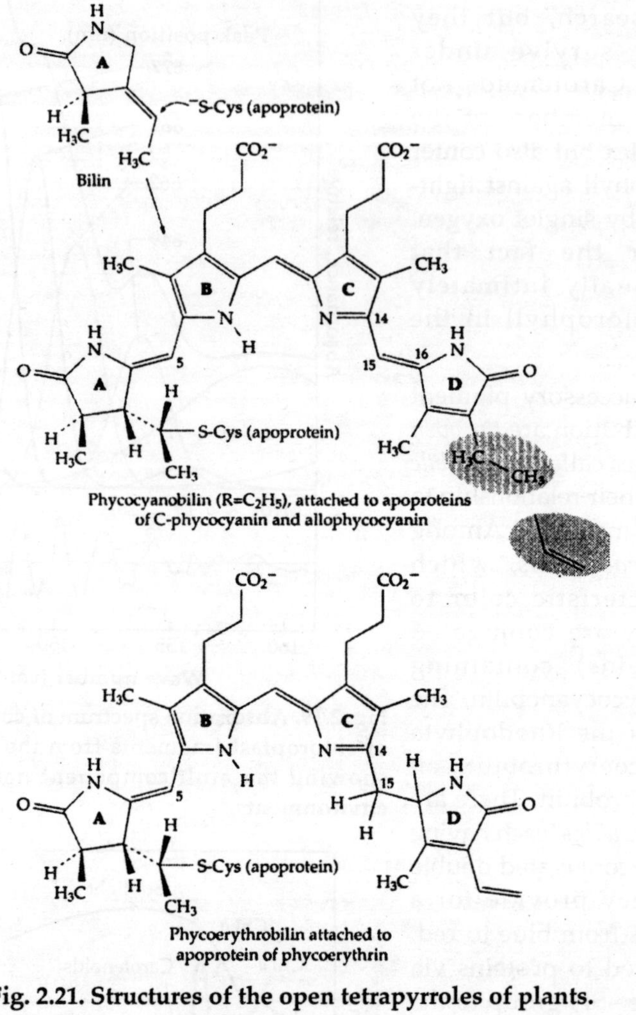

Phycocyanobilin (R=C_2H_5), attached to apoproteins of C-phycocyanin and allophycocyanin

Phycoerythrobilin attached to apoprotein of phycoerythrin

Fig. 2.21. Structures of the open tetrapyrroles of plants.

Table 2.3. The Common Bilin Pigments Present in Phycobiliniproteins[a]

Isomer	*Number of conjugated double bonds*	*Absorption maximum (nm) when conjugated to proteins*
Phycocyanobilin	8	~640
Phycobiliviolin	7	~590
Phycoerythrobilin	6	~550
Phycourobilin	5	~490

The bilins are derived biosynthetically rather directly from biliverdin IXα, whose formation is described in Fig. 2.24. The addition of an apoprotein –SH group to a carbon-carbon double bond of the bilin is catalyzed by a specific lyase. *Phytochrome* arises in a similar way as does the blue biliprotein *insecticyanin.*

3. The Light-Receiving Complexes

Irradiation of chloroplasts leads to easily measurable fluorescence from chlorophyll *a,* but no fluorescence is observed from chlorophyll *b* or from other forms of chlorophyll, carotenoids, or other pigments. It appears that the latter all serve as light-collecting or antenna pigments that efficiently transfer their energy to chlorophyll *a* at the reaction centers. As is evident from Fig. 2.22, the light-collecting pigments generally have higher energy absorption bands than do the reaction centers. Thus, a broad range of wave-lengths of light are absorbed by an organism, and energy from all of them is funneled into the reaction centers. The light-collecting pigments are bound to specific proteins, which are located close to the reaction centers and are arranged to provide efficient energy-transfer. Distances between adjacent pigment molecules vary from 1 to 7 nm.

Phycobilosomes

Algal and cyanobacterial phycocyanins and phycoerythrins are aggregated in special granules that are on the outsides of the photosynthetic membranes. The granules in the cyanobacteria are known as phycobilosomes. The $(\alpha\beta)_6$ hexamers form the disks of the phycobilosomes. These are held together by linker proteins, which fit asymmetrically into the central cavities.

The disks and linker proteins are assembled into rods which are joined to form the phycobilosomes. The latter are organized into closely packed parallel arrays on the surface of the photosynthetic membranes.

Purple Photosynthetic Bacteria

The reaction centers of *Rhodobacter spheroides, Rhodospirillum rubrum,* and related purple bacteria are embedded in the plasma membrane. Each center is surrounded by a ring of bacteriochlorophyll *a* molecules bound noncovalently to heterodimeric (αβ) protein subunits made up of ~52- to 54-residue chains. Each (αβ) dimer binds two molecules of BChl *a,* whose central Mg^{2+} ions are coordinated by conserved histidine imidazole groups, as well as a molecule of spirilloxanthin. About 15 – 17 αβ subunits form the ring, which is designated LH1. Most of these bacteria also have smaller rings, designated LH2, floating in the membrane near the LH1 complex. The LH2 rings consist of about nine αβ subunits with associated BChl *a* and carotenoid. Under some conditions a third complex LH3 may be formed. In *Rhodopseudomonas acidophila* nine of the 27 BChl *a* molecules absorb light maximally at ~800 nm and are designated *B800.* The other 18, designated *B800* absorb maximally at ~860 nm. The B850 BChl *a* molecules have direct contact with the chromophores of neighboring molecules, allowing for easy energy transfer. The B800 chromophores are more isolated.

Low-temperature (1.2 K) single-molecule spectroscopic techniques have been used to obtain the fluorescence-excitation spectra shown in Fig. 2.26C. For an ensemble of LH2 complexes (upper trace) the spectral absorption bands are broad, but for individual LH2 complexes structure can be seen clearly for the B800 chromophores but not for the B850 chromophores. This difference has been interpreted to mean that the excitation energy of an electronically excited B850 molecule is delocalized over the whole ring of 18 BChl *a* molecules as an *exciton.* This permits both fast and efficient energy transfers from B800 to B850 and from B850 of one LH2 ring to another or to an LH1 ring and to the reaction center. Energy transfer may occur by the Förster dipole-dipole mechanism. Many of the antenna are supported by binding through their Mg^{2+} ion to an imidazole group of a protein as can also be seen in reaction center chlorophylls. Hydrogen-bonding to the C13-oxo groups of the chlo-rophylls may also be possible. The orientations of the transition dipole moments of the chlorophyll molecules may be arranged to facilitate rapid energy transfer.

Green Sulfur and Nonsulfur Bacteria

In these organisms chlorophylls are present in rodlike particles of protein present within *chlorosomes,* baglike structures which may be as large as 100 × 260 nm and are attached to the inside of the cytoplasmic membranes, which contain the reaction centers. The over 10,000 light-collecting chlorophylls *c, d,* or *e,* which may be present per reaction center, allow these bacteria

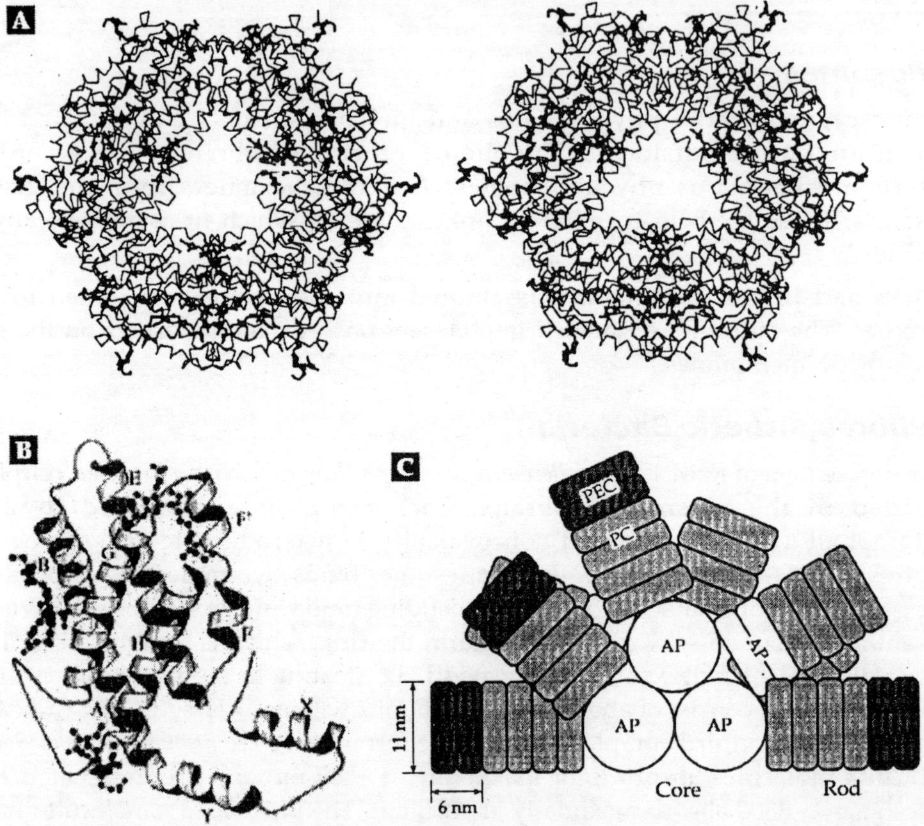

Fig. 2.22. (A) Stereoscopic view of a hexameric $(\alpha\beta)_3$ phycobiliprotein. (B) The β subunit of the complex with two molecules of bound phyco-erythrobilin and one of phycourobilin. From Chang *et al.* (C) Schematic representation of a phycobilosome of a strain of the cyanobacterium *Anabaena*. Each disk in the structure contains an $(\alpha\beta)_3$ phycobiliprotein. The circles marked AP are cross-sections of rods, each one composed of about four disks of allophycocyanin (AP). The projecting rods contain C-phycocyanin (PC) and phyco-erythrocyanin (PEC).

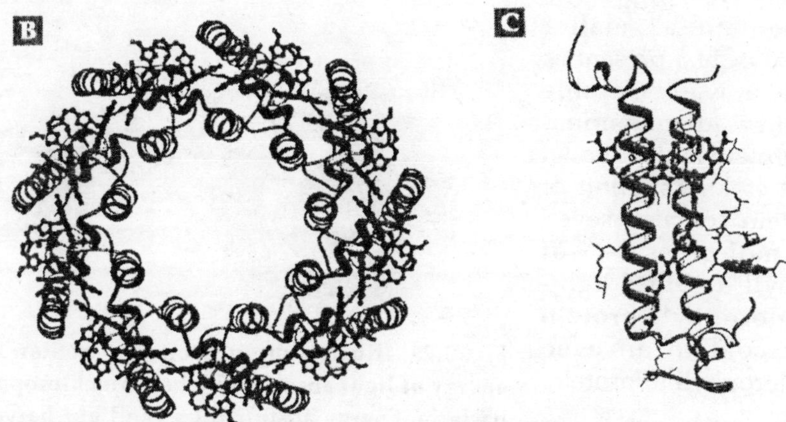

Fig. 2.23. (A) The van der Waals contact surface of the periplasmic face of the reaction center and surrounding light-harvesting complex LH1 of *Rhodobacter sphaeroides*. Made with VMD by Theoretical Biophysics Group, UIUC (B) Ribbon drawing of the structure of the circular light-harvesting complex LH2 of the purple photosynthetic bacterium *Rhodopseudomonas acidophila*. The tetrapyrrole rings of the 18 bacteriochlorophyll molecules are also shown. (C) Structure of one of the nine αβ protomers with three associated bacteriochlorophylls. One of these is near the top of the protein, and the other has its chromophoric group protruding at nearly a right angle on the right side. The complete phytyl side chains are also depicted in a stick representation. (B) and (C).

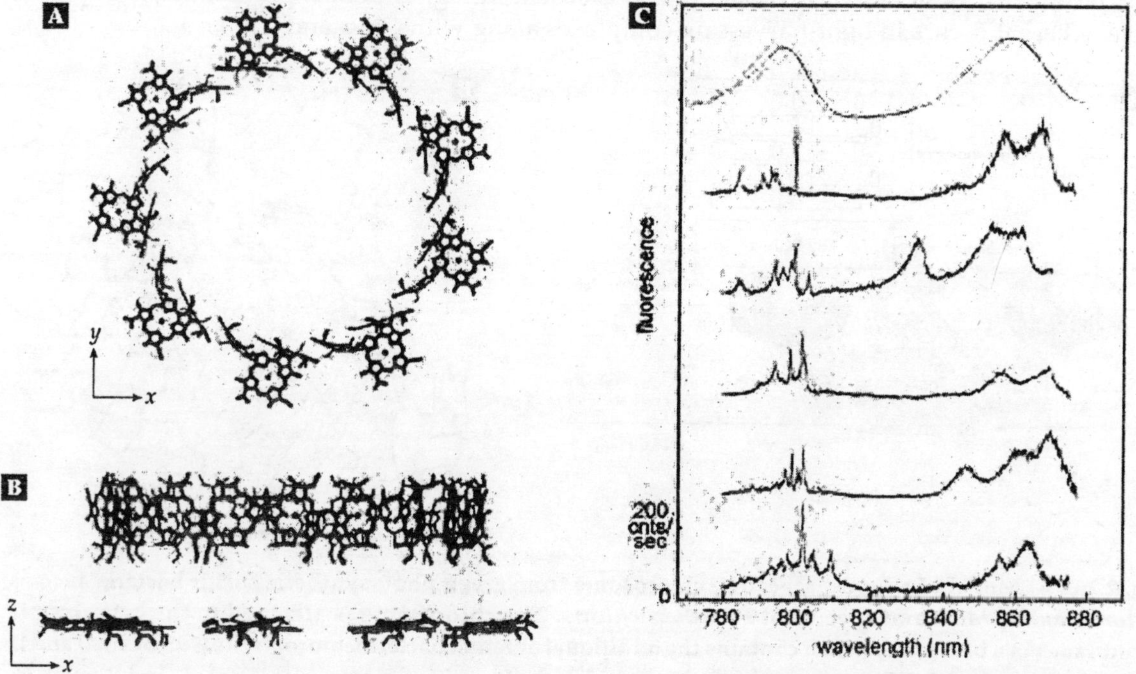

Fig. 2.24. (A), (B) Arrangement of bacteriochlorophyll chromophores in the cyclic LH2 array of *Rhodopseudomonas acidophila*. The B850 subunits are gray while the B800 subunits are black. (C) Fluorescence-excitation spectra. Top trace, for an ensemble of LH2 complexes, other traces, for several individual LH2 complexes at 1.2K. Fine structure is evident for the B800 but not for the B850 chromophores.

to live in extremely weak light. The bacteria also contain a smaller "antenna" of ~5000 BChl *a* present as a complex with a water-soluble protein. The three-dirnensional structure of this protein is also shown in Fig. 2.28. Each 45-kDa subunit of the trimeric protein contains seven embedded molecules of bacteriochlorophyll α. Other light-collecting chlorophyll-protein complexes may contain an even higher ratio of chlorophyll to protein.

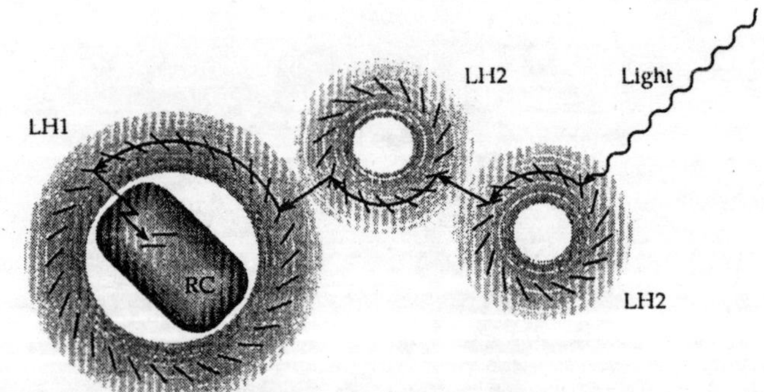

Fig. 2.25. Illustration of proposed exciton transfer of the energy of light absorbed by bacteriochlorophyll *a* of purple bacteria. Energy absorbed by the light harvesting complex LH2 is transferred in steps to another LH2, to LH1 and to the reaction center. The short lines within the circles represent the edges of the BChla chromophores.

Eukaryotic Plants and Cyanobacteria

Photosynthetic dinoflagellates, which make up much of the marine plankton, use both carotenoids and chlorophyll in light-harvesting complexes. The carotenoid *peridinin,* which absorbs blue-green in the 470- to 550-nm range, predominates. The LH complex of *Amphidinium carterae* consists of a 30.2-kDA protein that forms a cavity into which eight molecules of peridinin but only two of chlorophyll a (Chl *a)* and two molecules of a galactolipid are bound.

The allenic carotenoid *fucoxanthin,* which is absent in higher plants, predominates in brown algae, where it occurs in light-harvesting complexes along with Chl *a* and Chl *c*.

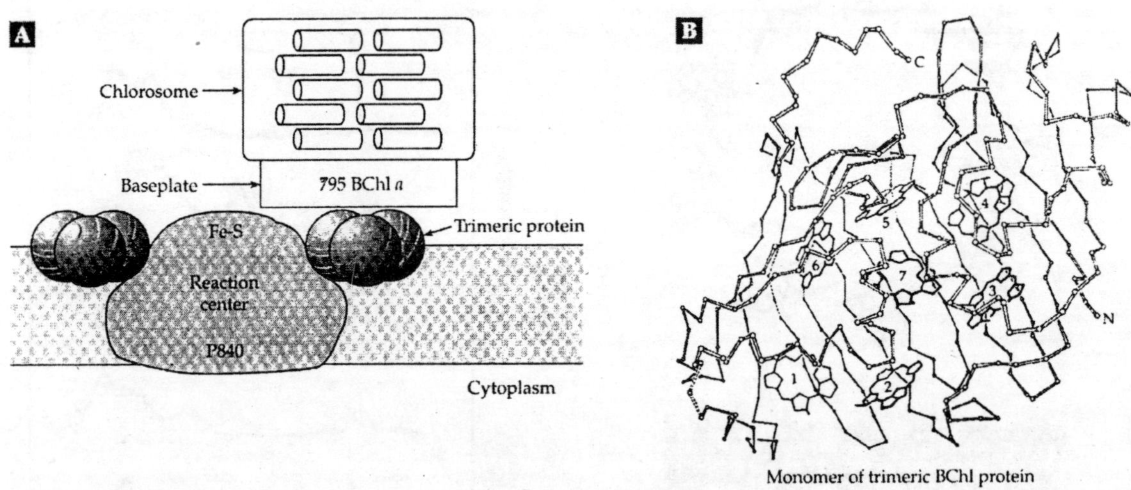

Fig. 2.26. (A) Model of a light-harvesting chlorosome from green photosynthetic sulfur bacteria such as *Chlorobium tepidum* and species of *Prosthecochloris.* The chlorosome is attached to the cytoplasmic membrane via a baseplate, which contains the additional antenna bacteriochlorophylls (795 BChl *a)* and is adjacent to the trimeric BChl protein shown in (B) and near the reaction center. After Li *et al.*, and Remigy *et. al.*, (B) Alpha carbon diagram of the polypeptide backbone and seven bound BChl *a* molecules in one subunit of the trimeric protein from the green photosynthetic bacterium *Prosthecochloris.* For clarity, the magnesium atoms, the chlorophyll ring substituents, and the phytyl chains, except for the first bond, are omitted. The direction of view is from the three-fold axis, which is horizontal, toward the exterior of the molecule.

A family of Chl *a*/*b* binding proteins are found in green plants. These have apparently evolved independently of chlorophyll-binding proteins of green bacteria. Quantitatively most important is the complex known as LHCII, the major Chl *a*/*b* protein associated with PSII and which may also provide energy to PSI. This one protein, whose structure is shown in Fig. 2.30, is thought to bind half of all of the chlorophyll in green plants. The protein is organized as trimers. Each 232-residue monomer binds 5 – 6 Chl *b*, 7 – 8 Chl *a*, ~ two molecules of lutein, and one of neoxanthin. LHCII also carries all four characteristic thylakoid lipids: mono- and digalactosyl diacylglycerols, phosphatidylglycerol, and sulfoquinovosyl diacylglycerol. The Chl *a* and Chl *b* molecules are in close contact (Fig. 2.30). Subpicosecond transient absorption spectroscopy indicates that half of the Chl *b* to Chl *a* energy transfers occur in < 0.2 ps. Notice the close association of the two luteins in Fig. 2.30 with the chlorophyll rings. The carotenoids are thought to quench chlorophyll triplet states to prevent formation of singlet oxygen.

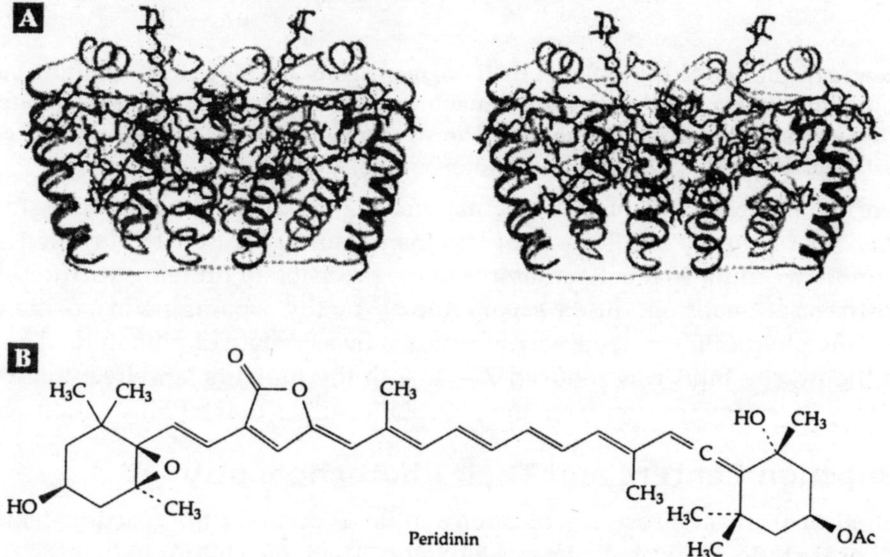

Fig. 2.27. (A) Stereoscopic drawing of light-harvesting complex from the dinoflagellate protozoan *Amphidinium carterae*. The central cavity contains eight molecules of peridinin, two of which can be seen protruding from the top. Deeply buried toward the bottom are two molecules of Chl *a*. Also present are two molecules of digalactosyl diacylglycerol.

PSII also contains several additional chlorophyll-binding proteins, designated CP24, CP26, CP29, CP43, CP47, etc. These lie on both sides of a pair of reaction-center cores. A large fraction of the LHCII complexes are separate from the reaction centers and are mobile, while a smaller fraction are bound to the outer ends of the core complex. The light-harvesting chlorophylls of PSI are arranged around the core. Some are bound in the N-terminal part of the major core subunits, the products of genes *PsaA* and *PsaB*. They bind ~90 Chl *a* and 14 β-carotene molecules per reaction center. In addition, the peripheral LHCI, composed of four different proteins arranged around the core, binds ~ 110 Chl *a* and Chl *b* and ~70 molecules of xanthophyll. Some species of cyanobacteria use antenna rings around their PSI trimers instead of phycobilinosomes.

In every case the light-harvesting complexes are arranged to allow rapid and very efficient transfer of electronic excitation energy from one chromophore to another and finally to the chromophores of the reaction center. The speed and efficiency appear to depend upon very rigid

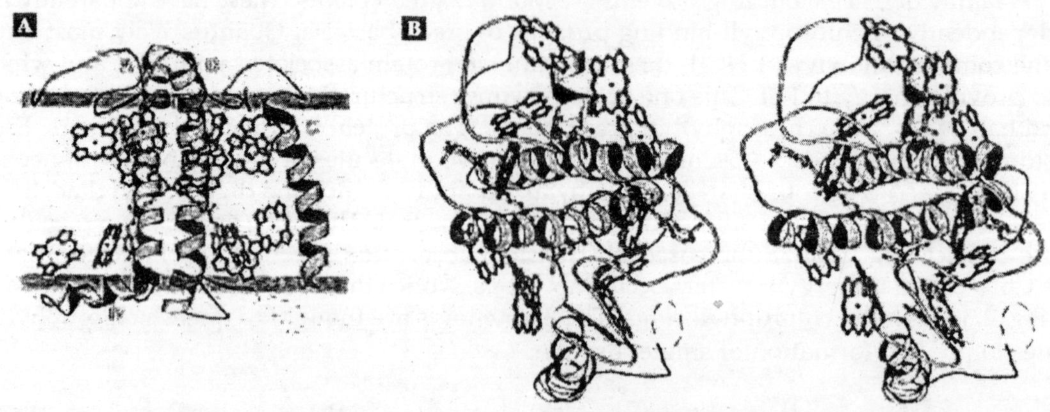

Fig. 2.28. Views of light-harvesting protein LHCII of green plants. (A) Side view indicating the approximate position in the lipid bilayer of the thylakoid membrane. Helices are labeled A–D. (B) Stereoscopic top view from the stromal side of the membrane. The structure, at 0.34 run resolution, was determined by electron crystallography on highly ordered two-dimensional crystals.

structures of the proteins and precise orientations of the bound chromophores to allow direct excitonic transfer of energy at distances of less than 2 nm or transfer by the Förster mechanism at distances not exceeding 10 nm. An example of the precision of protein structures was observed when a posttranslationally modified asparagine N^5-methylasparagine at position 72 of the β subunit of many phycobiliniproteins was substituted by aspartate or glutamine. The fluorescence lifetime of the nearby bilin was reduced 7 – 10% in the mutants, an effect that could cut the >95% efficiency of energy transfer from the phycobilinisomes to the PSII reaction center.

4. The Reaction Centers and Their Photochemistry

The initial or primary processes of photosynthesis occur in the reaction centers in which chlorophyll or bacteriochlorophyll absorbs a photon. Then, the chlorophyll, in its singlet excited state (Chl*), donates an electron to some acceptor A to form a radical A^- and to leave an oxidized chlorophyll Chl^+ radical

$$\text{Chl} \xrightarrow{h\nu} \text{Chl}^* \xrightarrow{A} \text{Chl}^+ + A^- \quad ...(2.31)$$

In the scheme of Fig. 2.18, acceptor A is Q_A for PSII and A_0 for PSI. The oxidized chlorophyll (Chl^+) quickly reacts further by receiving an electron from some donor.

The photooxidation of chlorophyll indicated in Eq. 2.31 is accompanied by bleaching in the principal light absorption band. However, since there is so much light-gathering chlorophyll for each reaction center, the effect is small. The study of the process has been aided greatly by preparation of isolated bacterial photochemical reaction centers.

Reaction Centers of Purple Bacteria

The exact composition varies, but the properties of reaction centers from several genera of purple bacteria are similar. In *Rhodopseudornonas viridis* there are three peptide chains designated H, M, and L (for heavy, medium and light) with molecular masses of 33, 28, and 24 kDa, respectively. Together with a 38-kDa tetraheme cytochrome (which is absent from isolated reaction centers of other species) they form a 1:1:1:1 complex. This constitutes reaction center P870. The three-dimensional structure of this entire complex has been determined to 0.23-nm resolution. In addition to the 1182 amino acid residues there are four molecules of bacteriochlorophyll (BChl), two of bacteriopheophytin (BPh), a molecule of menaquinone-9, an atom of nonheme iron, and four molecules of heme in the *c* type cytochrome. In 1984, when the structure was determined by Deisenhofer and Michel, this was the largest and most complex object whose atomic structure had been described. It was also one of the first known structures for a membrane protein. The accomplishment spurred an enormous rush of new photosynthesis research, only a tiny fraction of which can be mentioned here.

The reaction centers are embedded in the cytoplasmic membranes of the bacteria, with the bottom of the structure, protruding into the cytoplasm and the heme protein at the top projecting out into the periplasm which lies within infoldings of the plasma membrane. Subunits L and M each contain five ~4.0 nm long roughly parallel helices, which span the cytoplasmic membrane. Another membrane-spanning helix is contributed by subunit H, which is located mainly on the cytoplasmic side. An approximate twofold axis of symmetry relates subunits L and M and the molecules of bound chlorophyll and pheophytin.

Spectral measurements suggesting exciton splitting were among early observations that led to the conclusion that the bacteriochlorophyll involved in the initial photochemical process exists as a dimer or *special pair* a conclusion verified by the structure determination. The special pair of BChl *b* lies in the center of the helical bundle that is embedded in the membrane. Nearly perpendicular to the rings of the special pair are two more molecules of BChl *b*. The central magnesium atoms of all four bacteriochlorophylls are held by imidazole groups of histidine side chains. Below the chlorophylls are the two molecules of bacteriopheophytin and below them the nonheme iron and the menaquinone, the first quinone acceptor Q_A. It corresponds to Q_A of PSII.

Isolated reaction centers usually contain or will bind a second quinone, which may be ubiquinone-10 (Q_{10}) and which is usually designated Q_B. Its binding site is to the left of the nonheme iron in Fig. 2.31 in a position symmetrically related to that Q_A. The of reaction centers also contain a carotenoid 1, 2-dihydroneurosporene. The reaction centers of purple bacterium, *Rhodospirillum rubrum,* each contain one molecule of spirilloxanthin; a variety of carotenoids are present in other species. The reaction centers of a third purple bacterium, *Rhodobacter spheroies,* are closely similar in structure to that in Fig. 2.31 but lack the tetraheme. Reaction centers of these bacteria accept electrons directly from a soluble cytochrome *c*.

Reaction Center Kinetics

After an 0.8-ps or shorter flash of light the decay of the singlet excited state of the bacteriochlorophyll dimer in isolated reaction centers can be followed by loss of its characteristic fluorescence. The lifetime of this excited state in *R. sphaeroides* is only 4 ps indicating a rapid occurrence of the initial electron transfer of Eq. 2.31. A rise in absorbance at 1250 nm is interpreted as formation of the bacteriochlorophyll cation radical $BChl^+$ in the special pair. Other spectral changes support the formation of BPh^- as the first reduction product. However, this is thought to occur in two steps with the monomeric BChl receiving the electron in ~2.8 ps and passing it to the pheophytin in ~0.7 ps corresponding steps for PSII of green plants are also shown.

The quinone Q_A (the secondary acceptor) is next reduced by the BPh^- radical in ~200 ps with development of a characteristic EPR signal at g = 1.82. Over a much longer period of time

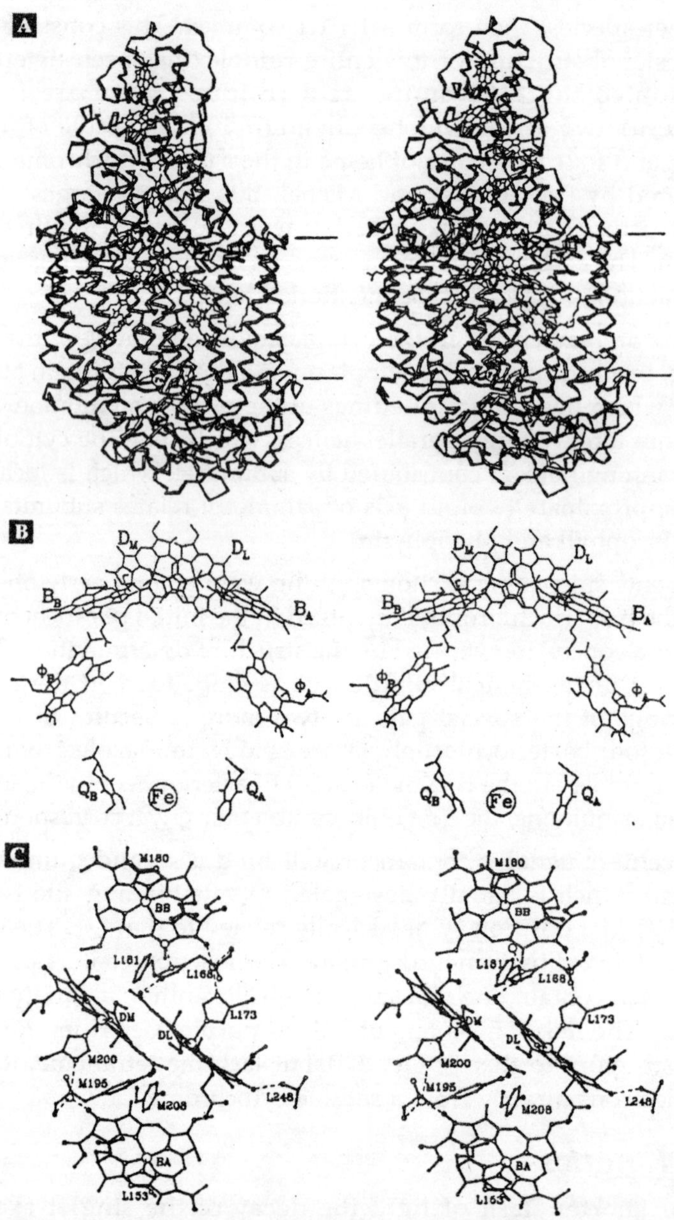

Fig. 2.29. (A) Stereoscopic ribbon drawing of the photosynthetic reaction center proteins of *Rhodopseudomonas viridis*. Bound chromophores are drawn as wire models. The H subunit is at the bottom; the L and M subunits are in the center. The upper globule is the cytochrome *c*. The view is toward the flat side of the L, M module with the L subunit toward the observer. (B) Stereo view of only the bound chromophores. The four heme groups Hel-He4, the bacteriochlorophylls (Bchl) and bacteriopheophytins (BPh), the quinones Q_A and Q_B, and iron (Fe) are shown. The four hemes of the cytochrome are not shown in (B). From Deisenhofer and Michel.[320] (C) Stereoscopic view of the Bchl *b* molecules along the local twofold axis. The special pair (D_M, D_L) is in the center with its tetrapyrrole rings almost perpendicular to the plane of the paper; the monomeric chlorophylls are labeled B_B and B_A. The four histidine ligands to the magnesium ions of the bacteriochlorophylls as well as two tyrosines (M195 and M208) and three water molecules (large circles) are also shown.

(~ 320 ns) an electron passes from the tetraheme cytochrome subunit to the Chl^+ radical in the special pair. The relatively slow rate of this reaction may be related to the fact that the bacteriochlorophyll of the special pair is 2.1 nm (center-to-center) from the nearest heme in the

	PSII (Green plants)		*Rhodopseudomonas Viridis*
	H_2O		
	↓		
	$(Mn)_4$ Manganese cluster		Cytochrome *c*
	↓		↓
	Yz(Tyr)		Fe (Tetraheme cytochrome *c*)
	↓		↓
	Chl_2 (P680)		$BChl_2$ Special pair cation radical (P870)
	↓		↓
	Chl_2^*		$BChl_2^*$ $h\nu$ Electronic excitation of $BChl_2$
~10 ps	↓	2.8 ns	↓
	$Chl_A(A_0)$		$BChl_A$
35-200 ps	↓	0.7 ns	↓
	$Phe(A_1)$		$BPhe_A$
	↓	200 ps	↓
	Q_A(Plastoquinone)		Q_A (Menaquinone)
	↓	24 μs	↓ 2e– in two steps
	Q_B (Plastoquinone)		Q_B (Ubiquinone)

cytochrome while BPh and Q_A are only 1.4 nm apart. Over a period of ~24 μs after formation of the radical anion Q_A^- an electron from the Q_A^- radical is passed to Q_B, a weakly bound ubiquinone-9, to form the Q_B^- radical. Upon absorption of a second photon by the special BChl pair another electron is passed through the chain to form Q_A^- Q_B^-. Uptake of two protons with transfer of the second electron from Q_A^- to Q_B^- yields the ubiquinol QH_2, which dissociates from its binding site (Q_B) into the ubiquinone pool dissolved in the lipids of the membrane bilayer.

Why is this multistep sequence of electron transfers necessary? A variety of techniques such as femtosecond IR and electronic spectroscopy, resonance Raman spectroscopy at low temperatures, and study of many mutants have been directed toward an answer to this question. It has been generally accepted that light energy absorbed by any one of the *"monomeric"* Chl or pheophytins in the reaction centers is funneled *"downhill"* to the special pair within 0.1 – 0.2 ps to generate *P**. The ultrafast ~3 ps electron transfer from *P** to the adjacent monomeric BChl or Chl is necessary to prevent loss of energy by fluorescence from *P**. The subsequent energetically downhill transfer to a pheophytin and on to Q_A prevents reverse electron transfer, which could also lead to fluorescence. Both the efficiency and the quantum yield are very high.

The rate of the ultrafast proton transfer becomes even higher at cryogenic temperature, suggesting quantum mechanical tunneling. The transfer is generally treated using Marcus theory, which indicates a very small reorganization energy for the process. Another aspect of the process is a possible coupling of a vibrational mode of the protein matrix to the electron transfer. Femtosecond near-IR spectra show low-frequency vibrational modes of the excited-state reaction center chromophores which may facilitate electron transfer. The transfer of an electron from the tetraheme cytochrome *c* of *R. viridis* or from the small cytochrome c_2 of *Rhodobacter spheroids to* Chl^+ of the special pair has similar characteristics but is slower than the initial electron transfer from *P**. On the other hand, electron transfers from Q_A^- to Q_B^- involve two distinct steps and coupled uptake of two protons. An unexplained fact is that photochemical electron transport through the reaction centers always occurs through the L-side to Q_A rather than the M-side. However, rapid electron transfer to the pheophytin on the B-side (M-side) has been observed following excitation with blue light. This may represent a photoprotective mechanism. In the PSI system, in which the two phylloquinones are tightly bound both the A-side and the B-side seem to function in electron transfer.

Table 2.4. Properties of Various Reaction Centers

	Rhodopseudomonas viridis (now Blasatochloris)	*Green plants and* PSII	*Cyanobacteria* PSI	*Green sulfur bacteria, heliobacteria*[a]
Subunits	L/M,H	D1/D2, CP43,CP47	PsaA/PsaB nine others	$(PscA)_2$
Masses, kDa	24/28, 33	38/39.4	83/83	65/65
Input	Cyt *c* (4 Fe)	H_2O, $(Mn)_4$, Y_z, Y_D	Cyt *f*, Plastocyanin or cyt *c*	Cyt c_1, c_2
Special pair	P870	P680 (BChl *b*)$_2$	P700 (Chl *a*)$_2$	P840 (Chl *a*)$_2$
Monomeric chlorophyll	$BChl_L$, $BChl_M$		A_0 Both Chl *a*	A_0 Both Chl *a*-like
Pheophytin	$BPhe_L$, $BPhe_M$		A_0 (Phe, Phe)	
Quinone	Q_B, Q_A (Ubiquinone, menaquinone-9)	Q_B, Q_A (Plastoquinone)	Q_B, Q_A (A_1) (Both phylloquinone in cyanobacteria plastoquinone in chloroplasts)	Q_B, Q_A (Both menaquinone-7)[b]
Iron	Fe^{3+}		F_X (Fe_4S_4)	F_X
Output	Ubiquinone Cyt bc_1, Cyt c_2	Ubiquinone Cyt b_6f	F_A, F_B, Fd NADP$^+$	F_A, F_B, Fd NAD$^+$

Cyclic Photophosphorylation in Purple Bacteria

QH_2 is eventually dehydrogenated in the cytochrome bc_l complex, and the electrons can be returned to the reaction center by the small soluble cytochrome c_2, where it reduces the bound tetraheme cytochrome or reacts directly with the special pair in *Rhodobacter spheroides*. The overall reaction provides for a cyclic photophosphorylation that pumps 3 – 4 H^+ across the membrane into the periplasmic space utilizing the energy of the two photoexcited electrons. These protons can pass back into the cytoplasm via ATP synthases located in the same membrane with their catalytic centers in the cytosol.

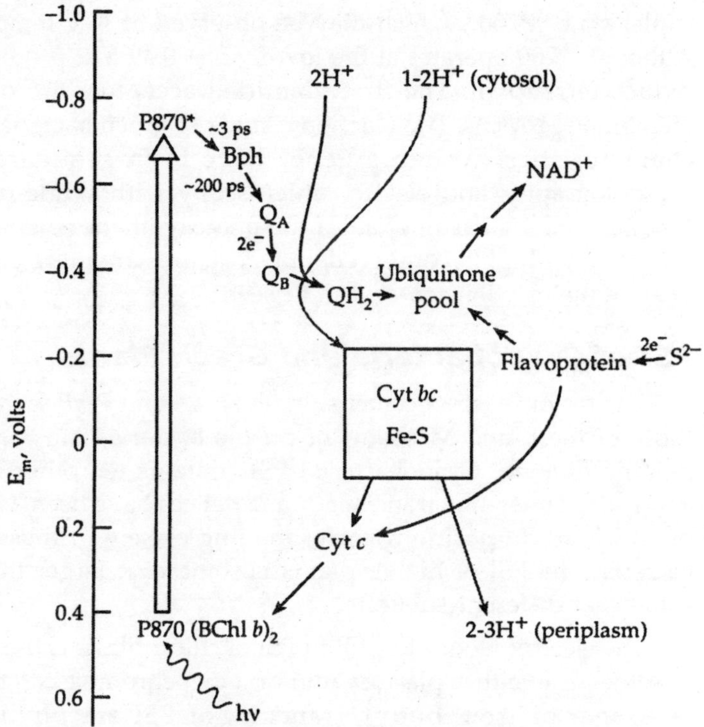

Fig. 2.30. Simplified diagram of cyclic electron flow in purple bacteria. Two protons from the cytoplasm bind to Q_B^{2-} in the reaction center to form QH_2 (ubiquinol), which diffuses into the ubiquinone pool. From there it is dehydrogenated by the cytochrome bc_1 complex with expulsion of two protons into the periplasm. A third and possibly a fourth proton may be pumped (green arrows) across the membrane, *e.g.*, via the Q cycle. The protons are returned to the cytoplasm through ATP synthase with formation of ATP. Some electrons may flow to the reaction centers from such reduced substrates as S^{2-} and some electrons may be removed to generate NADPH using reverse electron transport.

Comparison with Other Reaction Centers

Subunit L of the R. *viridis* reaction center was found to have a 25% sequence homology with a quinone-binding protein now known as D1, a component of the reaction center core of PSII of chloroplasts. This protein was identified as the specific target protein for inhibition by herbicides such as DCMU and atrazine. These compounds act as competitive inhibitors of quinone binding and bind in the Q_B site in the R. *viridis* reaction center. This fact, together with the discovery that the core of PSII of green plants consists of a heterodimer of the related polypeptides Dl and D2, suggested that ESII is very similar to the bacterial center of Fig. 2.31. Both use a quinone as primary acceptor. However, P680 operates at a more positive potential (E_m ~ 1 V), consistent with the fact that it must provide an oxidizing agent able to oxidize H_2O to O_2 ($E^{o\prime}$ = +0.82 V). Plastoquinone rather than menaquinone is the primary acceptor in PSII. A chlorophyll *a* dimer is apparently the initial electron donor. The methyl ester carbonyl groups on the edges of rings I of BChl may coordinate to other groups of the proteins. These ester groups are absent in Chl *a*.

The PSI reaction center (P700) of maize chloroplast was also found to contain a pair of homologous polypeptides with appropriately placed histidyl residues for chlorophyll binding. FTIR spectra also indicate the presence of two Chl *a* molecules. Small shifts in IR frequencies upon electronic excitation suggest that one chlorophyll (P_1) is hydrogen-bonded through its 109-ester and 9-oxo groups while the chlorophyll (P_2) is free. This may account for the low value of E_m. The charge on $P700^+$ appears to be carried in part on both chlorophylls. However, in the triplet state ^{3}P700*, which may be observed at low temperature, excitation is localized on P_1. Although P700 operates at the low E_m of + 0.49 V it produces a powerful reducing agent able to reduce ferredoxin. The first identified acceptors are other Fe-S centers present in integral membrane proteins. The reaction centers of green bacteria and PSI of cyanobacteria have similar characteristics. As more genes have been sequenced and X-ray diffraction, electron crystallography, and electron microscopy with single-particle averaging have advanced, the fundamental similarity of all of the photosynthetic centers has been confirmed. Gene sequences are often not highly conserved, but structures are more conserved. An evolutionary relationship of all of the reaction centers can be seen.

PSI of Cyanobacteria and Green Plants

The major reaction-center subunits *PsaA* and *PsaB* each have a C-terminal domain, resembling those of the L and M chains of purple bacteria, and a large N-terminal antenna-chlorophyll-binding domain. Cyanobacterial PSI contains ten other subunits, PsaC to PsaF, PsaI to PsaM, and PsaX. Thirty-one transmembrane helices have been assigned to the various subunits, several of which are in positions corresponding closely to those in the reaction centers of the purple bacteria. The PSI of higher plants is somewhat larger than that of cyanobacteria and contains somewhat different subunits.

The electron donor to Chl$^+$ in PSI of chloroplasts is the copper protein plastocyanin. However, in some algae either plastocyanin or a cytochrome *c* can serve, depending upon the availability of copper or iron. Both Q_A and Q_B of PSI are phylloquinone in cyanobacteria but are plastoquinone-9 in chloroplasts. Mutant cyanobacteria, in which the pathway of phylloquinone synthesis is blocked, incorporate plastoquinone-9 into the A-site. Plastoquinone has the structure with nine isoprenoid units in the side chain. Spinach chloroplasts also contain at least six other plastoquinones. Plastoquinones C, which are hydroxylated in side-chain positions, are widely distributed. In plastoquinones B these hydroxyl groups are acylated. Many other modifications exist including variations in the number of isoprene units in the side chains. There are about five molecules of plastoquinone for each reaction center, and plastoquinones may serve as a kind of electron buffer between the two photosynthetic systems.

Look at the Z scheme of Fig. 2.17. The lower end of each vertical arrow is located at an electrode midpoint potential E_m (or $E^{o\prime}$) for the couple P$^+$/P, *i.e.*, for a one-electron reduction of the Chl$^+$ or BChl$^+$ radical. The top of the arrow is at the estimated value of E_m for the excited state P*. It is more negative than the ground-state value of E_m by the energy (in electron volts) of the light absorbed. This is a little misleading for it is not commonly appreciated that light carries entropy as well as energy. An important consequence of this fact is that not all of the energy of sunlight could be harnessed for chemical work. Knox has calculated that at 700 nm at most 78% of the energy could be captured. See Parson for further discussion. Nevertheless, the photoexcited P700* with E_m = –1.26 V is able to reduce a series of membrane-bound Fe-S centers of $E^{o\prime}$ ~ –0.5 to –0.6 V. There are three of these designated F_X, F_A, and F_B. Center F_X is an Fe_4S_4 cluster located at almost the same position as the single Fe^{3+} of the *R. viridis* reaction center. F_A and F_B are also Fe_4S_4 clusters both of which are carried on the small 79-residue PsaC. This protein binds to the reaction center on the stromal side as shown in Fig. 2.33. Close to it are two other subunits, PsaD and PsaE, which appear to assist the docking of ferredoxin or flavodoxin to PsaC and cluster F_X. The photosynthetic centers of green photosynthetic sulfur bacteria also have centers F_X, F_A, and F_B.

The soluble electron carriers released from the reaction centers into the cytoplasm of bacteria or into the stroma of chloroplasts are reduced single-electron carriers. Bacterial ferredoxin with two Fe_4S_4 clusters is formed by bacteria if enough iron is present. In its absence flavodoxin which may carry either one or two electrons, is used. In chloroplasts the carrier is the soluble *chloroplast ferredxin,* which contains one Fe_2S_2 center. Reduced ferredoxin transfers electrons to $NADP^+$ via *ferredoxin: $NADP^+$ oxidoreductase,* a flavoprotein of known three-dimensional structure.

PSII and Formation of Oxygen

The structure of PSII has been difficult to determine directly, but its core has been modeled in atomic detail using bacterial reaction centers as a guide. More recently electron crystallography provided a three-dimensional image at 0.8 nm resolution. The resolution has been extended down to 4 nm by X-ray crystallography. The structure of the cyanobacterial PSII is very similar to that of green plants. PSII contains at least 17 protein subunits, all of which are encoded by chloroplast genes. The large structurally similar D1 and D2 form the core. They are encoded by genes *PsbA* and *PsbD* and have molecular masses of 38.0 and 39.4 kDa, respectively. Both Q_A and Q_B are plastoquinone. It is of historical interest that these cofactors were first designated simply as Q, not for quinone but for *quencher.* This is because Q_A apparently quenched the fluorescence of P680, the reaction-center chlorophyll *a.* If chloroplasts are irradiated with 650-nm light, PSII is activated but PSI is not. Under these conditions Q_A becomes reduced, and the fluorescence of Chl^+ increases, presumably because the electron acceptor Q_A is absent. If PSI is activated by addition of far-red light, Q_A remains more oxidized, and fluorescence is quenched by a mechanism that appears to depend upon one of the additional bound chlorophylls, as well as the chlorophyll pair Chl_2 and the 9-kDa cytochrome. The latter is an essential PSII subunit (*PsbE* gene), which forms a tight complex with the D1/D2 pair.

Other subunits in PII include the 56- and 50-kDa antenna proteins CP43 (*PsbC* gene) and CP47 (*PsbB* gene). Three *extrinsic proteins,* which bind to the lumenal side of thylakoid membrane, are the 33-kDa *manganese stabilizing protein* (*PsbO* gene), cytochrome c_{550}, and a 12-kDa subunit. The position of the larger two of these proteins are shown in the model in Fig. 2.34, which is based on 0.8-nm resolution data. These extrinsic proteins seem to function together to facilitate binding not only of Mn ions but also of Ca^{2+} and Cl^-, both of which are essential for O_2 evolution. Other smaller subunits are also present. Together with its antenna complexes PSII may form large supercomplexes with as many as 25 subunits. The distribution of complexes varies in the different regions of the thylakoid, *e.g.,* the stroma lamellae and grana stacks.

The four-electron dehydrogenation of two water molecules to give one molecule of O_2 by PSII is still not well understood. From experiments on oxygen evolution in the presence of repeated short flashes of light it was found that a four-quantum process is required. There must be some way of storing oxidizing equivalents until enough are present to snap together an oxygen molecule. There is abundant evidence that manganese is required for this process and that the oxidation of H_2O occurs on a cluster of four atoms of manganese.

The 33-kDa protein PsbO, which is present in all oxygen-forming photosynthetic organisms, is closely associated with the Mn_4 cluster. Removal of this protein leads to a gradual loss of two of the four Mn ions. The structure of the Mn_4 cluster is not yet certain, but on the basis of EXAFS spectroscopy a pair of di-μ-oxo bridged Mn dimers linked by carboxylates and with a fifth μ-oxo bridge as in Fig. 2.35 has been proposed. The distance between Mn ions in the di-μ-oxo-Mn_2 groups is ~0.27nm, and the planes of these groups are roughly parallel to the surface of the thylakoid membranes. The protein groups that bind the Mn atoms include carboxylates, and probably one or more histidine imidazole groups, perhaps of His 332 and His 337 of the D1 chain.

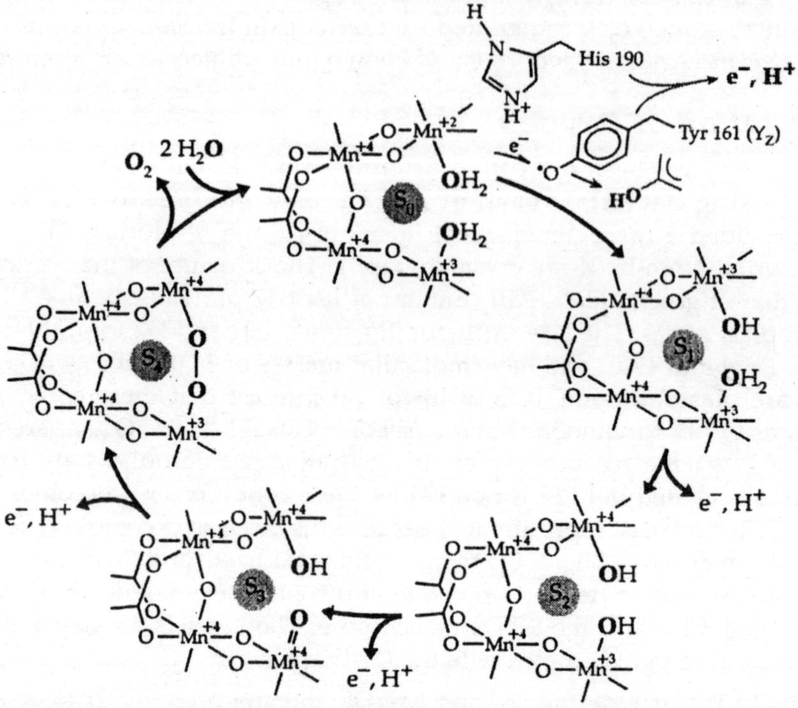

Fig. 2.31. Proposed sequence of S-states of the manganese cluster of photosystem II. The successive states as two molecules of H_2O (green oxygen atoms) are converted to O_2 is shown with the successive states $S_0 - S_4$ labeled. To save space and possible confusion tyrosine 161 (Y_Z) and the nearby His 190 are shown only by S_1. The Y_z radical is thought to remove a hydrogen atom or H^+ from one bound H_2O and an electron from one Mn ion at each of the four S-states $S_0 - S_3$ functioning in each case to eject a proton into the thylakoid lumen and to transfer an electron to P^+ of the reaction center. However, the exact sequence of e^- transfer and H^+ release may not be shown correctly.

The immediate donor of an electron to the reaction-center cation P^+ ($ChlZ^+$) of PSII was identified by EPR spectroscopy as a tyrosine radical. On the basis of directed mutations this tyrosine, which is usually designated Y_Z, is Tyr 161 of the Dl chain and is located ~1.2 nm from one of the chlorophylls of the $(Chl)_2$ pair. The two molecules in the pair are not in close contact, their central Mg atoms being ~1.0 nm apart. One of the two probably forms the $P688\cdot^+$ intermediate. Y_Z is also close to the Mn_4 cluster and to the imidazole group of His 190 of subunit D1. The $Y_Z^\bullet$ radical is able to accept an electron from the Mn_4 cluster within 30-1300 μs depending upon the oxidation state of the cluster. If a proton is transferred synchronously from a bound H_2O, a neutral –OH group will be created on Y_Z. The proton may then be transferred to His 190, which can eject the proton on its other nitrogen atom into the lumen. Alternatively $Y_Z^\bullet$ may accept an electron to become tyrosinate $-O^-$, which then donates an electron to P^+, while His 190 accepts a proton directly from a bound H_2O.

In a mechanism proposed by Hoganson and Babcock four successive transfers, each of one H^+ + one e^-, leads to a three-electron oxidation of Mn ions, *e.g.*, from the 2^+ and 3^+ oxidation states to all Mn^{4+}, and to joining of the two water oxygens to form a manganese peroxide linkage.

Oxidation of the peroxide dianion to O_2 by the adjoining Mn^{4+} and Mn^{3+} ions completes the cycle. This mechanism is hypothetical, and various alternatives have been presented. Most assume a structure similar to that shown in Fig. 2.35. Some are based on nonenzymatic model reactions. Chloride ions are essential to O_2 formation especially in going from state S_2 to S_3 and S_3 to S_0. This suggests that Cl^-may function in passing electrons between Mn ions. Calcium ions are also necessary, but if has been difficult to establish an exact function. A bicarbonate ion may also be an essential ligand in the Mn_4 cluster.

ATP Synthesis in Chloroplasts

The flow of electrons between PSII and PSI is of great importance for ATP formation. As previously mentioned, plastocyanin is usually the immediate donor to P700 and serves as a mobile carrier to bring electrons to this reaction center. In this function it is analogous to cytochrome *c* of mitochondrial membranes. The essentiality of plastocyanin was shown by study of copper-deficient *Scenedesmus.* The photoreduction of CO_2 by H_2 is impaired in these cells, but the Hill reaction occurs at a normal rate.

Like mitochondria, chloroplasts (when illuminated) pump protons across their membranes. However, while mitochondria pump protons to the outside, the protons accumulate on the inside of the thylakoids. The ATP synthase heads of coupling factor CF_1 are found on the outside of the thylakoids, facing the stromal matrix, while those of F_1 lie on the insides of mitochondrial membranes. However, the same mechanism of ATP formation is used in both chloroplasts and mitochondria.

The cleavage of water at PSII also occurs on the inside of the thylakoids. The splitting of one water molecule leaves two protons (one per electron) inside the thylakoids, while the electrons are "photoejected" through the lipid bilayer to acceptor Q_A on the outside. The chlorophyll in PSI is likewise in contact with the inside of the bilayer with acceptor A_0 on the outside. Since the conversion of NAD^+ to NADH on the outside generates a proton, the overall reaction would be the pumping of one and a half protons per electron passing through the Z scheme.

The pathways involved in cyclic photophosphorylation in chloroplasts are not yet established. Electrons probably flow from the Fe–S centers Fd_X, Fd_a, or Fd_b back to cytochrome b_{563} or to the PQ pool as is indicated by the dashed line in Fig. 2.18. Cyclic flow around PSII is also possible. The photophosphorylation of inorganic phosphate to pyrophosphate (PP_i) occurs in the *chromatophores* (vesicles derived from fragments of infolded photosynthetic membranes) from *Rhodospirillum rubrum.* The PP_i formed in this way may be used in a variety of energy-requiring reactions in these bacteria. An example is formation of NADH by reverse electron transport.

Protection of Chloroplasts Against Radiation and Oxygen

Carotenoids often act as accessory light-receiving pigments, but an additional function is protecting photosynthetic organisms against toxic effects of light. Carotenoid photoprotection has been demonstrated in photosynthetic bacteria and in the reaction centers and light-harvesting complexes (LHC) of green plants. Excited chlorophyll molecules can pass from the singlet ($^1Chl^*$) to the triplet ($^3Chl^*$) state by intersystem crossing. The triplet chlorophyll can then react with ordinary oxygen ($^3O_2$) to form singlet oxygen ($^1O_2$). The formation of $^3Chl^*$ is favored when the intensity of sunlight is high and energy is absorbed in the LHCs faster than it can be utilized in the reaction centers. Carotenoids are able to quench excitation of both $^3Chl^*$ and $^1O_2$. Strains of *Rhodobacter sphaeroides* that lack carotenoids rapidly form both $^3BChl^*$ and $BChl^+$ cation radicals in their LH1 and LH2 complexes. However, the presence of carotenoids quenches the triplet bacterio-chlorophyll effectively and no formation of $BChl^+$ radicals is seen. Another effect of excessive light energy is the reaction of O_2 with the highly reducing P700* of photoexcited PSI. This interaction can form triplet oxygen, which can react to generate superoxide anion radicals, H_2O_2, and hydroxyl radicals.

In higher plants the quenching of both Chl* and $^1O_2$ depends upon carotenoids and also upon the large transthylakoid membrane potential that is generated by high light intensities. High light intensity also induces a rapid reductive deoxygenation of epoxycarotenoids via the *xanthophyll* or *violaxanthincycle.* Epoxycarotenoids are found only in photosynthetic O_2 evolving organisms. Although occurring in response to light, the cyclic photodeoxygenation and reoxygenation is not a non-photochemical process. Violaxanthin contains the epoxy structure at both ends of the molecule. Reduction of one end produces *antheroxanthin* and of both ends *zeaxanthin.* These three Carotenoids are found in almost all higher green plants and algae. The deepoxidation is mediated by ascorbic acid, occurs in the lumen of the thylakoids, and is favored by the low pH developed during illumination. Epoxidation is catalyzed by a monooxygenase located on the stromal side. The significance of the xanthophyll cycle is puzzling. There may be specific binding sites, perhaps in the inner antenna complex on CP29 that bind zeaxanthin or antheroxanthin. This could alter the antenna structure to form an *"exciton trap"* in which the fluorescence lifetime would be decreased and excitation energy would be dissipated rapidly as heat. Zeaxanthin is also found in the macular area of the primate retina.

Recent studies, using an *Arabidopsis* mutant defective in the xanthophyll cycle, point to a chlorophyll-binding protein PsbS, which participates in nonphotochemical quenching at high light intensity. Another *Arabidopsis* protein, which is probably a blue light receptor, participates in an avoidance response by which chloroplasts move to the side wall to avoid strong light.

Yet another carotenoid function in PSII has been proposed. Under some conditions, when electron flow from Tyr_z is blocked, the bound monomeric chlorophyll Z (Chlz) acts as a secondary electron donor to $P680^+$. The cytochrome b_{559} subunits may have a similar function. Both cyt b_{559} the carotenoid are essential for assembly of PSII, and both may participate in a protective cycle. Chloroplasts generate both O_2 and powerful reducing materials such as the membrane-bound FeS centers of PSI, which may form superoxide ions by single-electron donation to O_2. Probably for this reason, chloroplasts are rich in superoxide dismutase which converts superoxide to O_2 and H_2O_2. The latter can diffuse into peroxisomes and react with catalase and peroxidases. It can also be reduced to H_2O within the chloroplasts by ascorbic acid and ascorbate peroxidase. The resulting dehydroascorbate can be reduced back to ascorbate by glutathione and dehydroascorbate reductase, in the following electron transfer sequence:

$$\text{NADPH} \rightarrow \text{G}-\text{S}-\text{S}-\text{G} \rightarrow \text{dehydroascorbate} \rightarrow H_2O_2 \qquad ...(2.33)$$

Under extreme conditions of excess light energy *photoinhibition* is observed as a result of damage to the PSII structure. The D1 polypeptide is cleaved, probably as a result of oxidation by $^1O_2$ and proteolysis. Damaged proteins are replaced and the PSII structure rebuilt, but the effect is a long-lasting decrease in photosynthetic efficiency. The cyanobacterium *Synechococcus* has three *PsbA* genes and resists UV-B radiation by exchanging a delicate Dl polypeptide with more resistant ones as necessary. Other adaptations to varying light-intensity involve movement of light-harvesting complexes from the thylakoid stacks, which contain much PSII, to the stroma lamellae, which contain more PSI. Some herbicides act by binding into the Q_B site in PSII. They may cause light-induced oxidative stress that kills the plant.

5. Control of Photosynthesis

The key reaction of the Calvin-Benson cycle of CO_2 reduction is the carboxylation of ribulose bisphosphate to form two molecules of 3-phosphoglycerate. The properties of ribulose bisphosphate carboxylase *rubisco,* which catalyzes this reaction, have already been discussed. It is controlled in part by CO_2 and by natural inhibitors but regulation of rubisco starts at the transcriptional level.

Light-induced Transcription

Plants depend upon light both as a source of energy and also for control of development. Many genes are activated by light in response to at least three groups of photoreceptors. These are *phytochromes* and the blue light responsive *cryptochromes* and the ultraviolet light *UV-B photoreceptors.* The synthesis of chlorophyll, of reaction center proteins, and of many enzymes are controlled by light-induced transcription. Among these processes are synthesis of both the large and small subunits of rubisco. The small subunits are synthesized in the cytoplasm in a precursor form. After illumination the concentration of the rubisco mRNA may be increased 100-fold. On the other hand, the large subunit of the carboxylase is encoded in chloroplast DNA, and stimulation of its synthesis by light appears to be at the translational level.

Light-induced Control via the Ferredoxin/Thioredoxin System

Rubisco is activated by CO_2 and by fructose 6-P and is inhibited by fructose 1, 6-P_2 whose accumulation is a signal to turn off the carboxylase. Conversely, fructose 6-*P* in high concentrations turns on the Calvin-Benson cycle. Like the reactions of gluconeogenesis photosynthetic CO_2 incorporation is dependent on the highly regulated fructose-1,6-bisphosphatase. In chloroplasts it is activated by light through

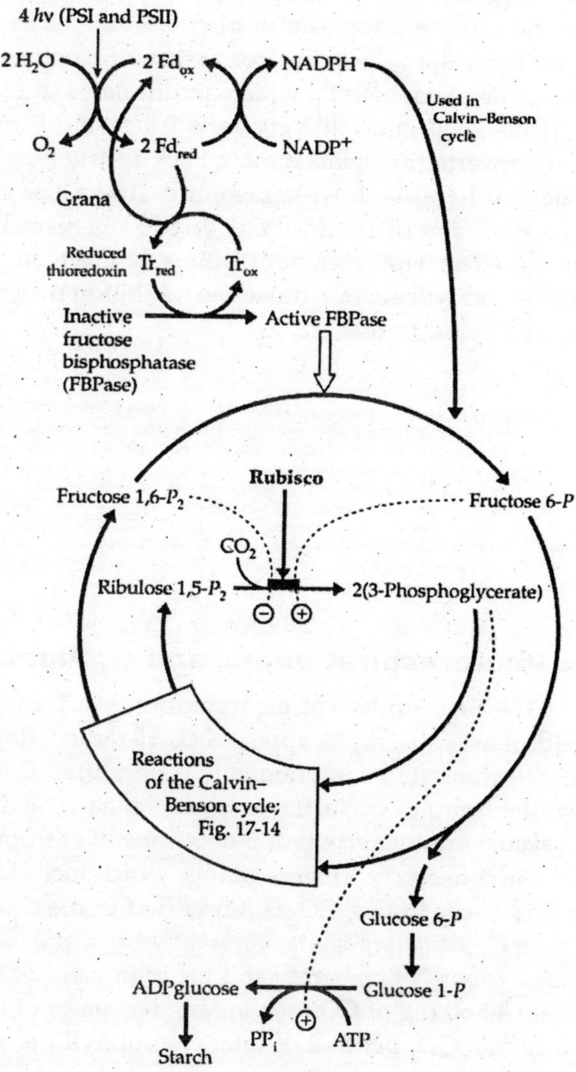

Fig. 2.32. Some control mechanisms for photosynthetic assimilation of carbon dioxide.

the mediation of reduced ferredoxin and thioredoxin. The small mobile thioredoxin is reduced to its dithiol form by reduced ferredoxin and then reduces one or more disulfide linkages in the fructose 1, 6-bisphosphatase to activate that enzyme. Other light-activated enzymes of the Calvin-Benson cycle include sedoheptulose-l, 7-bisphosphatase, the phosphoribulokinase that forms ribulose 1, 5-bisphos-phate and the $NADP^+$-dependent glyceraldehyde-3-phosphate dehydrogenase. $NADP^+$-dependent malate dehydrogenase, which has a major function in C_4 plants, is totally inactive in the dark but is activated by the ferredoxin-thioredoxin system in the light. The activity of the LHCII complex is also affected.

Another aspect of chloroplast metabolism is synthesis of starch. Formation of ADP-glucose from glucose 1-phosphate is induced by 3-phosphoglycerate, a *"feed-ahead"* type of regulation. Although fructose 2, 6-bisphosphate is absent from chloroplasts, it has an important regulatory function in the cytoplasm of plants as it does in animals. In the plant cytosol triose phosphates from the chloroplasts are converted to fructose 6-*P*, glucose 6-*P*, UDP-glucose, and sucrose. Inorganic phosphate P_i, which accumulates in plant vacuoles, also has a regulatory function. It activates the kinase that converts fructose 6-*P* to fructose 2, 6-P_2 and inhibits the phosphatase that converts the bisphosphate back to fructose 6-*P*. The accumulated fructose 2, 6-P_2 inhibits fructose-1, 6-bisphosphatase and slows the conversion of triose phosphates to sucrose. Accumulation of fructose 6-*P* due to decreased utilization for sucrose formation will have a similar effect. However, both 3-phosphoglycerate and dihydroxyacetone phosphate have opposite effects and will act to remove the inhibition by lowering the fructose 2, 6-P_2 level and to promote rapid sucrose formation.

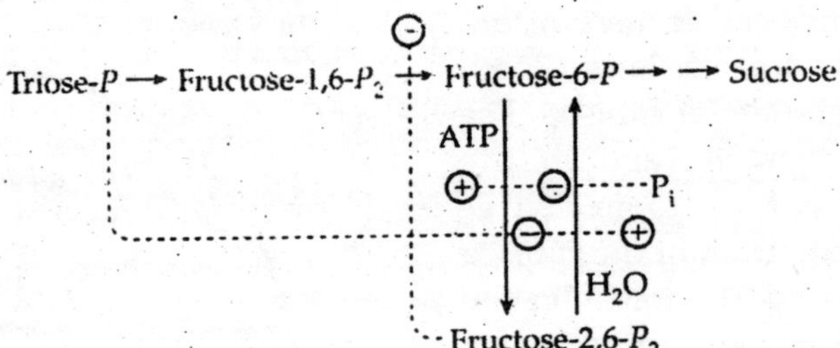

6. Photorespiration; C_3 and C_4 Plants

The first product of incorporation of CO_2 via the Calvin-Benson cycle is 3-phosphoglyceric acid. It was the rapid appearance of radioactivity from $^{14}CO_2$ in phosphoglycerate and other three-carbon (C_3) compounds that permitted Calvin and associates, using green algae, to work out the complex cycle. Green algae, spinach, and many common crop plants are often known as *C_3 plants.* Another group of plants, mostly of tropical origin and capable of extremely fast growth, behave differently. In these plants, which include sugar cane, maize, and crabgrass, radioactivity from ^{14}C-containing CO_2 is found first in the C_4 compounds oxaloacetate, malate, and aspartate. These *C_4 plants* are characterized by high efficiencies in photosynthesis, a fact that explains the rapid growth of crabgrass and the high yield of corn. Maximum rates of CO_2 incorporation may attain 40-60 mg of CO_2 per square decimeter of leaf surface per hour (~ 0.3 mmol CO_2 m^{-1} s^{-1} or ~0.10 mol CO_2 per mol of total chlorophyll per second), more than twice that for common crop plants.

Like all other organisms plants respire in the dark, but illumination of C_3 plants markedly increases the rate of oxygen utilization. This light-enhanced respiration (*photorespiration*) may attain 50% of the net rate of photosynthesis. Photorespiration prevents plants from achieving a maximum yield in photosynthesis. For this reason, its understanding and control assume great importance in agriculture. It is difficult to measure the rate of photorespiration, and the literature on the subject often refers instead to the *CO_2 compensation point*. This is the CO_2 concentration (at a given constant light intensity) at which photosynthetic assimilation and respiration balance. (Similarly; the *light compensation point* is the light intensity at which the rate of photosynthetic CO_2 incorporation and that of respiration exactly balance.) Normal air ha a CO_2 content of ~0.03% or 300 ppm. For common C_3 crop plants the CO_2 compensation point is ~40 – 60 ppm at 25°C. The C_4 plants are characterized by a much lower CO_2 compensation point, often less than 10 pprn. In strong sunlight the CO_2 level of air in a field of growing plants drops. Furthermore, as the temperature rises on a hot day, the CO_2 compensation point rises. The result is a serious decrease in efficiency of photosynthesis for the C_3 plants but not for the C_4 plants.

Metabolism of Glycolic Acid

The 2-carbon glycolic acid is formed in large quantities in the chloroplasts of C_3 plants and moves out into the cytosol. The major source of this acid is phosphoglycolate whose formation is catalyzed by rubisco in the chloroplasts through competition of O_2 for the CO_2 binding site of the enzyme. It is easy to understand why an increase in the O_2 pressure in air increases the CO_2 compensation point for a plant. Another less important source of glycolate is transketolase, which may yield glycolaldehyde as a side product. Glycolaldehyde can be oxidized readily to glycolate. Glycolate is metabolized rapidly, some in the chloroplasts, but most in the peroxisomes. There the flavoprotein glycolate oxidase converts it to glyoxylate with formation of H_2O_2. Some of the hydrogen peroxide formed may react nonenzymatically, decarboxylating glyoxylate to formate and CO_2, but most is probably destroyed by peroxidases or catalase. The latter enzyme is lacking in chloroplasts, one reason why oxidation of glycolate must occur in the peroxisomes.

Glyoxylate undergoes transamination to glycine, which can be oxidatively decarboxylated in the mitochondria. It can also be converted to serine, some of which returns to the peroxisomes to be oxidized to hydroxypyruvic acid and glyceric acid. The latter can be synthesized into glucose. The net result is the stimulation of a large amount of metabolism that ultimately produces CO_2 and apparently accounts for the light-induced respirations of plants. Because much of the glycine formed in the peroxisomes is oxidatively decarboxylated in the mitochondria, photorespiration also generates large amounts of NH_3. This is recycled into amino acids within the photosynthetic cells, an energy-requiring process.

Although much metabolism occurs as a result of photorespiration, it appears to waste energy rather than to provide energy to the plant. Why then don't plants avoid this process? Wouldn't a small change in the structure of ribulose bisphosphate carboxylase allow plants to avoid photorespiration and to grow more efficiently? The answer is not clear. It has been difficult to create such modified plants, and there is a possibility that they would not grow well. One theory is that photors piration protects plants when the CO_2 pressure is low and the absorbed light would damage the chloroplasts if there were not a way to utilize the accumulating reduced Fe-S proteins generated by PSI. Photorespiration provides a mechanism. Most efforts to breed plants with lower photorespiration rates or to inhibit it chemically have failed.

The C_4 Cycle for Concentration of Carbon Dioxide

The C_4 plants reduce their rate of photorespiration by using a CO_2 concentrating mechanism that enables them to avoid the competition from O_2. All species of C_4 plants have a characteristic

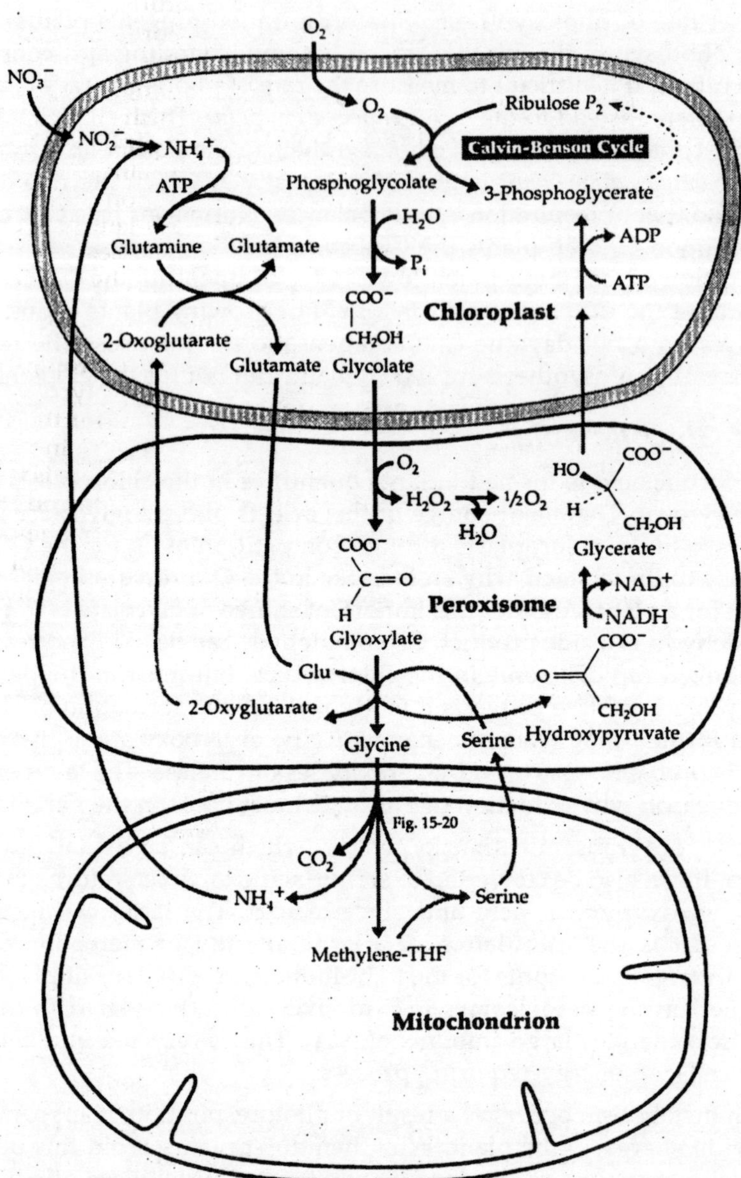

Fig. 2.33. Production of glycolate by chloroplasts and some pathways of its metabolism in peroxisomes and in mitochondria.

internal leaf anatomy in which a single dense layer of dark green cells surrounds the vascular bundles in the leaves. This *bundle sheath* is surrounded by a loosely packed layer of cells, the *mesophyll* an arrangement that is sometimes called the *"Kranz anatomy."* In C_4 plants there is a separation of the chemical reactions between the mesophyll and the bundle sheath cells. The incorporation of CO_2, as bicarbonate ion into oxaloacetate, occurs in the mesophyll cells, principally through the action of PEP carboxylase. Oxaloacetate is reduced to malate by light-generated NADPH. Alternatively, it undergoes transamination to aspartate. Both malate and aspartate then diffuse out of the mesophyll cells and into bundle sheath cells where the malate undergoes oxidative decarboxylation via the malic enzyme to pyruvate. Aspartate also can be converted to oxaloacetate, malate, and pyruvate in the same cells. The overall effect is to transport CO_2 from the mesophyll cells into the bundle sheath cells along with two reducing equivalents, which appear as NADPH following the action of the malic enzyme. The CO_2, the NADPH, and additional NADPH generated in the chloroplasts of the bundle sheath cells are then used in the Calvin-Benson cycle reactions to synthesize 3-phosphoglycerate and other materials. Of the CO_2 used in the bundle sheath cells, it is estimated that 85% comes via the C_4 cycle and only 15% enters by direct diffusion. The advantage to the cell is a higher CO_2 tension, less competition with O_2, and a marked reduction in photorespiration.

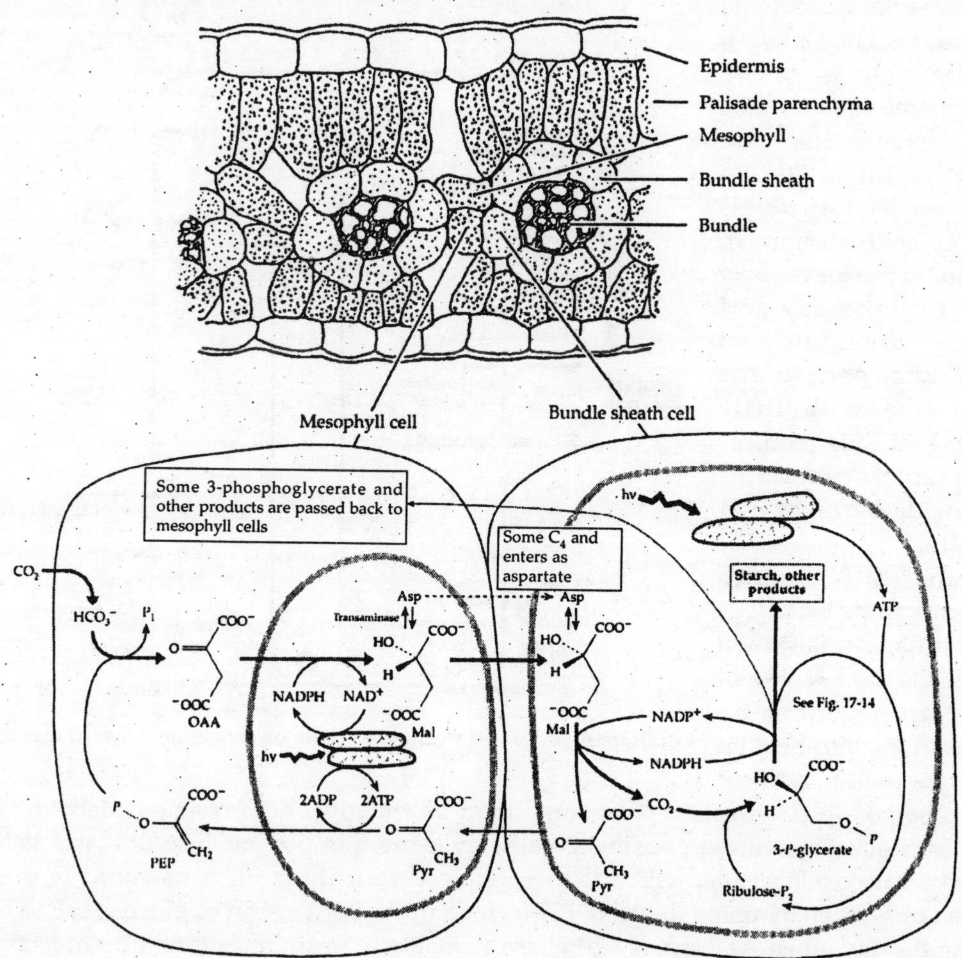

Fig. 2.34. The C_4 cycle for concentrating CO_2 in the C_4 plants.

The pyruvate produced in the bundle sheath cell is largely returned to the mesophyll cells where it is acted upon by *pyruvate-phosphate dikinase*. This phosphorylates pyruvate to PEP while splitting ATP to AMP and PP_i. The latter is in turn degraded to P_i. In effect, two high-energy linkages are cleaved for each molecule of pyruvate phosphorylated. Because of this extra energy need, it is thought that cyclic photophosphorylation is probably more important in the chloroplasts of the mesophyll cells than in the bundle sheath cells. It also accounts for the fact that C_4 plants are less efficient than C_3 plants under cool or shaded conditions. Other CO_2-concentrating mechanisms exist in plants. For example, cyanobacteria accumulate HCO_3^- ions in carboxysomes, polyhedral bodies to which rubisco adheres. An ABC type ATP-dependent transporter powers the bicarbonate accumulation.

Metabolism in the Family Crassulaceae

The crassulacean plants are a large group that includes many ornamental succulents such as *Sedum*. They have a remarkable metabolism by which large amounts of malic and isocitric acids are synthesized at night. During the day when photosynthesis occurs these acids disappear. The stomata in the leaves stay closed during the day and open only at night, an adaptation that permits the plants to live with little water. However, this requires that the plant accumulates carbon dioxide by night and incorporates it photosynthetically into organic compounds by day. A possible mechanism is shown in Fig. 2.39. On the left side of the figure are reactions by which starch can be broken down at night to PEP. While it would also be possible to produce that compound by the glycolysis reactions, labeling studies have indicated that the pentose phosphate pathway is more important. The PEP acts as the CO_2 acceptor to create oxaloacetate, which is then reduced to malic acid. A balanced fermentation reaction can be written by using the NADPH formed in the conversion of glucose 6-*P* to ribulose 5-*P*. During the day when ATP and NADPH are available in abundance from photoreactions, the conversions on the right side of the figure can take place. The initial step, the release of CO_2 from malic acid by the malic enzyme, is the same as that employed by C_4 plants. In this case, it is used to release the CO_2 stored by night, making it available for incorporation via the Calvin-Benson cycle. The remaining pyruvate is reconverted to starch.

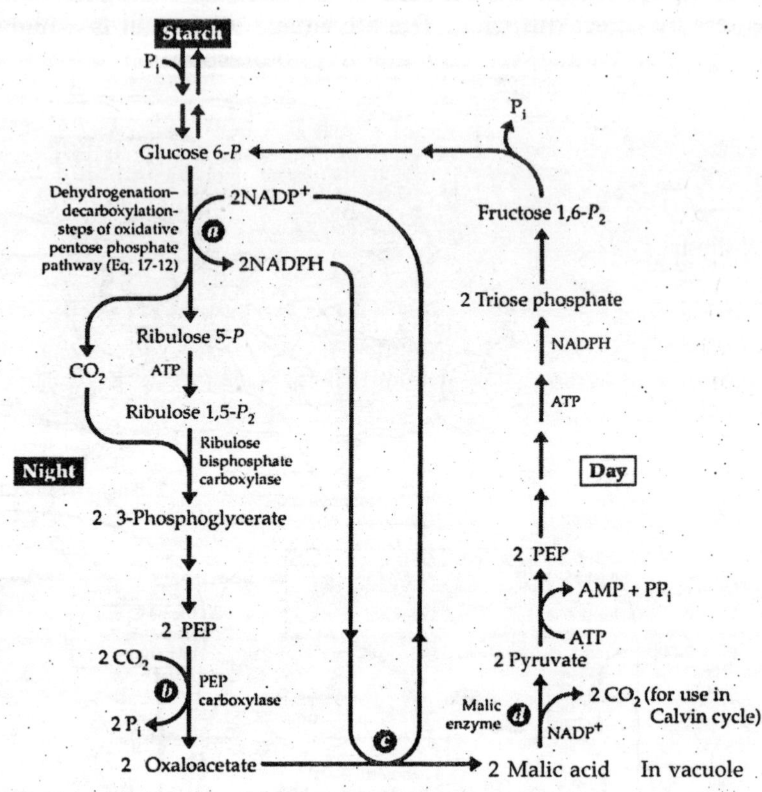

Fig. 2.35. A proposed night-day cycle of crassulacean acid metabolism.

$$C_6H_{10}O_5\text{ (starch)} + 2\,CO_2\text{ (g)} \rightarrow 2C_4H_4O_5^{2-}\text{ (malate)} + 4\,H^+\ \Delta G'(\text{pH7}) = -159\text{ kJmol}^{-1} \quad ...(2.35)$$

Many plants store substantial amounts of malate in their cytoplasm and in vacuoles. It apparently serves as a ready reserve for carbohydrate synthesis.

7. Photosynthetic Formation of Hydrogen

A system consisting of chloroplasts, ferredoxin, and hydrogenase has been used to generate H_2 photosynthetically. This may be a prototype of a method of solar energy generation for human use. Another photochemical hydrogen-generating system makes use of both the nitrogen-fixing heterocysts and photosynthetic vegative cells of the cyanobacterium *Anabaena cylindrica.* In this instance hydrogen production is accomplished by nitrogenase. Photogeneration of H_2 by bacteria is just one of many kinds of photometabolism observed among photosynthetic microorganisms.

F. VISION

The light receptors of the eye perform a very different function from those of chloroplasts. Visual receptors initiate nerve impulses, and their primary requirement is a high sensitivity. By the use of stacked membranes containing a high concentration of an intensely absorbing molecule the most sensitive visual receptors are able to trap nearly every photon that strikes them. The retina of the human eye contains more than 10^8 tightly packed receptor cells of two types. The *rod cells* are extremely sensitive. Used for night-time vision, they give a *"black and white picture"* and are concentrated around the periphery of the retina. The retina works as a coincidence detector. An ensemble of ~500 rods must register 5 – 7 isomerizations within a few tenths of a second in order to trigger a nerve impulse. The less sensitive *cone cells,* which are most abundant in the center of the retina, are of three types with different spectral sensitivities. They provide color vision.

The retinal receptors have a very active metabolism. Human rod cells may live and function for a hundred years. A self-renewal process leads to a casting off of the older membranous discs from the end of the rod and replacement by new discs at the end nearest to the nucleus. The rod outer segment is surrounded by a plasma membrane. Within the membrane but apparently not attached to it are ~500 stacked discs of ~2 μm diameter and with a repeat distance between centers of ~32 nm. Each disc is enclosed by a pair of membranes each ~7 nm thick with a very narrow space between them. From electron micrographs it appears that this space within the discs is sealed off at the edges. Somewhat larger spaces separate adjacent discs.

The membranes of the rod discs are ~60% protein and 40% lipid. About 80% of the protein is *rhodopsin* (visual purple), a lipoprotein that is insoluble in water but soluble in detergent solutions. Digitonin is widely used to disperse rhodopsin molecules because it causes no change in optical properties. In addition to rhodopsin, in the outer segment discs of frog retinal rods, there are ~65 molecules of phospholipid and smaller amounts of other materials for each molecule of rhodopsin. The cone cells have a similar architecture but have a different shape and contain different light receptors. The receptors in the cones are present in deep indentations of the plasma membrane rather than in discs within the cytoplasm.

1. Visual Pigments

The rod pigment rhodopsin is readily available from cattle retinas and has been studied for many years. It has a molecular mass of ~41 kDa of which ~2 kDa is contributed by two asparagine linked oligosaccharides. Both bovine and human rhodopsins consist of a 348-residue protein known as *opsin* to which is bound a molecule of vitamin A aldehyde, *retinal.* Human and bovine opsins are 93% homologous. A totally synthetic 1057 base pair gene for bovine rhodopsin was made by Khorana and associates. The gene was constructed with 28 unique sites for cleavage by specific restriction endonucleases. These have allowed easy specific mutation of the gene and production of a wide variety of mutant forms of opsin. Similar synthetic genes have been constructed for the three human cone pigments and for the related bacterial protein *bacteriorhodopsin.*

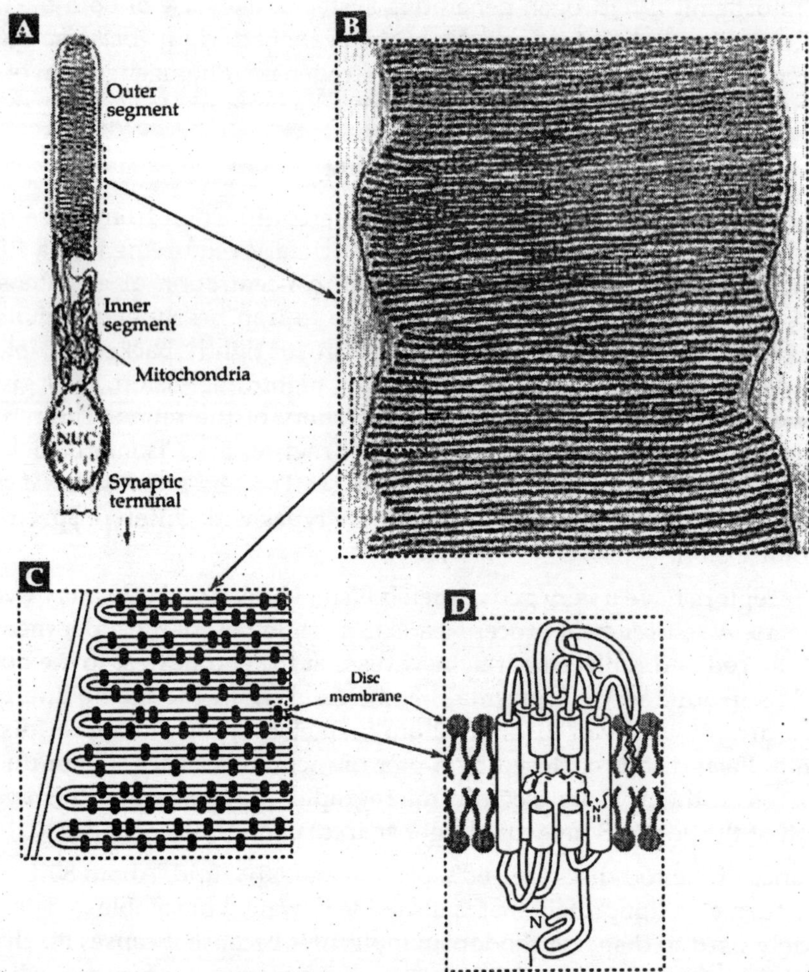

Fig. 2.36. (A) Diagram of a vertebrate rod cell. (B) Electron micrograph of a longitudinal section of the outer segment of a retinal rod of a rat. There are 600 – 2000 discs per rod and 2×10^4 and 8×10^5 rhodopsin molecules per disc. (C) Enlarged section; from Dratz and Hargrave. (D) Schematic drawing of rhodopsin. The two α helices in the front have been partly cut away to reveal the 11 *-cis* retinal in protonated. Schiff base linkage to lysine 296.

Transmembrane Structure

From its circular dichroism rhodopsin was estimated to be 60% helical, and its amino acid sequence suggested that it contains seven parallel membrane-spanning helices just as does bacteriorhodopsin Rhodopsin and other visual pigments are also members of the large family of *G-protein coupled receptors,* which includes the β_2 adrenergic receptor pictured. It has been hard to determine the three-dimensional structure of rhodopsin or other receptors of this family. However, their relationship to bacteriorhodopsin, whose structure was obtained in 1975 by

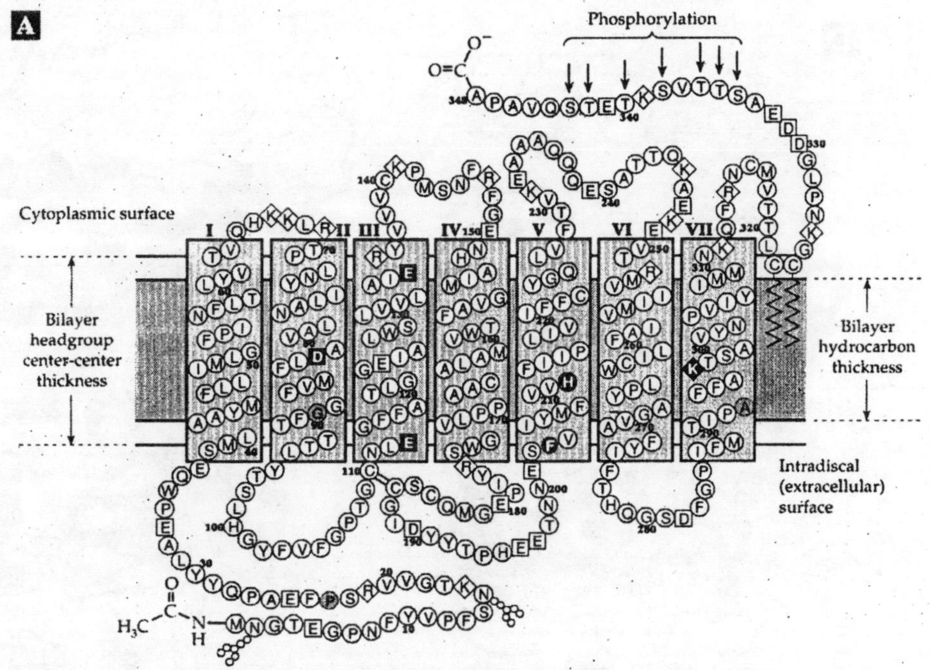

● A few residues essential to the mechanism of vision.

◍ Point mutations causing serious diseases. P23H is a frequent cause of retinitis pigmentosa. Mutations G90D and A292E cause congenital night blindness.

● The mutation C203R in the green-sensitive pigment causes red-green color blindness.

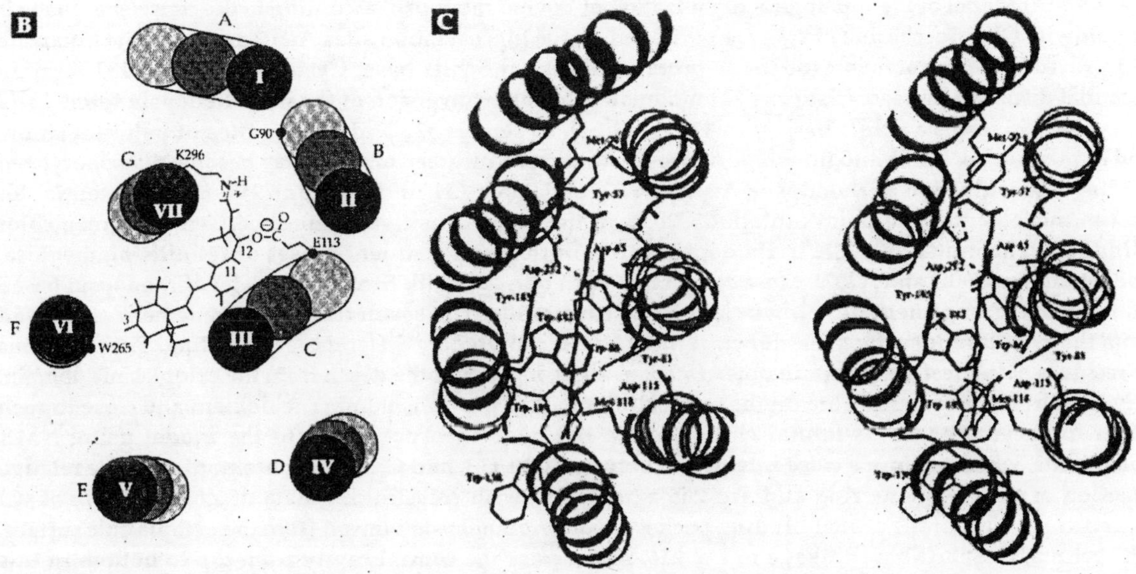

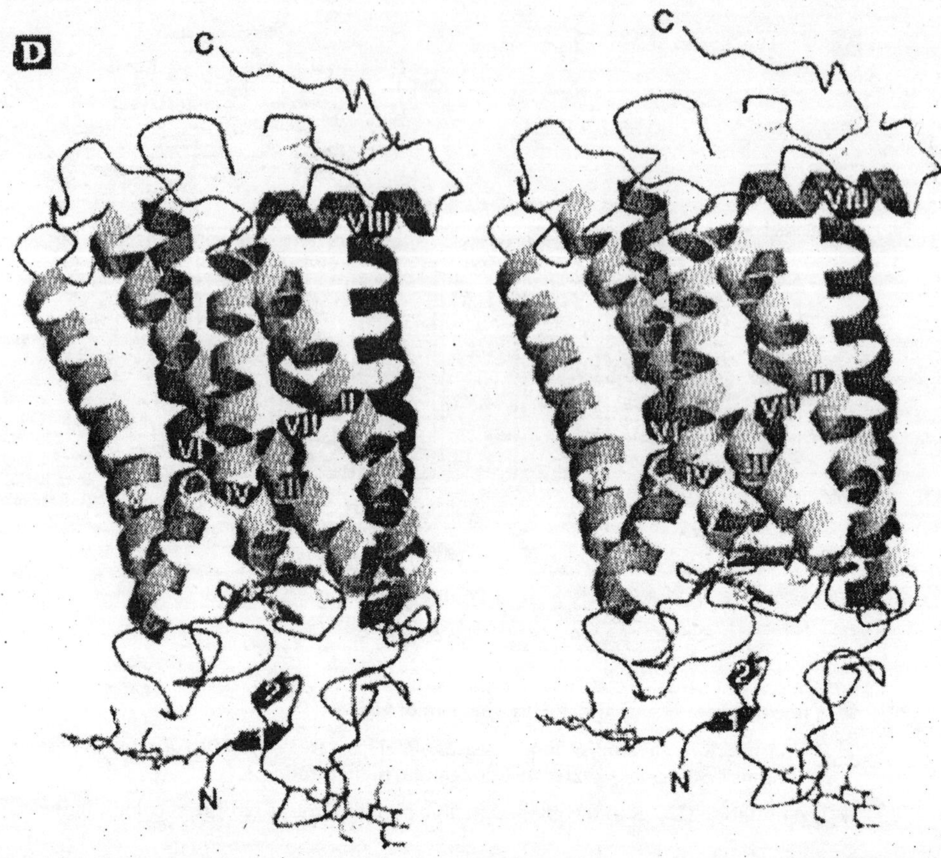

Fig. 2.37. (A) Model of the topological organization of bovine rhodopsin according to the consensus analysis of Baldwin. Oligosaccharide chains are portrayed on the glycosylation sites Asn 2 and Asn 15. Glutamate 113 provides the counter ion for the *N*-protonated retinal Schiff base. Cysteines 110 and 187 form an essential disulfide linkage. Histidine 211 modulates the interconversion of the metarhodopsin forms I and II. Lysine 296 forms a Schiff base with 11-*cis*-retinal. Cysteines 322 and 323 are sites of palmitoylation, and as many as six serine and threonine hydroxyls (indicated by green arrows) may become phosphorylated during desensitization of rhodopsin. Aspartate 83, glutamate 134, and histidine 211 may be essential for proton movements. The point mutation C203R in the green-sensitive, pigment causes red-green color blindness. The mutation P23H in rhodopsin is one of the most frequent causes of retinitis pigmentosa, and mutations G90D and A292E cause congenital night blindness. (B) Structural model of rhodopsin based on the helix arrangement of Baldwin and NMR constraints. The seven transmembrane helices, viewed from the cytoplasmic surface, are shown at three levels indicated by differences in shading. The α-carbons of residues of interest are shown in dots of various sizes, indicating the depth from the cytoplasmic domain. Gly 90, Glu 113, and Lys 296 are on the extracellular half of the transmembrane domain and close to each other in space. The 11-*cis*-retinal chromophore has been incorporated into the model using NMR constraints, which require a close interaction between Glu 113, and C_{12} of the chromophore. The relative position of the (β-ionone ring and Trp 265 agrees well with crosslinking data of Zhang *et. al.*, of (C) Stereoscopic view of the retinal-binding pocket of *bacteriorhodopsin* viewed from the cytoplasmic surface. The retinal, in Schiff base linkage to Lys 216, runs across the central cavity from top to bottom in this view. (D) Ribbon drawing of bovine rhodopsin (stereoview).

electron crystallography and more recently by X-ray crystallography at 0.15 nm resolution permitted modeling based on similarities among the proteins, results of electron crystallography and mass spectrometry have been combined with studies of mutant forms of rhodopsin and other visual pigments to provide the picture given in Fig. 2.41 A, B. Recently a higher resolution structure has been obtained by X-ray diffraction. Mutations have been introduced in every part of the rhodopsin molecule, and the effects on photoreception, protein stability, and other properties have been observed.

Some of the essential residues identified are indicated in Fig. 2.41 A, a schematic diagram showing, the seven helices and connecting loops. Fig. 2.41 B shows an end view of the helix bundle with its retinal prosthetic group buried in the interior of the protein. Rhodopsin is roughly cylindrical with a length of 6 – 6.5 nm and a diameter of ~2.8 nm and is embedded in the phospholipid bilayer with its long axis perpendicular to the membrane surface. The two oligosaccharide chains, which are attached near the N terminus, project into the intradiscal space on the side away from the cytoplasm. Palmitoyl groups on two cysteine side chains help to anchor the protein. Rhodopsin apparently exists in the membranes as monomers. From the composition it can be calculated that the average center-to-center distance of the cylindrical molecules must be ~5.6 nm.

The Visual Chromophores

Rhodopsin has been an object of scientific interest for over 100 years. Wald and associates established that rhodopsin contains 11-*cis*-retinal bound to the opsin in Schiff base linkage. When native rhodopsin is treated with sodium borohydride, little reduction is observed. However, after the protein is bleached by light, reduction of the Schiff base linkage becomes rapid, and the retinal is incorporated into a secondary amine, which was identified as arising from Lys 296.

In a crystal structure 11-*cis*-retinal has the 12-*s-cis* conformation shown at the top in Eq. 2.36 rather than the 12-*s-trans* conformation at the center and in which there is severe steric hindrance between the 10–H and 13–CH_3. Nevertheless, ^{1}H– and ^{13}C–NMR spectroscopy suggest that the retinal in rhodopsin is in a twisted 12-*s-trans* conformation. The Schiff base of 11-cis-retinal with N-butylamine has an absorption maximum at ~ 360 nm but *N*-protonation, as in the structure in Eq. 2.36, shifts the maximum to 440 nm with ε_{max} = 40,600 M^{-1} cm^{-1} (Fig. 2.42). This large shift in the wavelength of the absorption maximum (the *opsin shift*) indicates that binding to opsin stabilizes the light-excited state by almost 33 kJ/mol compared to that in the free *N*-protonated Schiff base. This is evidently the result of a fixed negative charge, that of Glu 113, which is near the polyene chain of the retinal and probably separated from it by a hydrogen-bonded water molecule.

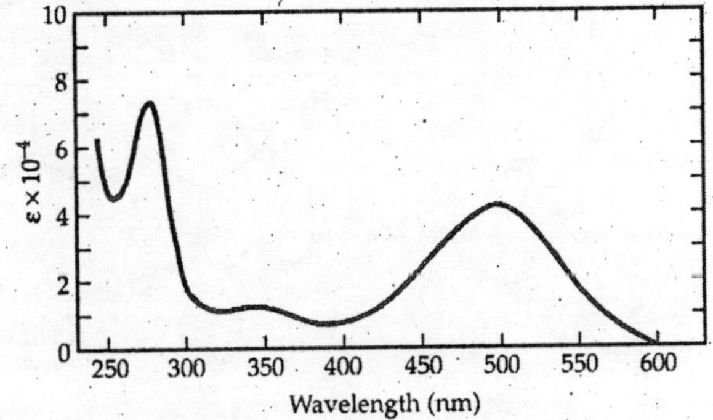

Fig. 2.38. Absorption spectrum of cattle rhodopsin in aqueous dispersion with a nonionic detergent.

Three types of cone cells in the human retina are needed for color vision. Four genes specify the proteins for rhodopsin and for related cone photoreceptors absorbing blue (~425 nm), green (~530 ran), and red (~560 nm) light. All of the cone opsins also bind 11-*cis*-retinal. The rhodopsin gene is located on human chromosome 3, while that of the blue pigment is found on chromosome 7. However, the green and red sensitive pigment genes, whose sequences are 96% identical, are close together on the q arm of the X-chromosome and near the gene for glucose-6-phosphate dehydrogenase. Examination of cloned DNA from persons with inherited red-green color-blindness shows that loss of a functional form of one of these genes is usually responsible for the problem. Among Caucasians 8% of males and 1% of females differ from the normal in their color vision. About 30% of affected males are *dichromats* and lack either the red-sensitive pigment (they are protanopes) or the green-sensitive pigment (*deuteranopes*). They usually have a partial gene deletion. The other 70% often have hybrid genes created by errors in recombination during cell division. A few deuteranopes have the point mutation C203R in the green pigment. John Dalton, of atomic theory fame, reasoned that his red-green confusion resulted from a blue tint in the vitreous humor of his eyes and ordered that they be dissected after his death (in 1884). There was no blue tint but DNA analysis performed more recently showed that Dalton was a deuteranope. Defects in the blue-sensitive receptor are relatively rare affecting about 1 in 500 persons, while only one person in 100,000 has a total lack of color discrimination.

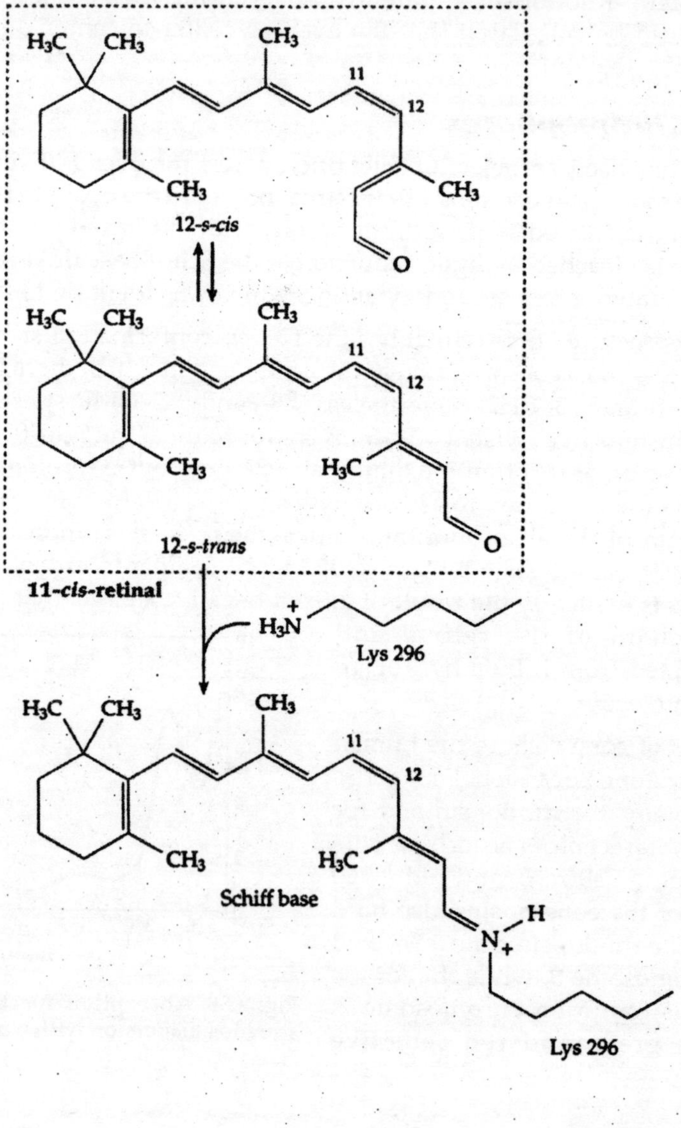

12-*s*-*cis*

12-*s*-*trans*

11-*cis*-retinal

Lys 296

Schiff base

Lys 296

All retinal-dependent visual pigments form Schiff bases with lysine side chains of the photoreceptor proteins. How can the same chromophore be *"tuned"* to absorb across the wavelength range of 360 to 635 nm? Modern techniques such as resonance Raman and FTIR spectroscopies and study of mutant forms have shown that interaction of the conjugated double bond system of the chromophores with immediately adjacent dipoles of side chain groups and peptide linkages is sufficient to account for the great variability in absorption maxima.

Visual pigments of many species have been investigated. Most vertebrate animals have a rhodopsinlike pigment plus a variable number of cone pigments. Mammals typically have only two, a short-wavelength pigment absorbing maximally in the ultraviolet, violet, or blue region and a long-wavelength pigment with maximum absorption in the green or red region. The bottlenose dolphin has only a rod pigment of λ_{max} 524 nm. In contrast, the chicken has four cone pigments with maximum absorption for violet, blue, green, and red. The red light receptor, called *iodopsin,* absorbs maximally at 571 nm. However, it binds chloride ions which induce an additional 40-nm red shift. The Cl^--binding site involves His 197 and Lys 200 which are present in an extracellular loop and quite far from the bound retinal. Human red and green color vision pigments, and also a green-sensitive pigment of the reptile *Gecko gecko,* undergo spectral shifts upon binding of Cl^- in the same site. However, rhodopsin and most other visual pigments do not share this behavior.

Fishes live in a variety of environments and have a diversity of visual pigments. Goldfish have genes for five opsins; one of which gives rise to an ultraviolet light receptor. They are also sensitive to polarized light. Related visual pigments occur throughout the animal kingdom. Even the eyespot of the alga *Chlamydomonas* contains rhodopsin with some sequence homology to invertebrate opsins. The pineal glands of chickens and probably of reptiles as well as those of fish also contain rhodopsinlike pigments. In a few freshwater marine species the visual pigments *(porphyropsins)* contain *3-dehydroretinal.* The peak positions of light absorption depend both upon the nature of the bound aldehyde and on the protein, the latter having the larger effect. Thus, retinal-based pigments absorb in the entire range 467 – 528 nm (18,900 – 21,400 cm^{-1}). The fruit fly, *Drosophila,* contains 3-hydroxy-11-*cis*-retinal in its rhodopsin and also contains other related photoreceptors.

2. The Light-induced Transformation

The retinal Schiff base chromophore is embedded in rhodopsin with its transition dipole moment parallel to the plane of the discs, *i.e.,* perpendicular to the direction of travel of the incoming photons. Absorption of a photon leads to a sequence of readily detectable spectral changes. The relaxation times indicated in Eq. 2.37 are for 20°C.

Rhodopsin (498 nm, 20.1×10^3 cm^{-1})

$h\nu \downarrow \tau < 6 \times 10^{-12}$ s

Bathorhodopsin (prelumirhodopsin; 543 nm, 18.4×10^3 cm^{-1})

$\downarrow \tau \sim 50$ ns

Lumirhodopsin (497 nm), 20.2×10^3 cm^{-1})

$\downarrow \tau \sim 50$ μs

Metarhodopsin I (478 nm, 20.9×10^3 cm^{-1})

$\downarrow\uparrow$ milliseconds

Metarhodopsin II (380 nm, 26.2×10^3 cm^{-1})

$\downarrow$ seconds

All-*trans*-retinal (387 nm, 25.8×10^3 cm^{-1}) + opsin ...(2.37)

The intermediate chemical species have been named bathorhodopsin, lumirhodopsin, and metarhodopsin I and II. At very low temperatures a transient form *photorhodopsin* with a wavelength maximum at ~580 nm may precede bathorhodopsin. Furthermore, nanosecond photolysis of rhodopsin has revealed a *blue-shifted intermediate* that follows bathorhodopsin within ~40 ns and decays into lumirhodopsin. The overall result is the light-induced isomerization of the bound 11-*cis*-retinal to all-*trans*-retinal and free opsin. The free opsin can then combine with a new molecule of 11-*cis*-retinal to complete the photochemical cycle.

What are the chemical structures of the intermediates in Eq. 2.37, and why are there so many of them? The answer to the last question is that the initial photochemical process is very fast. Subsequent conformational rearrangments and movement of protons are slower, occur in distinct steps, and give rise to the observed series of intermediates. To shed light on these processes many experiments have been done with analogs of retinal, often using very rapid spectroscopic techniques. These studies have shown that the isomerization of the Schiff base from 11-*cis* to all-*trans* occurs in the first very rapid step of Eq. 2.37. Constrained analogs of 11-*cis* retinal also combine with opsin to form rhodopsinlike molecules with absorption maxima near 500 nm.

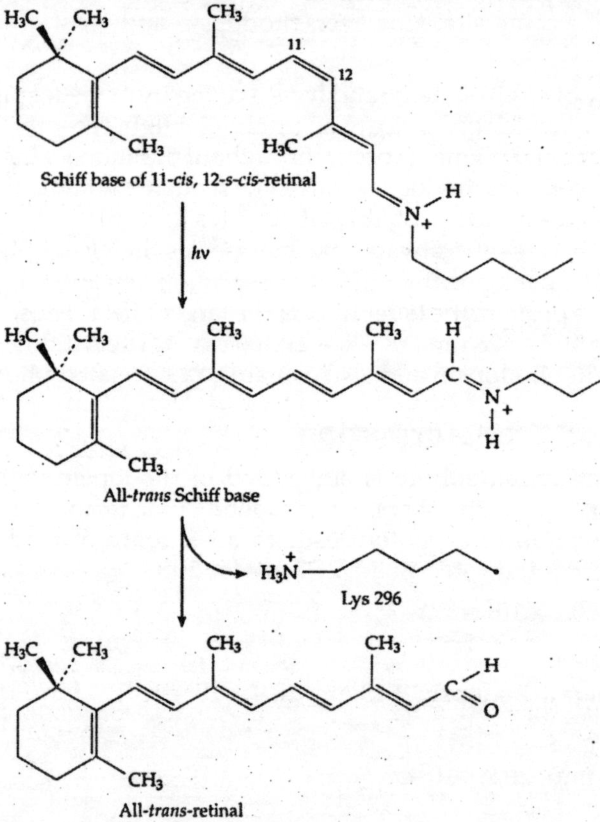

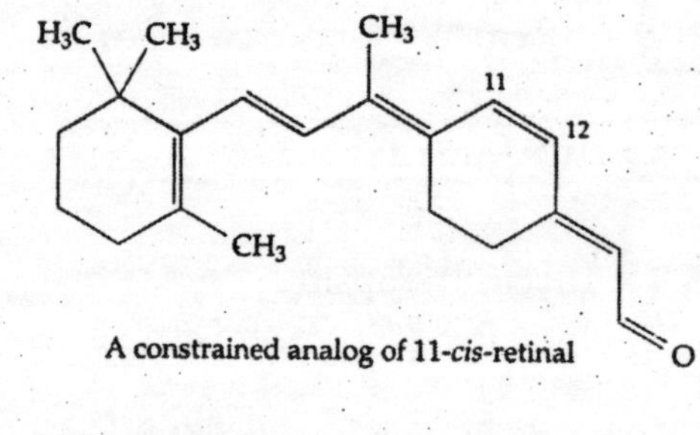

H_3C CH_3 CH_3 11 12 CH_3 H_3C H N^+

Schiff base of 11-*cis*, 12-*s*-*cis*-retinal

hν

H_3C CH_3 CH_3 CH_3 H C N^+ H CH_3

All-*trans* Schiff base

$H_3\overset{+}{N}$ Lys 296

H_3C CH_3 CH_3 CH_3 H C O CH_3

All-*trans*-retinal

H_3C CH_3 CH_3 11 12 CH_3 O

A constrained analog of 11-*cis*-retinal

However, most of these analogs cannot isomerize and illumination does not produce bathorhodopsinlike molecules.

In the photoexcited state the positive charge on the Schiff base is delocalized. For example, with some of the positive charge located on C5, rotation could occur around the more nearly single C11–C12 bond to give the all-*trans* isomer. However, it seems more likely that simultaneous rotation occurs around two connected single and double bonds. The mechanism of this photoisomerization, which is among the fastest known chemical reactions, is still being investigated. In native rhodopsin the conversion to bathorhodopsin occurs with a high quantum yield of 0.67 within 0.2 ps, a time comparable to the period of torsional vibrations of the retinal. This extreme speed suggests that the isomerization is a concerted process that is a *vibrationally coherent*. Vibrational motion in the electron excited state is utilized in the isomerization process.

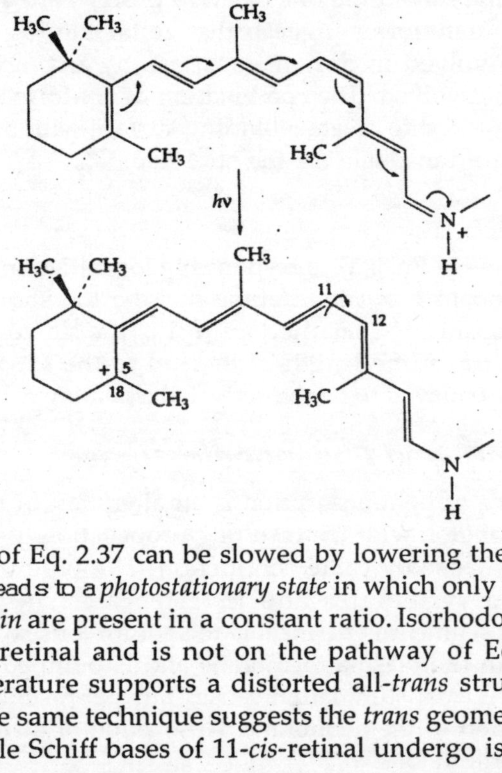

The reaction sequence of Eq. 2.37 can be slowed by lowering the temperature. Thus, at 70K illumination of rhodopsin leads to a *photostationary state* in which only rhodopsin, bathorhodopsin, and a third form, *isorhodopsin* are present in a constant ratio. Isorhodopsin (maximum absorption at 483 nm) contains 9-*cis*-retinal and is not on the pathway of Eq. 2.37. Resonance Raman spectroscopy at low temperature supports a distorted all-*trans* structure for the retinal Schiff base in bathorhodopsin. The same technique suggests the *trans* geometry of the C=N bond shown in Eqs. 2.38 and 2.39. Simple Schiff bases of 11-*cis*-retinal undergo isomerization just as rapidly as does rhodopsin.

Some step in the sequence of Eq. 2.37 must initiate a chemical cascade that sends a nerves impulse out of the rod cell. This is accomplished through a GTP-dependent G protein cascade. Light-activated rhodopsin initiates the cascade by interacting with the G protein *transducin*. Energy for the activation comes from the quantum of absorbed light. While the primary chemical reaction has long been recognized as the isomerization of the retinal Schiff base, is not obvious how this generates the signal for transducin to bind and become activated.

The seven helices of rhodopsin form a *"box"* around the bound retinal. The environment of the retinal is largely hydrophobic. However, there are also buried polar groups, some of which lie in conserved positions in more than 200 G-protein-coupled receptors and internal water molecules whose vibrational spectra can be detected. As in bacteriorhodopsin the buried polar groups and water molecules are doubtless hydrogen-bonded in an internal network. We can anticipate two effects of the isomerization reaction: (1) It will distort the shape of the box in which it occurs. (2) It will break some hydrogen bonds and allow new ones to form and may affect the balance of electrical charges within the protein. This in turn can lead to proton movements and alterations in the internal hydrogen-bonded network. Both of these anticipated effects have been observed.

Conformational changes induced in the rhodopsin protein by the isomerization of the retinal Schiff base include significant movement of the end of helix VI(F) at the cytosolic surface as well as smaller movements of other helices. Spectroscopic measurements indicate that the Schiff base nitrogen remains protonated in both the bathorhodopsin and metarhodopsin I forms and in metarhodopsin II, the first long-lived form in the sequence. It seems likely that the proton has jumped via a bridging water molecule and the E113 carboxylate to the external (intradiscal) surface of the molecule. At the same time one or more protons are apparently taken up on the cytoplasmic side. Study of mutant forms suggests that glutamate 134, near the cytosolic surface, and histidine 211 may be involved in proton transport. By analogy with bacteriorhodopsin, aspartate 83 is probably also involved. The combination of conformational change plus altered charge distribution may be needed to create a binding surface with a suitable shape and charge constellation to bind tightly to transducin for the next step.

3. The Nerve Impulse

Which of the intermediates in Eq. 2.37 is responsible for initiation of a nerve impulse? Some evidence favored metarhodopsin I, but its lifetime may be too short. On the other hand, the transformation of metarhodopsin I to metarhodopsin II is the slowest step that could trigger a nerve impulse, which must travel the length of the rod to the synapse in about one ms and metarhodopsin II is generally believed to be the activated signaling form of rhodopsin.

Transducin, Cyclic GMP, and Phosphodiesterase

The essential consequence of light absorption is an alteration in the membrane potential in the vicinity of the absorbed photon with the resulting propagation of a nerve impulse down the plasma membrane to the synapse by cable conduction. The type of potential change that is transmitted differs among vertebrates and invertebrates. In the case of mammalian photoreceptors the rod outer segment is permeable to sodium ions so that a large *dark current* of sodium ions flows in through the plasma membrane and is pumped out by sodium pumps in the inner portion of the cell. Visual stimulation causes this permeability to Na^+ to be decreased with an increase in polarization of the membrane. Absorption of a single photon by rhodopsin blocks the outflow of ~ 10^6 sodium ions.

At one time calcium ions seemed to be the logical internal messenger between rhodopsin and the plasma membrane. If light absorption opened channels from the internal space of the rod discs, calcium ions could be released and diffuse quickly to the plasma membrane and block the entrance of sodium ions. However, light *does not* increase the free [Ca^{2+}] in the cytoplasm but may decrease it from 500 nM to as low as 50 nM. Stryer suggested that the essential messenger is *cyclic GMP* and that a *decrease* in cGMP concentration initiates the nerve response. Cyclic GMP is apparently responsible for keeping the sodium ion channels open. Absorption of a photon in the rod disc produces activated rhodopsin R* (metarhodopsin II), which acts as an allosteric effector for the heterotrimeric G protein. Like proteins G_s and G_i of the adenylate cyclase system, transducin contains three subunits: α, 40 kDa, 350 residues; β, 36 kDa; and γ, ~8 kDa. In the resting state they are associated as $T_{\alpha\beta\gamma}$ with a molecule of GDP bound to the a subunit. When activated rhodopsin R* binds to transducin (step *a*) it catalyzes a rapid exchange of GTP for GDP (step *b*). This is followed by dissociation of $T_{\beta\gamma}$ from T_{α} GTP. The latter serves as an allosteric effector for a *cGMP phosphodiesterase* bound to the disc surface converting it to an active form (step *c*). The activated phosphodiesterase, an $\alpha\beta\gamma_2$ oligomer, hydrolyzes the cGMP (step *d*), reducing its concentration and thereby inhibiting the Na^+ outflow.

Because one molecule of activated cGMP phosphodiesterase can hydrolyze more than 10^5 molecules of cGMP per second the light response is highly amplified. There is also an earlier stage of amplification. Each molecule of light-activated rhodopsin (R*) is able to catalyze the exchange of GTP for GDP on hundreds of molecules of $T_{\alpha\beta\gamma}$ before R* passes on to other intermediates and releases all-*trans*-retinal from opsin (light green lines).

Rhodopsin Kinase, Recoverin, and Arrestin

Metarhodopsin II (R*) can become phosphorylated by rhodopsin kinase on as many as seven serine and threonine side chains on its cytoplasmic surface. The 45-kDa protein arrestin binds to such phosphorylated R* which is rapidly deactivated and desensitized so that it is less likely to be immediately reactivated. This is important in the adaptation of the eye to bright light. At the same time T_{α}2GTP is hydrolyzed back to T_{α}2GDP and reforms TαβγGDP, and guanylate cyclase regenerates the cGMP. At least four different arrestins are known. Some function in nonvisual tissues. In all cases they seem to serve as "uncouplers" of G protein-coupled receptors.

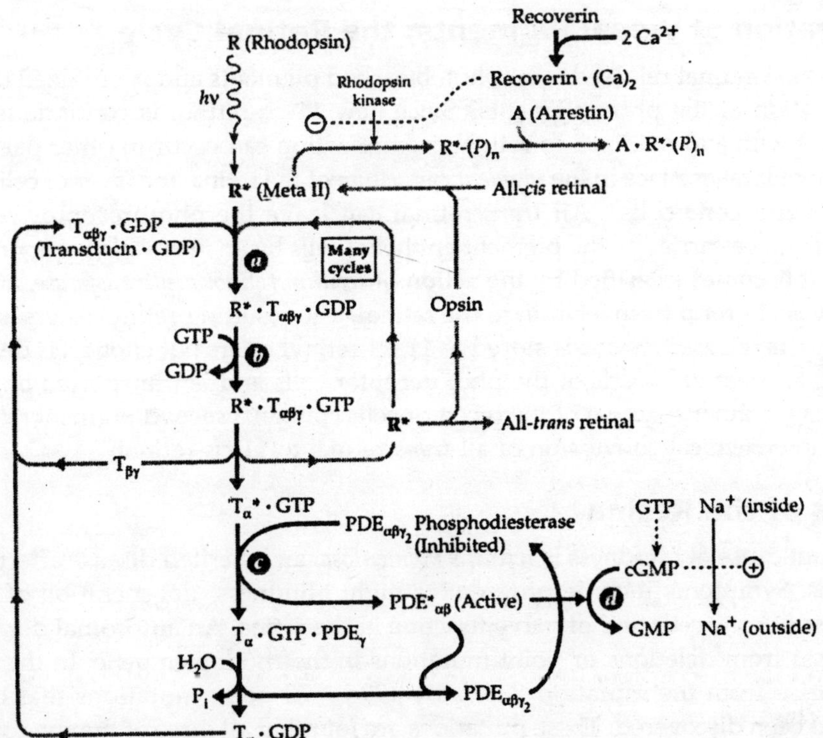

Fig. 2.39. The light-activated transducin cycle. In step *a* photoexcited rhodopsin (R*) binds the GDP complex of the heterotrimeric transducin ($T_{\alpha\beta\gamma}$). After GDP–GTP exchange (step *b*) the activated transducin T*•GTP reacts with the inhibited phosphodiesterase (PDEαβγ) to release the activated phosphodi-esterase ($PDE_{\alpha\beta}$).

Recovery of the inhibited rhodopsin, which occurs most rapidly in dim light, depends upon calcium ions. In dim light both Ca^{2+} and Na^+ enter the visual cells through the cGMP-controlled channels. At the same time Ca^{2+} flows out through a Na^+/Ca^{2+} exchanger. When the channels are blocked by cGMP formed in response to light, $[Ca^{2+}]$ falls as a result of continuing export by exchange with Na^+. In the dark the $[Ca^{2+}]$ concentration rises again and binds to a 23-kDa calcium sensor molecule called recoverin. Recoverin, like calmodulin, has four EF-hand Ca^{2+}-binding motifs and also an unexpected feature. A *myristoyl group* attached at the N terminus is bound into a deep hydrophobic pocket in the protein. However, when two Ca^{2+} ions bind, recoverin undergoes a conformational change that allows the myristoyl group to be extruded and to bind to a membrane surface. This allows recoverin to move out of the cytosol to the surface of the disc membrane where it binds to and inhibits the activity of rhodopsin kinase, increasing the sensitivity of photodetection.

Some Details About Cone Cells and Invertebrate Vision

The biochemistry of retinal cones is less well known but is similar to that of rod cells. Cone pigments are present in the plasma membrane rather than in isolated discs. Different α, β, and γ subunits of transducin are formed in rods and cones. Many differences are seen among various invertebrate visual systems. Inositol triphosphate (IP_3) and Ca^{2+} often serve as signals of photoexcitation. *G* proteins also play prominent roles.

4. Regeneration of Visual Pigments; the Retinal Cycle

How is all-*trans*-retinal released from photobleached pigments and isomerized to 11-*cis*-retinal for the regeneration of the photopigments? Since new 11-*cis*-retinol is continuously brought in from the bloodstream and oxidized to retinal, isomerization can occur in other parts of the body. However, much of it takes place in the *pigment epithelium* of the retina, the layer of cells immediately behind the rod and cone cells. All *trans*-retinal can leave the photoreceptor cells and, after reduction to retinal, be carried to the pigment epithelial cells by an *interphotoreceptor retinoid-binding* protein. There it becomes esterified by the action of *lecithin:retinol acyltransferase,* an enzyme that transfers a fatty acyl group from lecithin to the retinol. The resulting retinyl esters are isomerized, and 11-*cis*-retinol is released? Some is stored as 11-*cis*-retinyl esters but enough is dehydrogenated to 11-*cis*-retinal to meet the needs of the photoreceptor cells and is transported back to them. In the cephalopods the inner segment of the receptor cells contain a second pigment *retinochrome* that carries out a photochemical conversion of all-*trans*-retinal to 11-*cis*-retinal.

5. Diseases of the Retina

An important cause of blindness is *retinitis pigmentosa,* an inherited disease affecting about one in 3000 persons. Symptoms include progressive night blindness, degeneration of the rod cells, and gradual loss of cone cells and of nerve function in the retina. An autosomal dominant form of the disease arises from deletions or point mutations in the rhodopsin gene. In the United States 15% of cases arise from the mutation P23H. By 1996 ~ 70 point mutations that cause retinitis pigmentosa had been discovered. These mutations are found in all three of the rhodopsin domains: intradiscal, transmembrane, and cytosolic. Other rhodopsin point mutations such as G90D and A292E cause congenital night blindness. Retinitis pigmentosa also arises from defects in *peripherin-RDS,* a structural component of rod cells identified originally from the gene *rds* (retinal degeneration slow) of the mouse. Another form of congenital night blindness results from mutations in rhodopsin kinase. A dominant *rod-cone dystrophy* is caused by a defect in the photoreceptor guanylate cyclase. The most frequent cause of combined deafness and blindness in adults *(Usher syndrome)* is a defect in a cell adhesion molecule. In *choroidemia.* another X-linked form of retinitis pigmentosa, the choroid layer behind the pigment epithelium also deteriorates. A geranylgeranyl transferase specific for the Rab family of G proteins is defective.

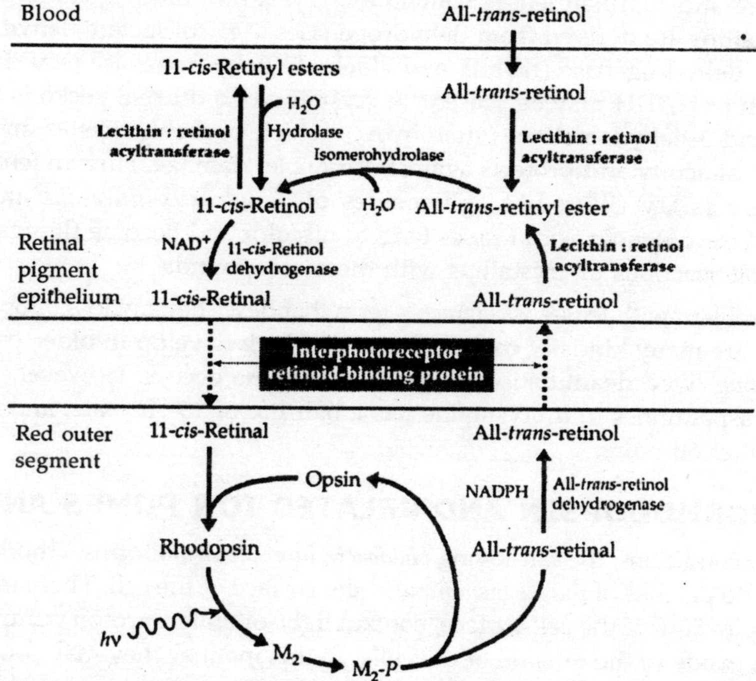

Fig. 2.40. Reactions of retinol and the retinal cycle of mammalian rod cells.

The most frequent cause of vision loss in the elderly is *macular degeneration.* Mild forms of the disease occur in nearly 30% of those over 75 years of age and more serious forms in 7% of that age group. There are many causes, some hereditary. Excessive accumulation of fluorescent lipofuchsin, perhaps arising in part from Schiff base formation between retinal and phosphatidylethanolamine, is sometimes observed.

6. Proteins of the Lens

The lens of the eye encloses cells that cannot be replaced and contains proteins that don't turn over and must last a lifetime. The fiber cells, which make up the bulk of the lens, have no nuclei. They elongate and stretch to cover the central nucleus, the original fetal lens, like the layers of an onion, the edges of the cells interdigitated with the next cell like a piece in a child's construction set. These cells are tightly packed with proteins in aggregates whose size is on the order of the wavelengths of light. The high concentration of proteins is needed to provide transparency and also a high refractive index. The membranes of the lens cells acquire increasing amounts of a 28-kDa *major intrinsic protein* as they age. Three classes of soluble lens proteins, called *crystalling,* are found in virtually all lenses. Alpha crystallins, which account for ~40% of the total soluble protein, are heterodimers of ~20-kDa subunits that associate into ~800-kDa complexes.

They have a chaperoninlike activity. Beta crystallin, which may account for ~35% of the protein, as well as the γ crystallins are β-sheet proteins with *"Greek key"* folding motifs.

In addition to the α, β, and γ crystallins many animals have recruited additional *"taxon-specific"* crystallins δ, ε, λ, etc., that have evolved from prexisting enzymes, chaperonins, or other proteins. For example, avian and reptilian lenses contain a δ crystallin homologous to argininosuccinate lyase. Many crystallins are derived from dehydrogenases, *e.g.*, for lactate dehydrogenase (duck), hydroxyacyl-CoA dehydrogenase (rabbit), and aldehyde dehydrogenase (squid and octopus). A high concentration of NADH may be present. A crystallin of a diurnal gecko is a retinol-binding protein with bound 3-dehydroretinal (vitamin A_2), which probably acts as an ultraviolet filter that improves visual acuity and protects against ultraviolet damage. Human lenses contain small molecules that act as UV filters, *e.g.*, glucosides of 3-*hydroxykynurenine* and 4-(2-amino-3-hydroxyphenyl)-4-oxobutanoic acid. Lenses tend to discolor and become fluorescent with age, in part by irreversible reactions of crystallins with these compounds.

A common problem with lenses is *cataract*, a term that describes any loss of opacity or excessive coloration. There are many kinds of cataract, most of which develop in older persons. Since lens proteins are so long-lived deamidation of some asparagine occurs. However, the reactions are slow. One of the asparagines in α crystalline has a half-life of 15-20 years, and some glutamines are undamaged after 60 years.

G. BACTERIORHODOPSIN AND RELATED ION PUMPS AND SENSORS

Under certain conditions, the salt-loving *Halobacterium salinarum* forms a rhodopsinlike protein, which it inserts into patches of *purple membrane* in the surface of the cell. These membranes, which may constitute up to 50% of the cell surface, contain light-operated proton pumps that translocate protons from the inside to the outside of the cells. In this manner they may provide energy for a variety of cell functions including ion transport and ATP synthesis. The 248-residue retinal-containing *bacteriorhodopsin* makes up 75% of the mass of the membrane. Its molecules aggregate into a two-dimensional crystalline array in the purple patches of the membrane. This allowed determination of the three-dimensional structure to 0.7 nm resolution in 1975 by electron microscopy and neutron diffraction. More recently the structure has been established at progressively higher resolution by electron crystallography and X-ray diffraction.

The most recent studies have been focused on determination of the structural alterations in the protein that accompany the proton pumping. A step-by-step picture is emerging. Internal changes in the retinal chromophore, movements of protons, and alterations in the shapes of some of the protein helices are involved. The surface loops have been studied both by electron crystallography and by atomic force microscopy. Each bacteriorhodopsin molecule is folded into seven closely packed α-helical segments which extend roughly perpendicular to the membrane. Although 100 residues shorter than rhodopsin, the folding pattern is very similar. The protein molecules form an extremely regular array with phospholipid molecules (mostly of phosphatidylglycerol) filling the spaces between them. The retinal is buried in the interior of the protein and is bound as an *N*-protonated Schiff base with the side chain of lysine 216.

The retinal in bacteriorhodopsin (bR) exists in both all-*trans* and 13-*cis* configurations. The *all-trans* form has an absorption maximum at 568 nm and the 13-*cis* at 548 nm. The two exist in a slow nonphotochemically mediated equilibrium in the dark. However, in the light the all-*trans* bR_{568} undergoes a rapid photochemical cycle of reactions which is presented in simplified form in Fig. 2.46 The subscript numbers designate the wavelength of maximum absorption. Many efforts have been made to determine the structures of the intermediates K, L, M, and O and to relate them to a mechanism of proton pumping. Both K and L contain 13-*cis* retinal. Therefore, as with rhodopsin the very first step is a photochemical isomerization. Intermediate M appears to contain a deprotonated Schiff base, but O is both *N*-protonated and again all-*trans*. It follows logically that the proton bound initially to the Schiff base is pumped out of the cell and is then replaced by a new proton in the O form. However, several questions must be answered if we are to understand this proton pump. Where in the sequence do proton transfers occur? How is the sequence driven by the absorbed light energy? Protons must enter the pump from the cytosol and exit on the exterior (periplasmic) side, flowing out against a concentration gradient. There must be a *"gating"* or *"switch"* mechanism that ensures that protons enter and leave the pump in the correct direction.

Some aspects of a possible mechanism for pumping the single proton bound to the retinal Schiff base are included in Fig. 2.46. In bR_{568} the Schiff base bound proton is H-bonded, via a water bridge, to the carboxylate group of Asp 85. The charge constellation in the interior of the protein, part of which is shown in this figure, is such that protonation of the Schiff base is stable and the pK_a of the protonated Schiff base is high, with estimates of 16 or above. One cause of the high pK_a is the presence of the nearby negative charges on D85 and D212. Absorption of light and isomerization of the retinal causes a downward movement of the $= NH^+-$ group of the Schiff base and facilitates movement of the Schiff base proton via the water molecule to the D85 carboxylate as indicated by the green arrows in Fig. 2-45B. Loss of the positive charge will instantly substantially raise the pK_a of D85 from a low value, while the loss of the negative charge will lower the pK_a of the Schiff base to closer to 7. The electrostatic interactions of the D212 and E194 carboxylates with the positive charge of R82 may also be altered. At some point in the sequence the interaction of R82 with the E194 carboxylate could cause E204, which is known by spectroscopic measurements to be protonated in the intermediate, to lose its proton to the outside. At some other point, perhaps between M_{412} and another intermediate, M_{408} a conformational switch must occur to limit flow of a proton back to E204 and to allow a new proton to enter from the cytosol. The D96 carboxylate is thought to accept this proton and to transfer it via a chain of water molecules or water molecules plus protein groups to the Schiff base, which may isomerize back to all *trans* in going from intermediate N to O. Isomerization to all-*trans* O is a slow step. Both O_{460} and bR_{568} are all-*trans* and have a 15-*trans* Schiff base linkage. There may be a difference in protein conformation with the chromophore being more distorted in O than in bR_{568}. Many recent studies have provided additional information.

Halorhodopsin

In addition to bacteriorhodopsin there are three other retinal-containing proteins in membranes of halobacteria. From mutant strains lacking bacteriorhodopsin the second protein, *halorhodopsin,* has been isolated. It acts as a light-driven *chloride ion pump,* transporting Cl^- from outside to inside. Potassium ions follow, and the pump provides a means for these bacteria to accumulate KCl to balance the high external osmotic pressure of the environment in which they live. The amino acid sequences of halorhodopsins from several species are very similar to those of bacteriorhodopsin as is the three-dimensional structure. However, the important proton-carrying residues D85 and D96 of bacteriorhodopsin are replaced by threonine and alanine, respectively, in halorhodopsin. Halorhodopsin (hR) undergoes a light-dependent cycle that

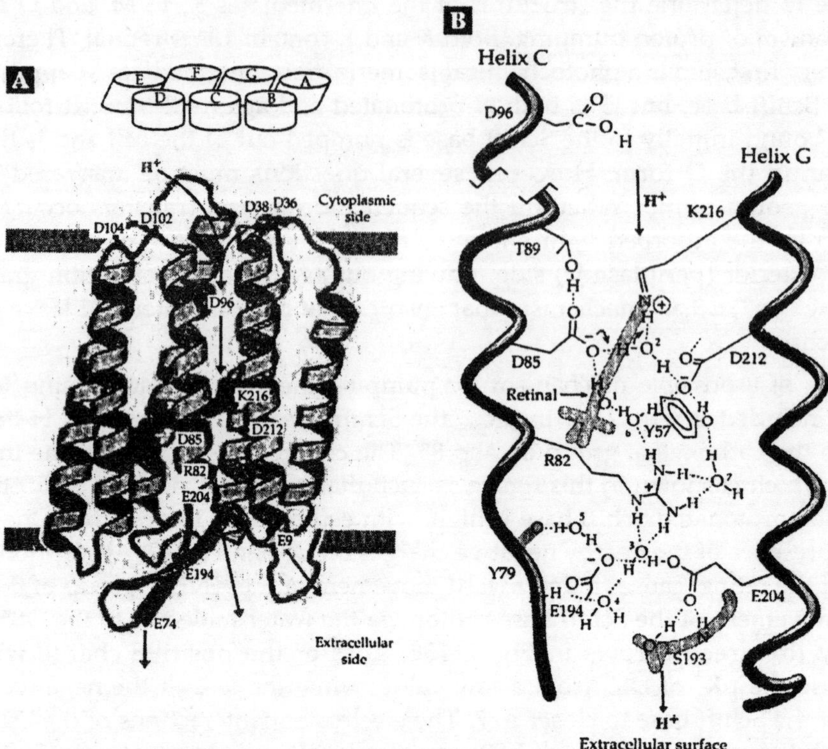

Fig. 2.41. (A) Some aspects of the structure of bacteriorhodopsin. Ribbon diagram with the retinal Schiff base in ball-and-stick representation. At the top the helices are labeled. The locations of aspartate, glutamate, and arginine residues that might carry protons during the proton pumping action are indicated. Retinal is shown attached to lysine 216. (B) drawing illustrating hydrogen-binding observed in the three-dimensional structure at 0.14 nm resolution. The side chains shown are those thought to be involved in proton transport and in a hydrogen-bonded network with bound water molecules, principally between helices C and G. The positions of many of the hydrogen atoms in this network have not been established. They have been placed in reasonable positions in this figure but may be quite mobile. For another view of the hydrogen-bonded network

involves an all-*trans* to 13-*cis* photoisomerization with some intermediates resembling those of the bacterior-hodopsin cycle.

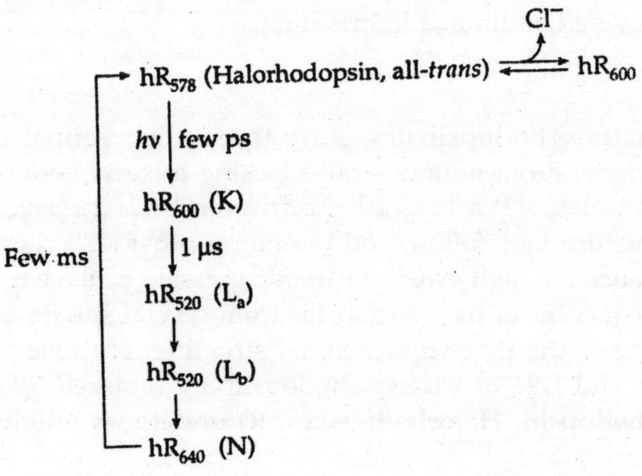

Sensory Rhodopsins

The third and fourth light-sensitive proteins of halobacteria are *sensoxy rhodopsins* (SR) that are used by the bacteria to control *phototaxis.* These bacteria swim toward long-wavelength light, the maximum in the action spectrum being at ~580 nm. They are repelled by blue or ultraviolet light, the maximum in the action spectrum being at ~370 nm. Evidently the bacteria can detect either a decrease with time in red light intensity or an increase with time in blue light intensity. Either is interpreted as unfavorable and causes the bacteria to tumble and move in a new direction. Sensory rhodopsin I (SRI) appears to be able to provide both light responses. Absorption of orange light by SRI_{587}, which contains all-*trans*-retinal, yields SRI_{373}, in which the retinal Schiff bases have been isomerized to 13-*cis* as in bacteriorhodopsin. The red light response is proportional to the fraction of SRI_{373} present. However, this is converted spontaneously back to SR_{587} within seconds. Nevertheless, photoexcitation of SRI_{373} with blue light causes a faster reconversion and induces swimming reversals, the repellent response. SRI exists in the bacterial membranes in a complex with a 57-kDa protein designated *halobacterial transducer I* (HtrT). which resembles bacterial chemotaxis receptors and is modulated by action of a methyltransferase. Interaction of SRI with HtrT depends upon a histidine residue, H166 of SRI. It may be part of a proton transfer pathway.

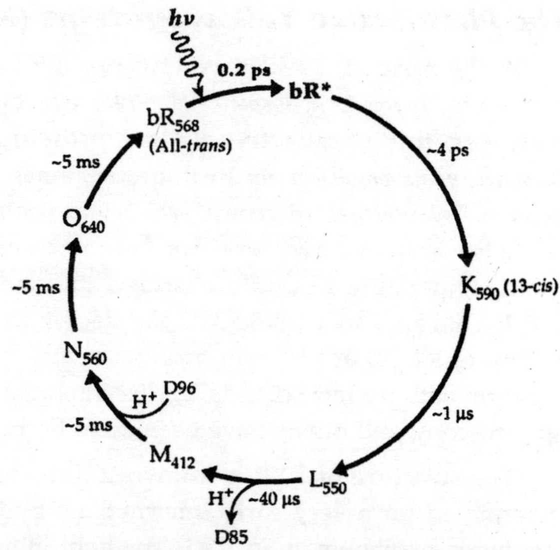

Fig. 2.42. The photoreaction cycle of bacteriorhodopsin. The subscript numbers indicate the wavelengths of maximum absorption of each intermediate and the approximate lifetimes are given by the arrows. Resting bacteriorhodopsin as well as intermediates J and O have all-*trans* retinal but K through N are thought to all be 13-*cis*. A proton is transferred from L to aspartate 85 and then to the exterior surface of the membrane. A proton is taken up from the exterior surface via aspartate 96 to form N.

Sensory rhodopsin II (SRII, also called phoborhodopsin) is specialized for repellant phototaxis.[591a] Blue light converts $SRII_{487}$ in < 1 ms to UV-absorbing $SRII_{360}$. It decays in ~ 100 ms to $SRII_{540}$ which reverts to the initial $SRII_{487}$ in ~ 0.5s. The cycle is accompanied by swimming reversals that result in a repellent effect of light. The three-dimensional structure is known.

Whereas retinal-based proton pumps all have the conserved residues D85 and D96 of bacteriorhodopsin, only the aspartate correspnding to D85 is conserved in the sensory rhodopsins. D96 is replaced by tyrosine or phenylalanine. In SRI D73 appears to be protonated and, therefore, does not form a counterion for the Schiff base iminium ion. However, in SRII D73 is apparently unprotonated and available to serve as a counterion and as a proton acceptor as in bacteriorhodopsin. There is also a corresponding aspartate (Asp 83) in rhodopsin. This suggests a common signaling mechanism for rhodopsin and the sensory rhodopsins. Finally, there are retinal-containing proteins in fungi and in algae. They may serve as blue light receptors.

The Photoactive Yellow Proteins (Xanthopsins)

A 124-residue, 14-kDa yellow protein isolated from a halophilic phototrophic purple bacterium, *Ectothiorhodospira halophila,* was at first thought to be a rhodopsin-type pigment. However, this photoactive yellow protein (PYP) carries a covalently bonded *coumaroyl (4-hydroxycinnamoyl) group* in thioester linkage, which is completely surrounded by the small protein. The coumaroyl group, which was probably derived from coumaroyl-CoA, is bound as a phenolate anion by hydrogen bonds to tyrosine and glutamate side chains. After a laser flash at room temperature a readily observed intermediate I_1 (also called pR) absorbing maximally at 460-465 nm appears within ~3 ns and decays within a few milliseconds to a bleached intermediate I_2 (also called pB or pM) with maximum absorption at ~355 nm. This returns to the original 446-nm form within a few seconds. Earlier intermediates I_0 and I_0^* have been identified by pico-second spectroscopy, and others have been identified at low temperatures.

The structure of PYP is known to 0.1 nm resolution Fig. 2.47. Structures have also been determined for a very early intermediate by trapping at –100°C and for I_1 (pR). The cofactor structures are shown in Eq. 2.41. The light-induced step is apparently the *cis-trans* isomerization, and changes in hydrogen-bonding follow. The hydrogen bond between the phenolate ion of the coumaroyl group and glutamate 46 appears to break, and E46 may donate a proton to the phenolate group to form the 337-nm chromophore of I_1. The signaling mechanism may be similar to that in sensory rhodopsins.

Stentorin

A protein with a bound chromophore called stentorin mediates the light-avoidance response of the protozoan *Stentor.* Stentorin, which is found in pigment granules in the cell surface, is a derivative of *hypericin,* a plant compound with antidepressant activity and the active ingredient in the herb St. John's wort.

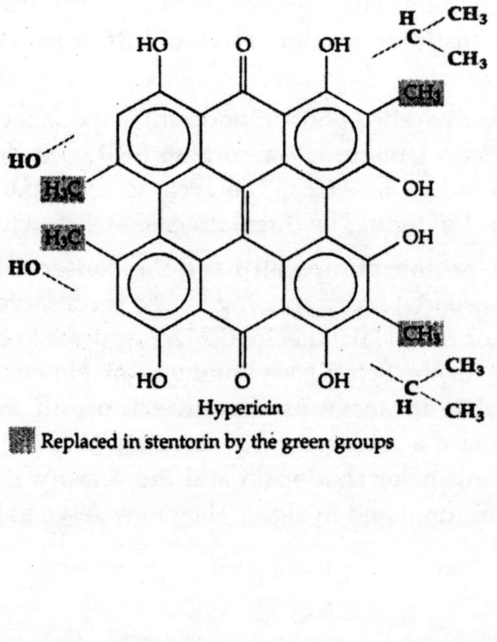

Hypericin

Replaced in stentorin by the green groups

Stentorin is covalently bonded to a 16-kDa protein in an acid-abile linkage. Its photocycle is not well investigated, but it is thought to initiate a response via proton transfer. However, stentorin proteins apparently do not belong to the bacteriorhodopsin family.

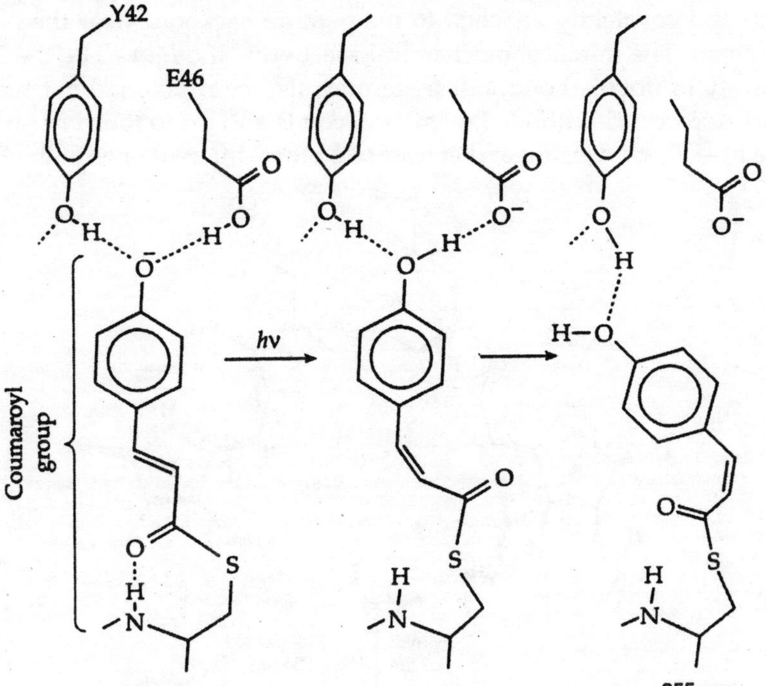

Y42
E46
Coumaroyl group
hν
446 nm
355 nm

Hypericin and related compounds have also aroused interest because of their antiviral and antitumor activities. Hypericin is a strong photoactivator which produces singlet oxygen with a quantum efficiency of 0.73. However, antiviral activity may involve a radical mechanism. Hypericin is attractive as a possible agent for photodynamic therapy. It can also receive energy from photoexcited firefly luciferin. A proposed application is to incorporate the gene for the enzyme luciferase of the firefly luminescence system into DNA from the virus HIV. This DNA could be used to promote synthesis of luciferase only in virus-infected cells. Addition of the nontoxic hypericin would lead to photoactivation of hypericin only in virus-infected cells, where the luciferin-luciferase complex would act as a *"molecular flashlight"* to activate the hypericin and destroy the cell.

H. PHYTOCHROME

In 1951, it was discovered that a flash of red light (maximum activity at 660 nm) during an otherwise dark period promoted a variety of responses in plants. These included flowering, germination of seeds (*e.g.*, those of lettuce), and the expansion of leaves in dark-grown pea seedlings. Interestingly, the effect of the short flash of red light could be *completely reversed* if followed by a flash of *far-red light* (730 nm). This discovery led to the isolation, in 1959, of the chromoprotein phytochrome, a kind of molecular switch that initiates a whole series of far-reaching effects in plants. The phototransformation is completely reversible and the switch can be thrown in one direction or the other many times in rapid succession by light flashes.

Green plants have a family of phytochrornes. There are five genes for the ~125-kDa chains of about 1100 residues each in *Arabidopsis,* and the corres spending phytochromes A – E each appear to have distinct functions. The chromophore is an open tetrapyrrole closely related to phycocyanobilin and covalently attached to the peptide backbone near the N terminus through a cysteine side chain. The initial photochemical reaction is thought to be the Z → *E* isomerizion around the C15 – C16 double bond, but there may also be rotation about the C14 – C15 single bond. The initial step occurs within a few microseconds and up to four intermediate species have been seen in the $P_r \rightarrow P_{fr}$ conversion and at least two other different ones in the $P_{fr} \rightarrow P_r$ conversion.

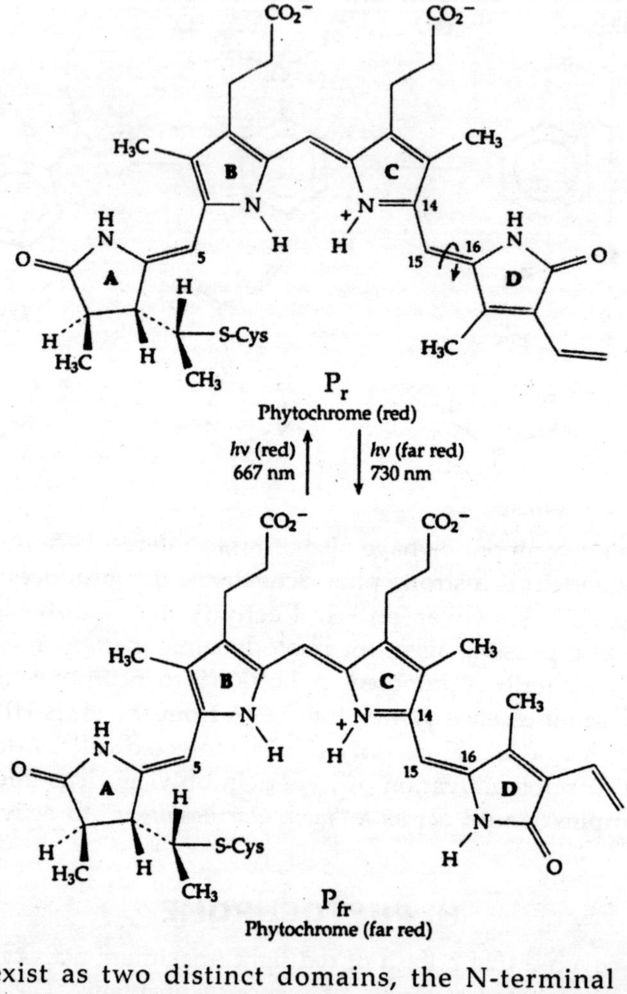

P_r
Phytochrome (red)

hv (red) 667 nm ↑ ↓ hv (far red) 730 nm

P_{fr}
Phytochrome (far red)

Phytochromes exist as two distinct domains, the N-terminal domain bearing the chromophore. However, the first 65 residues at the N terminus as well as the C-terminal domain probably interact with other proteins to transmit a signal. The slow responses to phytochrome are thought to involve regulation of transcription. Thus, the synthesis of mRNA molecules specific for the small subunit of ribulose bisphosphate carboxylase and for the chlorophyll *a*/*b* binding protein increases in response to formation of P_{fr}. These responses are quite rapid occurring within 15 – 30 min. Another response to P_{fr} is a decrease in the amount of the specific mRNA for phytochrome itself. That is, light induces a decrease in the concentration of this light-sensing molecule thereby decreasing the sensitivity of the system.

Phytochrome is found not only in higher plants but also in algae, where it controls the movement of chloroplasts, and also in cyanobacteria. Cyanobacterial phytochromes contain histidine kinase domains, which may function in a two-component system with a response regulator similar to protein CheY of the chemotaxis system in *E. coli.* Some nonphotosynthetic bacteria also use bacteriophytochromes for light sensing. In some cases biliverdin is the chromophore.

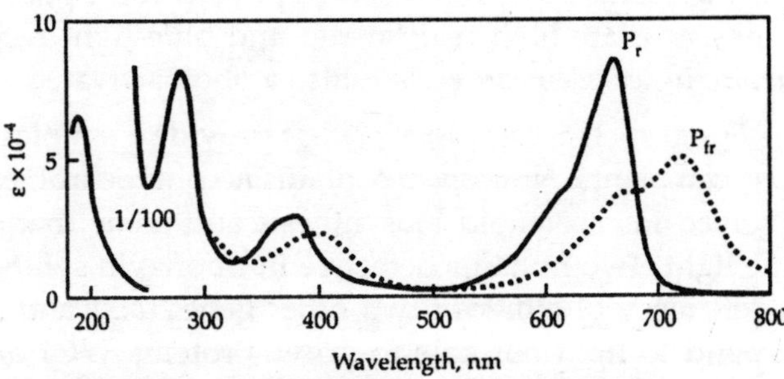

Fig. 2.43. Absorption spectra of red (P_r) and far-red (P_{fr}) forms ofpurified oat phytochrome following saturating irradiations with red and far-red light.

Phytochromes of higher plants also have histidine kinase-like N-terminal domains. Searches for associated signaling proteins have revealed a phytochrome-interacting factor in *Arabidopsis.* A possible partner for phytochrome B, it is a nuclear helix-loop-helix protein that may be a transcription factor. Phytochrome A may signal via a WD-repeat protein to control morphogenesis.

One response under phytochrome control is the closing of leaflets of *Mimosa* at the onset of darkness. The response occurs within 5 min, too short a time to be the result of transcriptional control. This and the finding that some phytochrome is tightly bound to membranes have led to the proposal that one primary effect of phytochrome is to alter membrane properties. It is not certain whether it is P_r or P_{fr} that is active in causing a response, but P_{fr} seems to be the most likely candidate for the "active" form. According to one suggestion, phytochrome in plastid membranes may mediate the release of gibberelins stored within the plastids.

I. SOME BLUE LIGHT RESPONSES

Numerous biological responses to light of wave-length 400 – 500 nm are known. These include phototropism in higher plants, the phototaxis of *Euglena,* and photorepair of DNA. On the basis of action spectra both carotenoids and flavins.were long ago proposed as photoreceptors. The action spectrum for opening of the stomates in coleoptiles matches the absorption spectrum of zeaxanthin. On the other hand, genetic evidence has strengthened the view that a flavin acts as the photoreceptor in the fungus *Phycomyces.* Recently compelling evidence for a flavoprotein receptor for phototropism in *Arabidopsis thaliana* has been obtained. Deficiency of a gene called *nphl* (nonphototropic hypocotyl 1) is associated with loss of blue light-dependent phosphorylation of a 120-kDa protein. This protein was identified as the product of *nphl* gene. The *nphl* protein is a soluble autophosphorylating Ser/Thr protein kinase with an N-terminal flavin-binding region. It apparently binds FMN and is a photoreceptor for phototropism in higher plants.

The complexity of the action spectra suggested the existence of more than one receptor. In higher plants there are not only blue light receptors but also violet receptors and phytochrome. In addition to nphl and a related protein NPL1. *Arabidopsis* employs two cryptochromes and *phototropins.* These are also riboflavin 5'-phosphate (FMN)-dependent proteins. The action of light apparently causes addition of a highly conserved cysteinyl –SH group to the C4a position of the flavin. Phytochrome absorbs blue and ultraviolet light to some extent as well as red or far-red. This adds considerable complexity to the interpretation of light responses in plants. The fern

Adiantum contains a protein with a phytochrome photosensory domain fused to an NPH1 structure. It may mediate both real far-red and blue-light responses. The protist *Euglena* makes use of a *photoactivated adenylate cyclase,* also a photoactivated enzyme, in a light avoidance response.

Fungi such as *Neurospora crassa* provide a simpler system for study of blue-light signaling than do green plants. *Neurospora* contains no phytochromes. However, numerous genes including some involved in carotenoid biosynthesis and some that control the circadian cycle are regulated by blue light. Two mutants defective in riboflavin synthesis show a reduced sensitivity to blue light. A deficiency of either of two other genes *wc*-1 and *wc*-2 results in "blind" *Neurospora* unable to respond to light but able to grow. Proteins WC1 and WC2 are probably transcription factors, which act as a heterodimer. WC1, which contains bound FAD, is the photorecetor. Recently a rhodopsinlike protein NOP-1 of *Neurospora* has been identified.

1. Cryptochromes

The elusive nature of the principal blue-light receptor in plants gave rise to the name cryptochrome. The gene for a cryptochrome in *Arabidopsis thaliana* was isolated by gene tagging and was cloned. It is surprisingly similar in sequence to the gene for the well-known *DNA photolyase.* It was soon recognized that cryptochromes, like photolyases, carry a bound flavin and also an antenna chromophore. The latter is probably a 5, 10-methylenetet-rahydrofolate, as in plant photolyases. It is a better light absorber than the flavin and passes electronic excitation to the flavin.

Two cryptochrome genes, *cry*1 and *cny*2, are present in *Arabidopsis.* The encoded proteins affect many aspects of plant growth. The cry-1 protein, together with NPH1, has a role in controlling phototropisin while cryptochrome cry-2 affects flowering time, apparently via antagonistic signals from cry-2 and phytochrome B. cry-1 is also involved in controlling the daily rhythm of the plant, the *circadian cycle.* The circadian clock, provides the organisms an oscillator with a period of about 24 hours. However, the oscillator must be *entrained* by the daylight cycle so that it remains in proper synchrony. The nature of the light signal and the mechanism of the entrainment are being investigated in many different organisms from fungi to human beings. In *Arabidopsis* the cycle is controlled by phytochromes A and B and by the cryptochrome cry-1.

Cryptochrome genes have been found in many organisms. In the fly *Drosophila* cryptochrome appears to interact directly with the clock proteins that control the circadian cycle. Most important are products of two genes *per* (period) and *tim* (timeless). They are helix-loop-helix DNA binding proteins that form heterodimers, are translocated to the nucleus, and repress their own transcription. Morning light leads to a rapid disappearance of the TIM protein. The cryptochrome CRY appears to react directly with TIM to inactivate it. However, details remain to be learned. The circadian clock mechanism appears to be universal and the cryptochrome-2 *(mcry*2 gene) appears to function in the mouse. A human cDNA clone was found to have a 48% identity with a relative of cryptochromes, the *(6–4) photolyase* of *Drosophila.* A second related human gene has also been found. The protein products of these two genes *(hcry*1, *hcry*2) lack photolyase activity. They too may encode cryptochromes.

Where in the body is the light sensed for entraining the circadian cycle? Genes for CRY1 and CRY2 are specifically expressed in ganglion cells of the retina in mice. Severing of the optic nerve destroys both vision and light entrainment of mammals. However, in mice with the retinal degeneration (*rd*) syndrome all rod cells and virtually all cone cells are destroyed but the circadian rhythm is normal. Furthermore, many blind persons with no conscious perception of light have normal light entrainment of their circadian cycle. For these reasons the ganglion cells of the retina, which are close to the location of the master circadian clock in the *suprachiasmatic nucleus* of the brain, are the most probable light sensory cells for the cycle. Recent evidence points to a retinal-based photoreceptor, *melanopsin*. However, vitamin A-deficient mice still display a normal circadian response.

2. Photolyases

A curious discovery was made many years ago. Bacteria given a lethal dose of ultraviolet radiation can often be saved by irradiating with visible or near ultraviolet light. This *photoreactivation,* which permits many bacteria to survive, results from the action of a *DNA photolyase,* which often absorbs light maximally around 380 nm and carries out a photochemical reversal of Eq. 2.26, cutting the pyrimidine-pyrimidine covalent bonds of thymine dimers in DNA. The enzyme is present in cells in such small amounts, only 10 – 20 molecules per cell, that it was difficult to investigate until the gene had been cloned. The significance cannot be doubted, for photoreactivation enzymes appear to be found in most organisms including some mammals. However, there is some doubt about the presence of a photolyase in the human body.

The *E. coli* DNA photolyase contains a blue flavin radical that arises from *FAD* and absorbs maximally at 580 nm. The enzyme also contains a second chromophore in the form of bound 5,10-methenyltetrahydrofolylpolyglutamate with 3 – 6 γ-glutamyl residues. The pterin coenzyme binds near the N terminus in a domain with an α/β folding pattern, while the FAD binds into a larger mostly helical domain. The pterin cofactor is not essential for repair activity, and it is generally agreed that because of its high molar extinction coefficient it acts as an effective *antenna.* It transfers energy in a nonradiative fashion to the $FADH^-$ anion located ~3 nm away.

The enzyme as isolated is in a stable blue radical form which must undergo a one-electron light-induced reduction to the anion $FADH^-$ before becoming active. A nearby indole ring of Trp may donate the electron and be reoxidized by a tyrosyl ring. The $FADH^-$ donates an electron to the pyrimidine dimer, initiating the sequence of radical reactions which cleaves both pyrimidine-pyrimidine bonds in the photodimers.

The structures of all of the photolyases are thought to resemble that in Fig. 2.49. However, in one large group, which includes methanogenic bacteria, *8-hydroxy-5-deazariboflavin* acts as the antenna chromophore. Another light-induced defect in DNA is the so-called 6 – 4 photoproduct, a different pyrimidine dimer. The 6 – 4 dimers are normally removed in most organisms by efficient *excision repair*. However, a 6 – 4 photolyase was discovered in both *Arabidopsis* and *Drosophila* and has also been found in *Xenopus* and the rattlesnake. It has a structure similar to that of the *E. coli* photolyase and presumably acts by a related mechanism that uses the light-excited reduced flavin in an electron donation and return cycle. A homolog of the *Drosophila* 6 – 4 photolyase gene has been found in human cells, but there is uncertainty about its function. Is it really a photolyase or is it a cryptochrome involved in the circadian cycle?

J. BIOLUMINESCENCE

The emission of visible light by living beings is one of the most fascinating of natural phenomena. Luminescent bacteria, glowing toadstools, protozoa that can light up ocean waves, luminous clams, fantastically illuminated railroad worms, and fireflies have all been the objects of the biochemists, curiosity. The chemical problem is an interesting one. The firefly's light with a wavelength of 560 nm (17,900 cm^{-1}) has an energy of 214 kJ/einstein. What kind of chemical reaction can lead to an energy yield that high? It is far too great to be provided by the splitting of ATP. Even the oxidation of NADH by oxygen would barely provide the necessary energy.

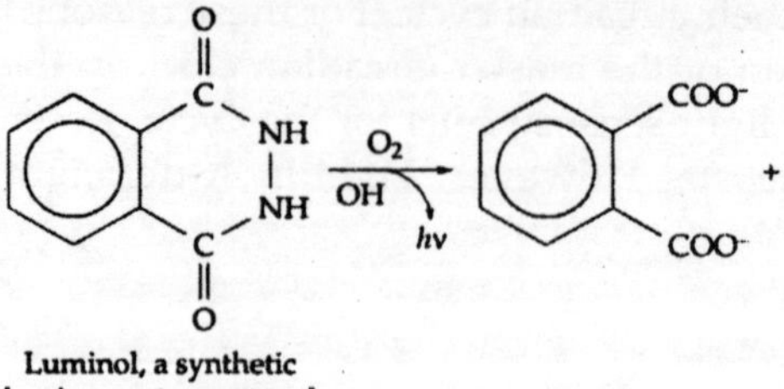

Luminol, a synthetic luminescent compound

A clue comes from the fact that chemiluminescence is very common when O_2 is used as an oxidant in nonenzymatic processes. The slow oxidation of alcohols, aldehydes, and many nitrogen compounds is accompanied by emission of light becomes CO_2, while the ring becomes oxidized.

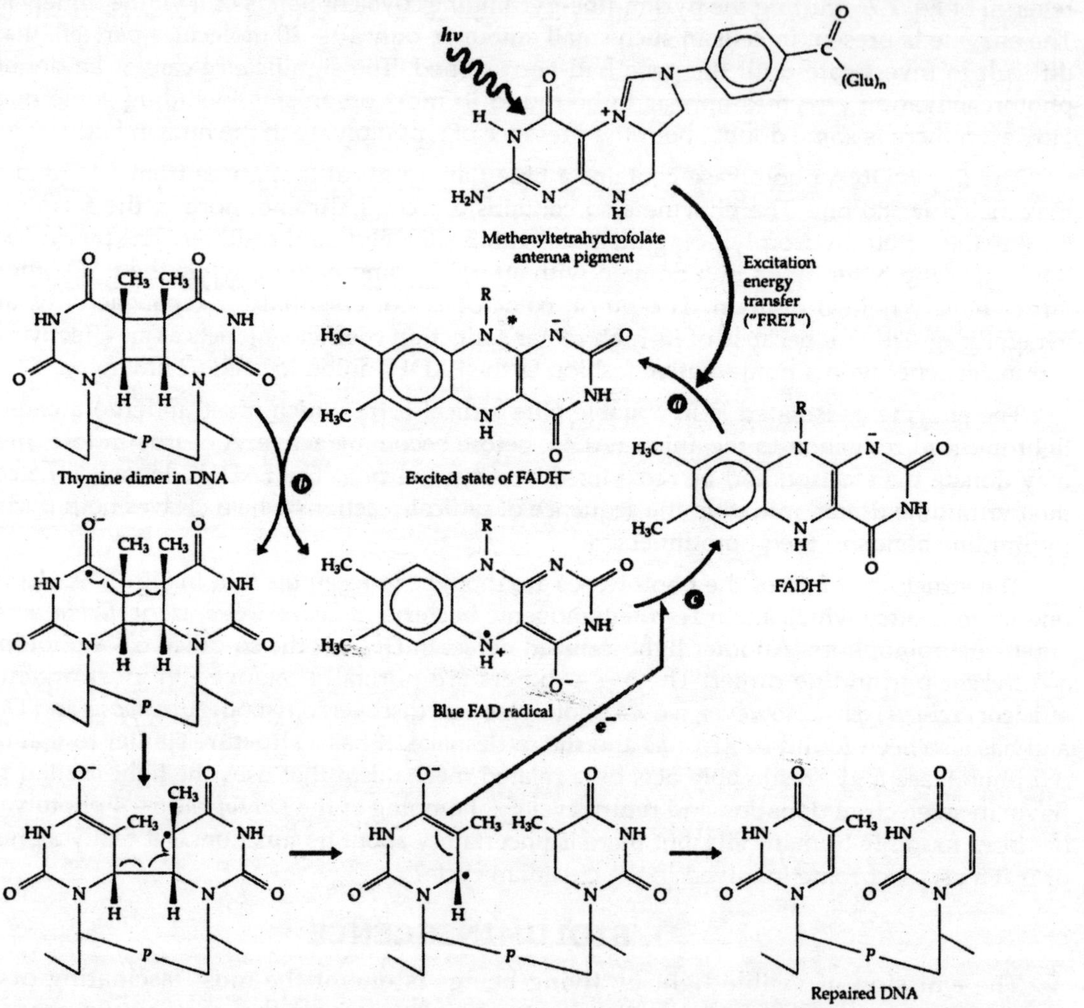

Fig. 2.44. Proposed mechanism of DNA photoylase action. The antenna pigment (top) absorbs a photon, then transfers energy to the anion of bound $FADH_2$ (step *a)* which donates an electron to the thymine dinner in DNA (step *b*). After free-radical-driven steps an electron is returned to the blue FAD radical (step *c*).

visible to the eye. Chemiluminescence is especially pronounced in those reactions that are thought to occur by radical mechanisms. The recombination of free radicals provides enough energy to permit the release of visible light. Cleavage of a peroxide linkage, *e.g.,* in a dioxetane is often involved. For example, the reaction of Eq. 2.44 is used in a sensitive light-detected assay for alkaline phosphatase.

In view of these facts it is perhaps not so surprising that many organisms have mastered the ability to channel the energy released in an oxygenation reaction into light emission. Attempts to extract luminous materials from organisms date from the last century when the French physiologist, DuBois, in 1887 prepared both a cold-water extract and a hot-water extract of luminous clams. He showed that the material in the cold-water extract, which he named *luciferase,* caused emission of light when a heat-stable material (which he called *luciferin*) present in hot-water extract was added. These names have been retained and are now used in a general way. Thus, the luciferins are a family of compounds whose structures have been determined for a number of species. Firefly luciferin is a carboxylic acid, but it must be activated in an ATP-requiring reaction to give *luciferyl adenylate,* whose structure is shown in the figure. The latter emits light in the presence of O_2 and luciferase. It can be seen that the original carboxyl group In addition, the acyl adenylate linkage is broken. In the *"sea pansy" Renilla reniforniis* (a coelenterate) the luciferin has quite a different structure. However, the reaction with O_2 to produce CO_2 and an oxidized product causes the light emission, just as in the firefly. The luciferin of *Renilla,* which is called *coelenterazine* is also found in the jelly-fish *Aequorea,* the shrimp *Oplaphorus,* the "firefly squid" *Watasenia scintillans,* and other luminous organisms.

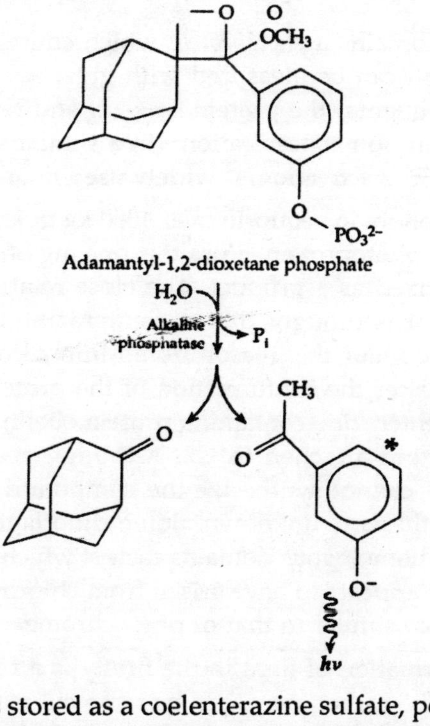

Adamantyl-1,2-dioxetane phosphate

In *Renilla* the coelenterazine is stored as a coelenterazine sulfate, possibly having the structure shown. To convert this storage form to the active luciferin the sulfo group is transferred onto adenosine 3′,5′-bisphosphate to form 3′-phosphoadenosine 5′-phosphosulfate, the reverse of step *d* of Eq. 2.38. The luciferin of the ostracod crustacean *Vargula hilgendorfii* has a structure close to that from *Renilla*. In *Vargula* (formerly *Cypridina)* the luciferin and luciferase are produced in separate glands and are secreted into the surrounding water where they mix and produce a bright cloud of light.

In most mechanisms suggested for luciferase action O_2 reacts at the carbon atom that becomes the carbonyl group in the product to form an intermediate peroxide. In the case of *Renilla* luciferin this can easily be visualized as a result of flow of electrons (perhaps one at a time) from the pyrazine nitrogen into the O_2. According to one proposal, the peroxide group that is formed adds to the carbonyl to form a four-membered dioxetane for coelenterazine peroxide. The latter opens, as indicated by the arrows, to give the products. This theory was tested using $^{18}O_2$. In the case of *Vargula* luciferin the expected incorporation of one atom of ^{18}O into CO_2 was observed, but with firefly and *Renilla* luciferins no ^{18}O entered the CO_2. In these two cases, a somewhat different mechanism may hold.

The jellyfish *Aequorea* contains a *photoprotein* which emits light only when calcium ions are present. Since light emission can be measured with great sensitivity (modern photomultipliers can be used to count light quanta) the protein *aequorin* and related photoproteins are used as a sensitive indicator of calcium ion concentration. (In a similar way the firefly luciferin-luciferase system, which requires ATP for activation, is widely used in an assay for ATP.)

To identify the chromophore in aequorin over 4000 kg of jellyfish were used to obtain 125 mg of electro-phoretically pure photoprotein. From this one mg of a chromophoric substance AF-350 was isolated and characterized as a product. The close relationship to the *Renilla* and *Vargula* luciferins is obvious, and it is thought that coelenterazine is present in aequorin and other photoproteins as a per-oxide. For this reason no additional oxygen is needed to complete the reaction when Ca^{2+} acts to alter the conformation of the protein. The structure of a photoactive intermediate from the coelenterazine-containing protein obelin (from *Obelia longissima*), however, shows only one oxygen atom attached to C2. Although coelenterazine is utilized by many cnidarians they apparently cannot synthesize the compound but must obtain it through their diet. The source of biosynthesis is unknown. Some dinoflagellates emit light from a 137-kDa luciferin that contains three homologous domains each of which binds a molecule of a tetrapyrrole. From its structure the latter appears to have arisen from chlorophyll whose ring has been opened to give a structure somewhat similar to that of phytochrome.

The first step in the formation of light in the firefly is a reaction with ATP to form luciferyl adenylate. The proton on the carbon may then be removed making use of the electron accepting properties of the adjacent ring system and carbonyl group before addition of the O_2. The reactions should be compared to those catalyzed by oxygenases. The large 62-kDa firefly luciferase has a two-domain structure that suggests that domain movement may be essential to bring reactants

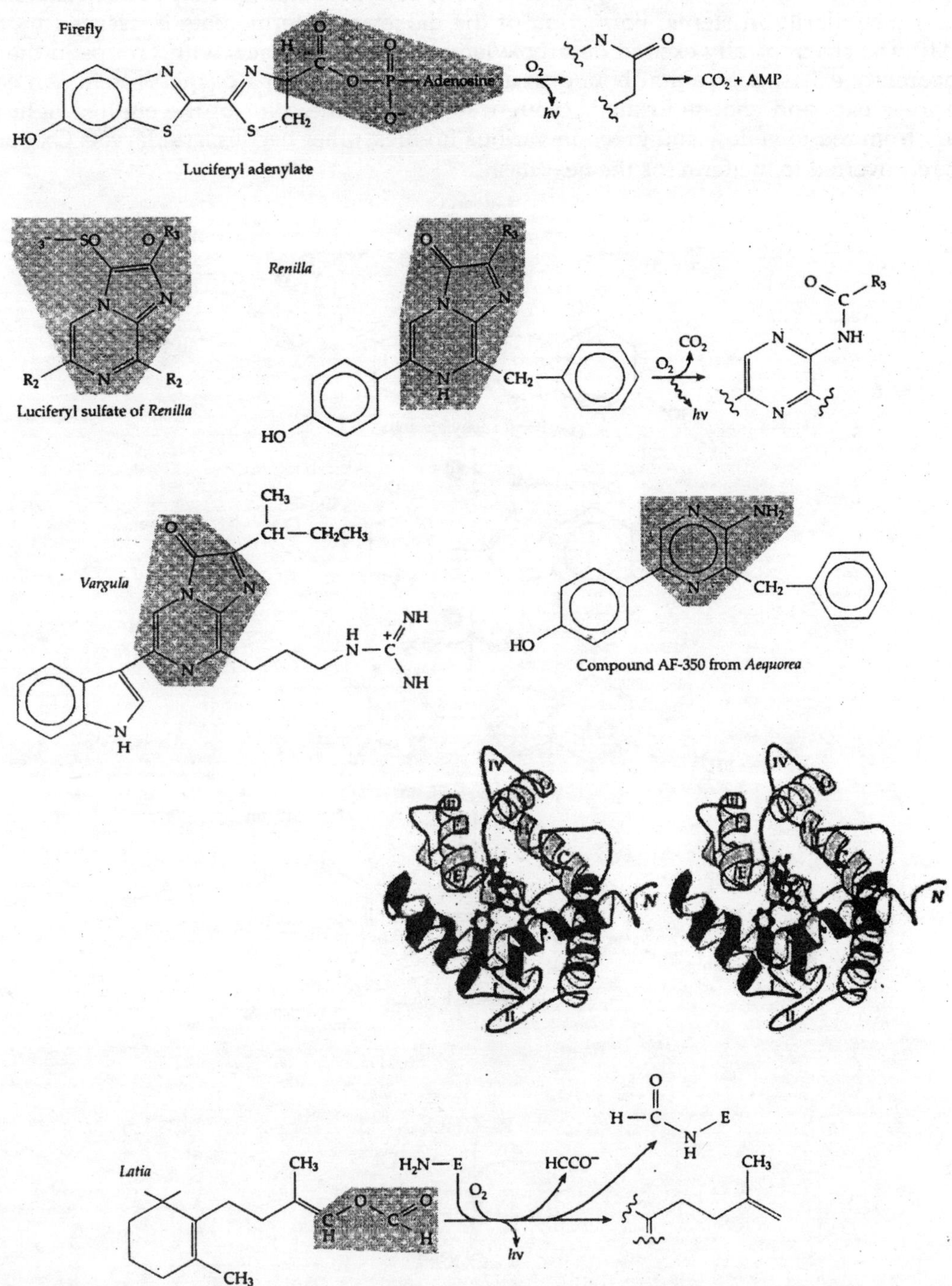

Fig. 2.45. Structures of luciferins from several luminous organisms. The forms shown are the *"activated"* molecules ready to react with O_2. However, compound AF-350 is a breakdown product of the Ca^{2+}-activated luminous protein aequorin.

together. The structure is homologous to those of acyl-CoA ligases and peptide synthetases which share a similarity in step *a*. Formation of the dioxetane intermediate is assisted by the loss of AMP. The electronically excited decarboxylation product interacts with groups in the protein. It apparently exists as an anion bound to acidic and basic groups of the protein. An equilibrium between oxo- and enolate forms is thought to regulate the color of the emitted light which can vary from red to yellow and green in various fireflies, other beetles, and larvae: Oxyluciferin can be reconverted to luciferin for the next flash.

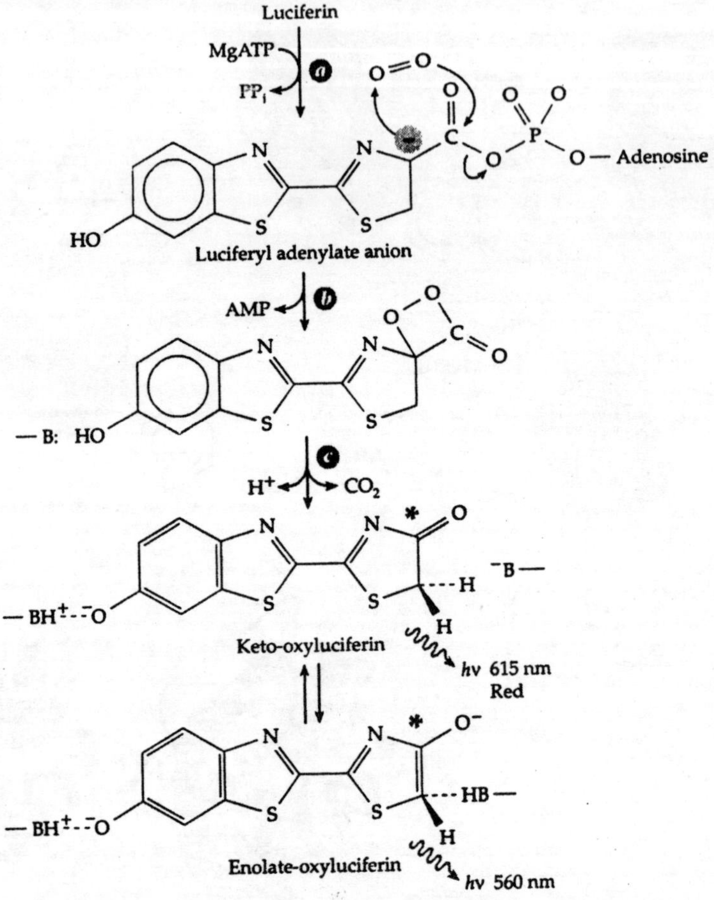

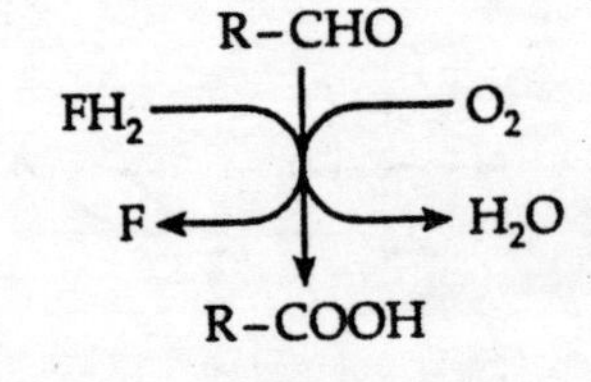

A very different light-producing reaction is used by the limpet *Latia*. The luciferin is an unusual terpene derivative that lacks any chromophore suitable for light emission. Evidently oxidation of this luciferin causes electronic excitation of some other molecule, presumably a *"purple protein"* which is also needed for luminescence A complex of luciferin plus the purple protein is believed to react with the luciferase. It is thought that the formyl group is released from its enolic ester linkage in the luciferin. A Schiff base of the resulting aldehyde may form with the enzyme and react with oxygen. Yet another type of luciferin is found in dinoflagellates.

Luminescent bacteria all appear to obtain light from a riboflavin-5′-phosphate dependent oxygenase, which converts a long-chain aldehyde (usually *n*-tetradecanal) to a carboxylic acid. Here FH_2 is the riboflavin 5′-P, which is thought to be supplied by a flavin reductase.

Bacterial luciferases are αβ heterodimers with subunit masses of ~40(α) and 35 (β) kDa. In *Vibrio harveyi* these are encoded by the *lux A* and *lux B* genes. At least five other genes are essential for light production including two regulatory genes. The tetradecanal and other long-chain aldehydes are supplied by reduction of the corresponding acyl-CoA. A special thioesterase releases a myristoyl group from an acyl carrier protein, diverting it for luminescence in *V. harveyi.* There is good evidence from ^{13}C NMR and electronic spectra for an enzyme-bound reduced flavin hydroperoxide. While this hydroperoxide can decompose slowly to flavin and H_2O_2 in the dark, it can also carry out the oxidation of the aldehyde with emission of light. The luminescent-emission spectrum resembles the fluorescence spectrum of the 4a –OH adduct which is probably the light-emitting species.

Cells of *Vibrio fischeri,* from the light organ of the fish *Monocentrus japonicus,* emit light only in dense cultures where a chemical inducer identified as N-(2-oxocaproyl) homoserine lactone accumulates.

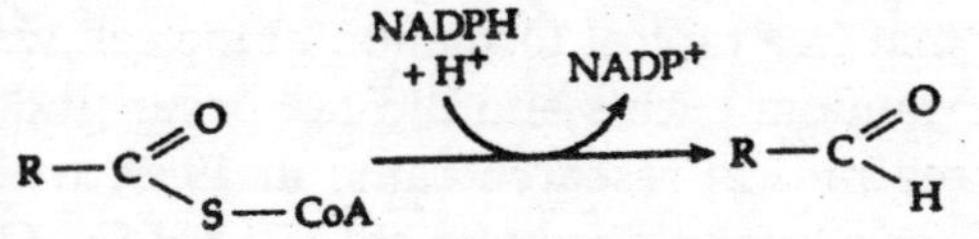

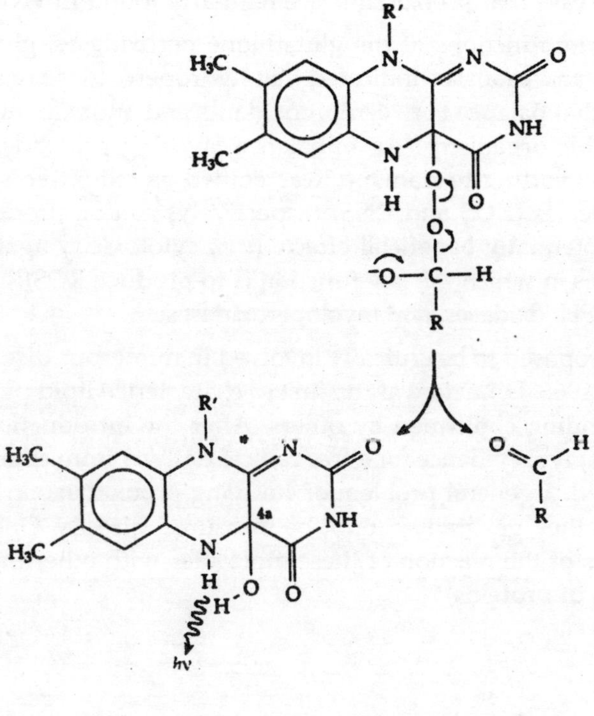

$$R-C(=O)-S-CoA \xrightarrow[]{NADPH + H^+ \;\; NADP^+} R-C(=O)H$$

CHAPTER 3

Hepatic Stress

Early free radical researchers were for the most part chemists (*e.g.*, Henry John Horstman Fenton of the "Fenton" reaction. However, free radical formation was considered for the most part an ex vivo phenomenon. Gershman and Gilbert first proposed that radiation-induced cell killing might be mediated by oxygen free radicals, based on the observation that the pattern of toxicity of radiation was quite similar to that hyperoxygen in vivo. In the same year, Harman published the "free radical hypothesis of aging". This strength of this hypothesis can be demonstrated by the simple fact that it has endured for almost 50 yr.

It was generally thought that free radical formation in the cell was limited to either external stimuli (such as radiation), or random events and did not occur under "normal" circumstances. A major paradigm shift in free radical research came in 1969, when Fridovich and McCord described the function of SOD as a catalytic reducer of O_2^- to H_2O_2. On the basis of these results, the concept that oxidants are produced by the cell under normal conditions gained hold. The discovery of enzymes that normally produce prooxidants (in contrast to electron leakage from other enzyme systems), such as NAD(P)H oxidases, xanthine oxidases, and myeloperoxidases, further strengthened the case that prooxidants are regularly found in vivo.

Discoveries of catalytic functions of the glutathione peroxidases, glutaredoxin peroxidases, thioredoxin peroxidases, and catalases indicated that hydroperoxides are also kept in close check in vivo. It became clear that balance between prooxidants and antioxidants is critical for survival and functioning of aerobic organisms. An imbalance favoring prooxidants and/or disfavoring antioxidants, potentially leading to damage, was coined as oxidative stress by Sies. Reactive oxygen and nitrogen species (ROS and RNS, respectively) can be products of normal cellular metabolism and have potentially beneficial effects (*e.g.*, cytotoxicity against invading bacteria); indeed, there are enzymes in which the key function is to produce ROS/RNS, such as nitric oxide synthases (NOS), NA(D)PH, oxidases, and myeloperoxidases.

Oxidative stress is proposed to be critically involved in numerous diseases, including alcohol-induced liver disease. In 1966, DiLuzio was the first to characterize lipid peroxidation after chronic exposure to alcohol a finding confirmed by others. After the introduction of the spin trapping technique, in vitro and in vivo, evidence for a free radical derived from ethanol, the α-hydroxyethyl free radical was presented. A general problem of studying prooxidants in vivo is that, because of their inherent reactivity, they generally cannot be measured directly. One is therefore often left with measuring products of the reaction of these molecules with other molecules (*e.g.*, oxidative or nitrative modification of proteins).

This indirect detection of "footprints" of ROS/NOS makes it difficult to conclusively identify the parent oxidant. However, coupled with in vivo modulation of free radical production, one can obtain relatively clear data supporting the involvement of specific free radical species involvement in alcohol-induced tissue damage. This chapter will summarize some techniques that, have been used in the study of oxidative stress in alcohol-induced liver injury. Immunohistochemical detection of 4OH-nonenal- and 3-nitrotyrosine-modified proteins will be detailed.

In the past, immunohistochemical techniques have served largely as qualitative measures of changes but have the advantage of illuminating the location and zonation of the haptene of interest. However, with the advent of digital photography and image analysis, one can now quantitate treatment-related changes in the tissue of interest. Other advantages of these techniques are that they are readily performed in most basic research laboratories and do not require pretreatment of the animals to attain a signal. The latter point implies that historical samples may be analyzed.

Although the methods described herein are based liver, these immunohistochemical techniques have been employed successfully in most tissue types with minor modifications. Although immunohistochemical analysis of ROS/RNS-modified proteins is a useful technique, careful attention must be paid to minimize the effect of variability and artifacts. Obvious possible sources of in vivo variability (*e.g.*, gender, age, dose, etc) should clearly be avoided. Furthermore, variability caused by the processing and staining of the tissue itself should also be minimized. This ex vivo variability is attributable predominantly to the staining itself; it is therefore imperative that all samples to be analyzed are stained simultaneously to avoid interassay inconsistencies. Reasonable measures should also be taken to avoid all possible sources of ex vivo variability, including harvesting and processing of the tissue for immunohistochemistry.

MATERIALS

All solutions should be of analytical grade or purer. Nanopure water is used for preparing all solutions. Glassware should be freed of possible detergent residue by prerinsing with nanopure water.

Generation of an In Vivo Positive Control for Staining

1. Laboratory safety equipment (gloves, mask, protective eyewear, etc.).
2. Lipopolysaccharide (LPS) from *Escherichia coli* (O55:B5; Sigma).
3. Sterile saline: 0.9 % sodium chloride.
4. Needles for injection.
5. Borosilicate glass tubes (12 × 75 mm; Fisher).

Harvesting of Tissues for Immunohistochemistry

1. Laboratory safety equipment (gloves, mask, protective eyewear, etc.).
2. Rodent anesthetic (*e.g.*, sodium pentobarbital, ketamine/xylazine, etc.).
3. Tools for harvesting tissue.
4. Sterile saline.
5. 10% neutral-buffered formalin (Fisher).
6. 10-mL syringes with plastic infusion needles (16 gage or larger).

7. Razor blades.
8. Plastic tissue cassettes.

Immunohistochemistry

1. Laboratory safety equipment (gloves, mask, protective eyewear, etc.).
2. Ventilated work area (*e.g.*, chemical fume hood).
3. Magnetic stir plate capable of maintaining a slow stir rate (<100rpm) and a 1-inch magnetic stir bar (Fisher).
4. >3X glass staining dishes (45x75x32 mm; "Coplin jars") and slide holders (Wheaton).
5. Xylene (Fisher) or less toxic alternative (*e.g.*, CitriSolv; Fisher).
6. 100% and 95% ethanol (Aaper).
7. Phosphate-buffered saline with 0.5% Tween-20 (PBS) (10x stock) 1.37*M* NaCl, 27 *mM* KC1, 100mM Na_2HPO_4, 18mM KH_2PO_4 (adjust to pH 7.4), and 5% Tween-20.
   (*a*) Autoclave before storage at room temperature. Sodium azide (0.02%) may be added to the stock to help prevent contamination. Prepare working solution by dilution of one part with nine parts water.
8. 10 mM sodium citrate (pH 6).
9. 30% hydrogen peroxide solution (Fisher).
10. Bovine serum albumin, fraction V (BSA; Sigma).
11. Clear nail polish or Dako Pen (Dako).
12. Slide rack (Fisher). A slide storage box (Fisher) works quite well for this purpose.
13. Rabbit polyclonal anti-4OH-nonenal antibody (Alpha Diagnostic); Rabbit polyclonal anti-nitrotyrosine antibody (Upstate).
14. En Vision, HRP-based staining system (Dako). It is not required, but this system usually gives a good signal to noise ratio and contains both the HRP-conjugated secondary antibody and DAB solutions. Alternatively, HRP-conjugated goat antirabbit antibodies (Vector) and diaminobenzidine (DAB) solution (Dako) may be purchased separately.
15. Aqueous hematoxylin (Dako).
16. Mounting media (Permount; Fisher) and coverslips (25 × 50 × 0.15 mm; Fisher).

Digital Image Acquisition and Analysis System

1. Microscope: Eclipse E600 with a 10 × ocular Plan Fluor series objectives (4 ×, 10×, 20×, 40x and 100 ×) (Nikon)
2. Camera: Spot RT slider BW/color digital CCD camera (Diagnostic Instruments)
3. Imaging software: MetaMorph image-acquisition and analysis software (Molecular Devices).

METHODS

Both staining protocols are relatively similar and have shared methods. The in vivo methods require prior approval by the local Institutional Animal Care and Use Committee (IACUC).

Generation of an In Vivo-Positive Control for 4OH-Nonenal and 3-Nitrotyrosine

It is important to have a positive in vivo control for the staining. In our experience, injections of hepatotoxic doses of LPS generate suffiecient signals of both 4OH-nonenal and 3-nitrotyrosine to serve as positive controls in Fig. 3.1 contains representative photomicrographs (10 × 10 magnification) of a liver from a mouse 24 h after injection with LPS (10mg/kg intraperitoneally) and immunostained for 4OH-nonenal. The supplier also recommends additional positive and negative controls for 3-nitrotyrosine staining. These can be generated by following manufacturer instructions.

1. Prepare an LPS-injectable solution in sterile saline (injection volume 1-2mL/ kg). Solution must be made in borosilicate glass tubes (LPS binds to plastic and Teflon). *Note:* LPS is highly toxic care should be taken when handling this compound to avoid exposure.
2. Inject the rodent (rat or mouse) with LPS intraperitoneally. This laboratory uses a dose of 10 mg/kg LPS (intraperitoneally) for both species. However, because of batch-to-batch variability in LPS potency, the dose should be titrated in the individual laboratory to achieve serum alanine aminotransferase (ALT) activity of 200–300 IU/L 24 h after LPS injection.
3. Harvest liver 24 h after LPS injection.

Harvesting Tissue for Immunohistochemistry

1. Anesthetize the rodent according to your institution's IACUC-approved protocol.
2. Perform a laparotomy to fully-expose the liver.
3. Exsanguinate the animal by cutting the aorta just before entering the heart.
4. Flush the liver/animal (Optional). Flush the liver with a 10-mL syringe filled with saline by inserting into the portal vein just below the liver. Follow with a 10-mL syringe filled with 10% neutral buffered formalin. This not only helps minimize possible ex vivo haptene formation but also helps retain the architecture of the organ. If other organs are also to be harvested, exsanguination, flushing, and fixing of the whole animal may be achieved by cardiac puncture and perfusion. A larger syringe (*e.g.*, 50 mL) will be required.
5. Cut thin (2- to 4-mm thick) transverse sections of liver (or other organ of interest).
6. Place in tissue cassettes (labeled with pencil) and fix at 4°C for >48 h.
7. Paraffin-embed tissue, section (5- to 6-µm thick), and mount on microscope slides.

Deparaffinization and Rehydration of Tissue Slides

At least one sample should be processed in duplicate in this procedure for primary antibody control. Normally, the positive in vivo control is selected for this control.

1. Slide racks for the Coplin jars have 10 slots. They may hold up to 20 slides if placed "back-to-back" (*i.e.*, tissue side facing out). The jar should be filled with liquid to completely cover the slides (~225 mL). With a stir bar in the bottom of the jar, it is placed on stir plate (set to the lowest RPM). Incubate in each of the following solutions for 5 min :

   (*a*) Xylene (or alternate solution); repeat once

   (*b*) 100% ethanol

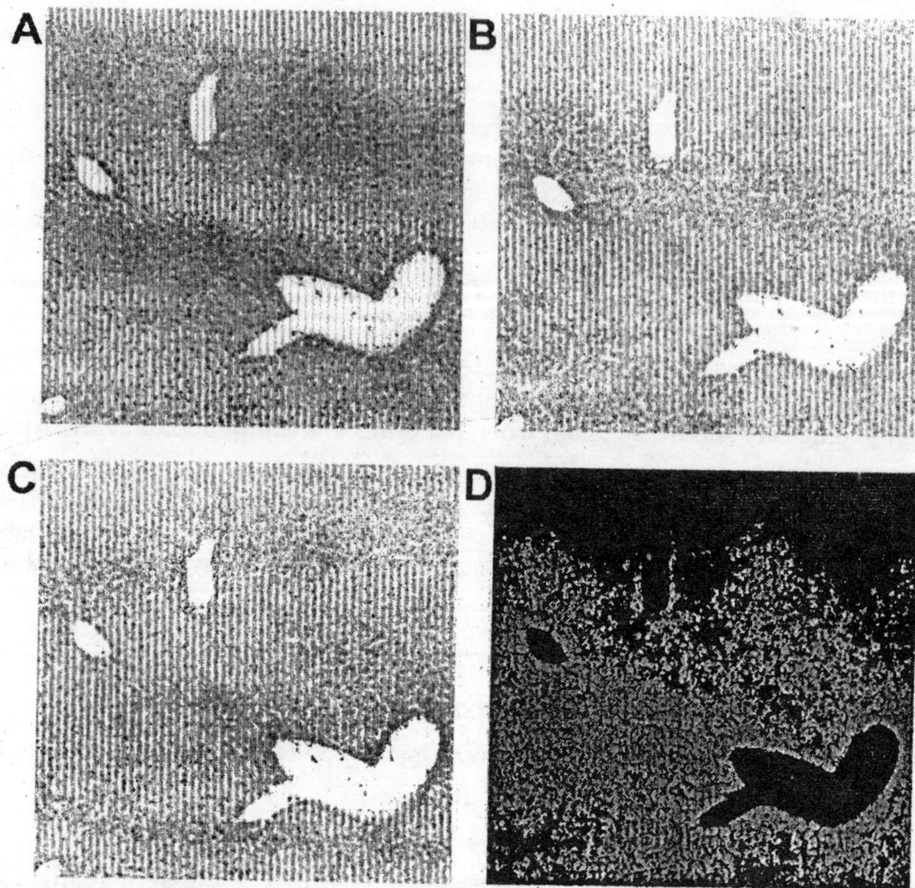

Fig. 3.1. Effect of LPS on 4OH-nonenal adduct accumulation in mouse liver as determined by quantitative immunohistochemistry. Representative photomicrographs are shown depicting 40H-nonenal immunostaining of a liver section (100x) of a mouse 24 h after injection with LPS (10 mg/kg intraperitoneally). The same microscope region is shown before (A) and after (B) white balancing of the image-analysis software. Note that the acellular space in (A) is light yellow instead of white. (C) is the same as (B), but after the blue exposure is increased 10%. Note that the contrast between the DAB chromogen (brown) and the hematoxylin counterstain (blue) is enhanced by digitally increasing the blue exposure. (D) shows the microscope region from (C), determined by the computer to be within the threshold range for positive staining.

(*c*) 95% ethanol

(*d*) 70% ethanol

(*e*) 30% ethanol

(*f*) Nanopure water

2. Create a buffer dam around tissue (optional). While carefully avoiding drying of the section (do one slide at a time), paint a circle around the tissue section with the Dako pen (or nail polish), carefully avoiding touching the tissue with the solution (>3 mm away from the section). This buffer dam decreases the likelihood of the sample drying-out during antibody incubations. Place samples in a new Coplin jar containing fresh water.
3. Perform antigen demasking (optional). This procedure is strongly recommended for 3-nitrotyrosine immunohistochemistry.

(*a*) Place slides in Coplin jar containing citric acid solution.

(*b*) Remove metal handle from slide rack. Place jar in a glass dish of water and microwave on high for 2.5 min (700-watt oven). Warning, buffer will get hot; avoid scalding.

(*c*) Let cool for 3 min. Refill Coplin jar with citrate buffer, if needed.

(*d*) Repeat steps b-c three times. Let cool in citrate buffer for 20 min.

(*e*) Rinse sections in Coplin jar containing nanopure water for 5 min. Repeat once.

4. Rinse the samples in a Coplin jar containing PBS-Tween for 5 min. Repeat once.

Peroxidase and Protein Block of Tissue Sections

1. Make 250 mL of 3% hydrogen peroxide solution diluted in PBS-Tween. Pour into Coplin jar.
2. Incubate the slides for 5 min in peroxide solution.
3. Rinse the samples in a Coplin jar containing fresh PBS-Tween for 5 min. Repeat once.
4. Make PBS-Tween containing 1% BSA (PBA). You will need approx 250 μL per slide.
5. Remove the slides from the Coplin jar one at a time. Pour off excess solution and pipette PBA into buffer dam around tissue. Place on slide rack.
6. Incubate slides for 1 h at room temperature. Keep rack covered and hydrated by placing wet paper towels in the bottom of the rack. Add fresh solution as needed to avoid sample drying.
7. Proceed immediately to immunohistochemistry.

Immunohistochemical Detection of Haptene Signals

1. While samples are blocking, prepare the primary antibody solution by diluting the antibody in PBA (~250 μL/slide). Historically, a 1:500 and 1:50 dilution of the 4OH-nonenal and 3-nitrotyrosine antibodies, respectively, were effective. Dilutions may vary on antibody batch.
2. Shake off excess PBA from slides one at a time and replace with primary anti-body solution. The primary antibody control slide will receive PBA without antibody.
3. Incubate on slide rack (covered and hydrated) for 30 min (4OH-nonenal) or an hour (3-nitrotyrosine). Incubation times may need to be altered to improve signal to noise ratio. Rinse the samples in a Coplin jar containing fresh PBS-Tween for 5 min. Repeat once. Shake off excess PBS-Tween from slides one at a time and replace with secondary antibody solution. Place on slide rack and incubate for 15 min (if using the Dako EnVision kit). Dilutions and incubation times for secondary antibodies from other, suppliers should be determined by manufacturer's suggestions.
4. Rinse the samples in a Coplin jar containing fresh PBS-Tween for 5 min. Repeat once.
5. Prepare DAB working solution according to manufacturer's instructions. Place a piece of white paper in the bottom of the slide rack. If the slide rack does not have a cover, make a cover with aluminum foil (DAB is light sensitive).
6. Apply DAB solution to each slide. Cover slides and incubate for 8 – 10 min. Use the in vivo positive control to determine the length of staining. When the brown DAB precipitate

appears to not increase in this slide, stop the incubation. Transfer slides to a Coplin jar containing nanopure water.

7. Rinse Coplin jar by placing under a gentle tap of water for 6 min.
8. Transfer slide rack to a Coplin jar containing hematoxylin. Incubate in solution for 30 seconds while gently moving the slide rack up and down.
9. Return slide rack to Coplin jar containing water. Rinse under tap for 6 min.

Dehydration and Coverslipping of Slides

1. Incubate the slide rack in each of the following solutions for 3-5 min.
   (*a*) Nanopure water
   (*b*) 30% ethanol 100% ethanol
   (*c*) 70% ethanol 95% ethanol
   (*d*) 95% ethanol 70% ethanol
   (*e*) 100% ethanol
   (*f*) Xylene (or alternate solution); repeat once
2. Keep the slides in the second xylene solution until ready to be coverslipped. Place a thin bead of Permount on one long edge of a coverslip. Gently role the slide onto the coverslip. Capillary action will pull the pieces of glass together. Avoid bubbles over the tissue sections. Should bubbles occur wash off coverslip with xylene solution and recoverslip. Let the slides dry in the fume hood for at least 24 h.

Image Analysis of Immunohistochemistry

An absolutely critical point in performing quantitative image-analysis of immunohistochemistry is to ensure that all samples to be compared are analyzed simultaneously. It is recommended that before performing image analysis that the slides be critically analyzed under the microscope for any variation in chromogen signal not owing to the in vivo condition. Furthermore, image-analysis should be viewed as a means to quantitate differences between treatment groups, not to create.

Specifically, the differences should be readily observable with the eye under the microscope. For microheterogeneous tissues, such as the liver, a low power magnification (*e.g.*, 100 ×, ocular and objective) is usually sufficient for analysis. For macrohetergeneous tissues (*e.g.*, specific brain regions), a higher magnification may be required to focus on the area of interest. Should staining be located in a specific area within a tissue (for example, the perivenous area of the liver lobule), higher magnification photomicrographs may be also justified. It is recommended, how-ever, that initial attempts be performed at lower magnifications, as this reduces the risk of investigator bias in acquiring the pictures. It is also a good idea to blind analysis of the tissues to reduce the risk of investigator bias.

1. Power-up microscope, camera and image-analysis computer. Turn-on the microscope light source to ~80% of full brightness and allow it to sit for at least 20 min. Most bright-field bulbs are unstable light emitters until they are sufficiently warmed-up. A simple way to validate that the bulb is stable is to take five consecutive photos of the same microscope region and validate that they are consistent in brightness, contrast and color.

2. While the system is warming-up, ensure that the microscope is optimized for taking pictures at the selected magnification (*e.g.*, adjusting the diaphragm height and width). These techniques are routine to experienced microscopists.
3. Make sure that the image analysis software is set to manually determine exposure settings. Otherwise, the software may adjust the exposure between pictures, causing variations in staining color and intensity.
4. Using the image-analysis software, calibrate the exposure settings for the micro-scope field. First, move the microscope stage to a region of the slide that does not contain the tissue (but does have the coverslip) and defocus the image. Microscopes often have a gradient of light intensity across the field. Most image-analysis programs have a function to account for this gradient and normalize this gradient in captured photomicrographs (in MetaMorph, the command is "get flatfield". Next, in the same area, determine optimal exposure settings for the digital camera (*i.e.*, "white balance"). Fig. 3.1A,B compare the same microscope field before (A) and after (B) appropriate white balancing.
5. After performing flatfield and white balance commands, the photomicrographs captured by the software should be similar to those seen with eye. Validate this by capturing a representative image and comparing the digital picture with the image seen with the eye. If necessary, manually adjust the RGB exposure settings.
6. Enhance contrast with digital filtering (optional). Image analysis is much more sensitive to contrast than the human eye. Immunostaining may look very clear by manual analysis but may be fraught with noise during digital analysis. This noise can be alleviated by enhancing the digital contrast. For example, slightly increasing the blue exposure on the RGB setting will enhance the contrast between the DAB chromogen and the hematoxylin counterstain. However, care should be taken if this filtering approach is performed that artifactual signals are not introduced. Fig. 3.1C is the same microscope region as Ib, but after the blue exposure has been increased 10%.
7. Once steps 1 through 6 are completed, it is imperative that the neither the microscope nor the exposure settings are altered during the image capture phase. Select and capture 10 microscope fields from each section (assuming 10x10 magnification). The regions should be randomly selected with the caveat that staining artifacts (*e.g.*, edges, folds, tears, etc.) should be avoided. Morphologic variations found in the tissue (*e.g.*, very large vessels in the liver lobule) may also be avoided. However, due diligence should be paid to avoid investigator bias in selecting regions. Should a higher magnification be used to focus on a specific region of the tissue, it is recommended that the number of pictures captured be increased.
8. Repeat this step for all tissue sections to be analyzed.
9. Select an image file with strong positive staining for the haptene. Use this image to determine the color, intensity, and saturation ranges for the image-analysis (called "threshold" in MetaMorph). It is recommended that a "conservative" range be initially attempted to reduce noise in the sections Fig. 3.1D shows the microscope region from Fig. 3.1C determined by the computer to be within the threshold range for positive staining.
10. Pool the summary area data from each tissue section to determine the percentage of microscope field positive for the haptene signal. Use this value to determine group means.

In Vivo Modulation of Free Radical Production Using Genetically Modified Mice

The inherent reactivity of ROS/RNS makes it very difficult to directly measure them under in vivo conditions. Even electron spin resonance spectroscopic analysis of free radicals usually requires a spin-trap to stabilize the radical signal long enough for detection. Therefore, although assessment of changes in ROS/RNS by the techniques described above are useful, these 'footprints' do not give any information on the type(s) and source(s) of reactive species produced that caused these changes.

To date, most work has employed chemical inhibitors of suspected prooxidant enzymes. The effect of these drugs in vivo may not necessarily be specific to inhibition of the enzyme in question. Therefore, caution should be used when basing conclusions on work with pharmacologic inhibitors. One advance towards this end in the alcohol field has been the ability to address key questions at the molecular level by using genetically-altered mice (*e.g.*, knockout/transgenic). Such strategies have their own potential pitfalls, such as induction of compensatory mechanisms in knockout mice, which may cause false-positive or negative results.

NOTES

1. To decrease ex vivo variability, our group always harvests sections for immunohistochemistry from the same liver lobes (left lateral and right medial) and the same location with the lobe (transverse section from the middle of the lobe) to avoid any potential geographic variation in the haptene signals.
2. There are many possible iterations of this system, but the emphasis should be on a system that is able to take high-quality and –resolution brightfield images. Important features of the system include flat-field correction, white balance, and manual control of red/green/ blue (RGB) exposures.
3. Windows-compatible personal computer. Most new PCs will have enough speed and RAM to handle MetaMorph. A large hard drive is recommended to accommodate the large microscope files.
4. Microscope sections must be of consistent high quality (e.g., few if any folds or tears). Should the laboratory lack a practiced hand at cutting and mounting tissues, it is highly recommended that the tissues be prepared and processed by a professional histopathology service.
5. A number of newer and more rapid methods to deparaffinize and rehydrate microscope slides are available that use less reagent volumes than Coplin jars. However, it has been the experience of this author that deparaffinization/rehydration can be inconsistent and incomplete with these alternate approaches. These techniques are therefore not recommended for quantitative immunohistochemistry.
6. The Dako En Vision kit supplies a peroxidase block reagent. However, more consistent blocking has been achieved by our group using freshly diluted concentrated hydrogen peroxide in a Coplin jar.
7. A critical component of quantitative immunohistochemistry is consistency, to avoid intra-assay variability. Careful attention should be paid so that the incubation steps for each sample are similar, especially in steps that require you to work with one sample at a time. A simple solution is to measure the length of time required to add reagent to each sample.

8. The following are general guidelines for using genetically altered mice in experiments. These guidelines are derived from the Banbury conference on genetic background in mice and have been adopted by many scientific journals:
   - (*a*) Genetic background of mouse strain must be clearly reported.
   - (*b*) Common strains of mice should be used to aide reproducibility by other laboratories.
   - (*c*) The genetically modified strain should be backcrossed into a common back-ground strain.
   - (*d*) The strain should be backcrossed into the wild-type strain at least four times to avoid genetic variation at other sites besides the locus of interest.
   - (*e*) Genetically modified line should not be maintained by crossing homozygotes, but rather maintained by backcrossing heterzygotes against the wild-type strain. This avoids the potential of genetic drift of the modified strain away from the background strain.

Results with modified strain should be validated with multiple generations. Further, weight-of-evidence can be increased by comparing pharmacologic targeting of the same protein with the knockout strain.

CHAPTER 4 Cytometry Examination

Chronic alcohol consumption is associated with several abnormalities in the immune system, including increased frequency and severity of certain infections. In addition, chronic alcohol consumption is associated with an increased risk for cancers of various organs. Natural killer (NK) cells are immune surveillance cells that inhibit the development and growth of certain tumors. A closely related lymphocyte subset, the so-called NKT cell, is also believed to serve a regulatory function in various diseases. There is a large body of literature examining the effect of alcohol on NK cell function. There is no consensus on the effect of alcohol on NK cells, and the relationship seems to differ depending on whether one examines acute or chronic alcohol consumption and whether alcoholic liver disease is present.

There is very little literature examining the effect of chronic alcohol consumption on NKT cells. This chapter is intended to detail the methodology used to examine splenic and hepatic NK and NKT cells in mice and is not intended as a comprehensive review of these two cell subsets. NK cells are part of the innate immune system and thus do not require any specific sensitization for the expression of their function. NK cells are lymphocytes that do not belong to the T- or B-cell lineages; they are often referred to as large granular lymphocytes, although not all NK cells are large. NK cells have several functions, most of which can be assigned to their ability to secrete cytokines or their ability to exert cytotoxicity.

NK cells are a major source of interferon gamma (IFN-γ). NK cells are also able to kill certain tumor target cells directly with no need for prior sensitization. Indeed, it is this killing ability that earned NK cells their name. NKT cells are a newly defined lymphocyte subset that shares characteristics of NK cells and T cells. NKT cells express certain markers unique to NK cells but also express some T-cell markers, including CD3 and a restricted pattern of TCR expression. NKT cells also are a major source of cytokines, including IFN-γ, and IL-4. NKT cells, however, are quite heterogeneous.

Some NKT cells subsets are thymically derived whereas others arise from an extrathymic pathway. Most, although not all, NKT cell subsets recognize glycolipids and are thus unique among the various lymphocyte lineages. A recently described cell, termed interferon-producing killer dendritic cells (IKDCs), shares phenotypic and functional qualities with both NK cells and DCs and can be identified by surface expression of both CDllc and CD49b. IKDCs resemble plasmacytoid DCs in that they express both B220 and moderate levels of MHC II but may be differentiated from this DC subset by their lack of Ly-6g. These cells are beyond the scope of this chapter and will not be covered. This chapter will detail the methodology for the isolation of NK and NKT cells from murine spleen and liver as well as the methodology employed to examine their number and function.

Examination of NK and NKT cells can be readily and easily accomplished if one pays attention to some of the pitfalls detailed in this chapter. We hope that the methods described below will simplify the task of investigators seeking to examine NK and NKT cells both of which are proving to be of vital importance in health and disease.

MATERIALS

Purification of NK and NKT from Murine Spleens and Livers

1. Minimal Essential Media (Gibco cat. no. 11095-080), Hank's Balanced Salt Solution (Gibco cat. no. 14175-095), and RPMI 1640 (Gibco cat. no. 11875-093). RPMI1640 is supplemented (cRPMI-5) with penicillin-streptomycin (final concentration 100U/mL and 100μmg/mL, respectively), L-glutamine (final concentration 2mM), gentamicin (final concentration 50μg/mL), 2-mercaptoethanol (2-ME) (final concentration 5×10^{-5} *M)*, and 5% heat-inactivated fetal calf serum. The above ingredients are filter-sterilized (0.2 μm) before use, stored at 4°C, and used within 30 d.
2. 60 × 15-mm sterile tissue culture dishes (Falcon #35-3002).
3. Tissue grinder (Thomas Scientific cat. no. 3431-D70).
4. 100-μm cell strainer (Falcon #352360).
5. Barrel from a 10-mL syringe (BD cat. no. 309604).
6. 0.83% NH_4Cl in 1 mM Trizma Base (RBC lysis buffer).
7. 40% Percoll (Sigma cat. no. 4937) made isotonic via addition of 10 × PBS containing 100U/mL heparin. For optimal separation of murine lymphocytes, osmolality should be adjusted to 330 mOsm/Kg H_2O.

Flow Cytometry Examination of Surface and Cytoplasmic Markers

1. Single-cell suspension of splenocytes or hepatic mononuclear cells (as described below).
2. Staining buffer: Hank's Balanced Salt Solution (Gibco cat. no. 14175-095) with 1 % bovine serum albumin (Sigma cat. no. 7030) and 0.1% sodium azide. pH 7.2-7.4.
3. 96-well round-bottom tissue culture plate (Corning Costar #3799).
4. 24-well flat-bottom tissue culture plate (Corning Costar #3524).
5. 5 mL polystyrene 12 × 75-mm tubes (Falcon #352054).
6. Monoclonal antibodies 2.4G2 (α-CD16/32) unlabeled, prepared from hybridoma culture supernatant (ATCC cat. no. HB-197), PK136 (α-NKl.l), PE and APC (BD Biosciences cat. no. 553165 and 550627), DX5 (α-CD49b) APC (eBio-science cat. no. 17-5971), XMG1.2 (α-IFN-γ) PE (BD Biosciences cat. no. 554412), and C363.29b (α-CD3ε) FITC (Beckman Coulter cat. no. 73191).
7. Murine CDld-PBS57 (a-GalCer analog) APC (NIH Tetramer Facility, Atlanta, GA).
8. Phorbol 12-myristate 13-acetate (Sigma; cat. no. P8139) and Ionomycin (Sigma; cat. no. 13909).
9. BD Cytofix/Cytoperm Plus Golgi Stop (BD Biosciences cat. no. 554715).
10. LincoPlex Mouse Cytokine Panel (Linco Research cat. no. MCYTO-70K-PMX16).

Cytotoxicity Assays

1. RPMI 1640 (Gibco cat. no. 11875-093) supplemented with penicillin-streptomycin (final concentration 100U/mL and 100mg/mL, respectively) and 5% heat-inactivated calf serum (Assay Medium)
2. 96 well Styrene V-bottom plates (Fisher Scientific cat. no. 14-245-72).
3. 2% acetic acid Chromium-51 Sodium Chromate radionuclide (^{51}Cr; Perkin Elmer cat. no. NEZ03001).

METHODS

Purification of NK and NKT from Spleen and Liver

The following should be performed in a Tissue Culture laminar flow hood using aseptic technique.

Isolation of Splenocytes

1. Remove spleens and place in 10mL of MEM.
2. Gently grind spleens with homogenizer and centrifuge cell suspension for 5 min at 200 g to pellet the cells.
3. Remove supernatant and resuspend cells in 2 mL of RBC lysis buffer for 45 s to lyse red blood cells. Immediately bring volume to 10 mL with cRPMI-5 and flash spin to remove aggregates.
4. Centrifuge cell suspension for 5 min at 200 g and resuspend cells in 10 mL of CPMI-5 to wash the cell pellet.
5. Centrifuge cell suspension for 5 min at 200 g and resuspend cells in 10 mL of cRPMI-5 for counting.

Isolation of Hepatic Mononuclear Cells

1. Perfuse liver via the portal vein with 10 mL of cold HBSS. Alternatively, if serum or PBMC are desired, thorough bleeding of the mouse prior to sacrifice will adequately remove peripheral blood from the organ.
2. Mince liver and press through a 100 μm cell strainer using the barrel of a 10 mL syringe.
3. Centrifuge cell suspension for 5 min at 200 g and resuspend cells in 20 mL of HBSS to wash the cell pellet.
4. Repeat wash then centrifuge cell suspension at 30 *g* for 5 min to pellet hepatocytes.
5. Transfer supernatant to a new 50-mL conical tube and centrifuge for 5 min at 200 g.
6. Resuspend cell pellet in 20 mL of 40% isotonic Percoll containing 100 U/mL heparin. Centrifuge for 20 min at 800 g.
7. Wash the cell pellet with HBSS and proceed with RBC lysis as detailed in the splenocyte isolation protocol.

Expected recoveries vary as a function of age, strain and immune status of the mice used, but typical yields range from 7 – 10 × 10^7 splenocytes and 2 – 5 × 10^6 hepatic mononuclear cells per mouse. With proper technique cell viabilities should exceed 90%.

Enrichment of NK and NKT Cells from Splenocytes and Hepatic Mononuclear Cells

Splenic NK and NKT are relatively rare populations comprising roughly 3 – 10% and 1 – 2% of lymphocytes in normal animals, respectively. Although their frequency is increased in preparations of hepatic mononuclear cells (10 – 15% NK and 15 – 30% NKT), enrichment of these cells is often required for functional analyses. The enrichment of NK cells has been facilitated by the availability of antibodies coupled to magnetic particles available from a variety of sources including Miltenyi Biotec (Auburn, CA) and Stem Cell Technologies (Vancouver, BC). Two general strategies are employed, each offering distinct advantages. Positive selection using a magnetic particle-coupled monoclonal antibody directed against a cell-surface marker specifically expressed by the cell type of interest (in this case CD49b or NK1.1) usually offers greater purity and yield.

The concern with positive selection is that engagement of a surface marker might alter the activation status of the cells. Negative selection, in which a cocktail of monoclonal antibodies are used to deplete unwanted cell types (B, T, DC and macrophages; *see Note* 2), will leave the investigator with unmanipulated NK cells. Depending upon the antibodies used, NKT cells may be the major contaminant in NK-cell preparations. One can use the differential expression of CDS to further separate these two populations. Using CD5 rather than CD3 will avoid potential activation of NKT cells via crosslinking of the CD3 complex. In situations where highly purified (>99%) populations are desired (*e.g.,* RT-PCR or western blot), sorting by flow cytometry is the method of choice. One can maximize cell recovery, and minimize the time needed on the sorter, by pre-enriching the cells of import by paramagnetic bead selection.

SURFACE AND CYTOPLASMIC MARKERS FOR NK AND NKT CELLS

Flow Cytometry Staining for Cell-Surface Molecules

1. Dilute cells to 1×10^7/mL with staining buffer and add 100μL to desired wells of 96-well tissue culture plate.
2. Add 100 μL of staining buffer to each well, mix gently, and centrifuge at 200 g for 30 s to pellet the cells.
3. Remove supernatant by quickly inverting and flicking plate.
4. Add 10 μL of 2.4G2 [1 mg/mL] to block Fc receptor-mediated binding and 20μL of each fluorochrome-conjugated antibody. Optimal concentrations for staining should be empirically determined. Incubate 10 mins at room temperature. If CD Id tetramer staining is being performed, increase incubation to 30 min.
5. Wash cells twice with staining buffer and resuspend in 200 μL of staining buffer. Transfer to 12 × 75 tubes and bring final volume to 500 μL with staining buffer. Analyze immediately on flow cytometer. Detector voltages should be set with unstained cells and compensation values set with cells stained with a single fluorochrome.

Intracellular Staining for IFN-γ

1. In a 24-well flat-bottom plate, add 5×10^6 cells to 2mL of cRPMI-5 . Add Golgi Stop (monensin) to a final concentration of 2μ*M*. Incubate alone (for unstimulated control) or with PMA [10ng/mL] and ionomycin [200ng/mL] for 5 h at 37°C in a 5% CO_2-humidified incubator. Another alternative for stimulation is to use YAC-1, the standard mouse NK target cell, for stimulation. Lymphocytes should be cultured with YAC-1 at a ratio of 10:1.

2. Harvest cells, wash and count then proceed with cell surface staining as detailed previously, followed by intracellular staining as described in the BD Cytofix/Cytoperm Kit manual.
3. Flow cytometry has proved to be an invaluable tool for dissection of the immune system and NK and NKT cell research is no exception. The choice of fluorochromes for particular applications will vary depending upon the instrumentation available to the investigator, but the methods are routine and conjugated monoclonal antibodies are widely available from a variety of vendors.

NK and NKT cells can be identified by their expression of several surface and cytoplasmic markers. The principal marker for NK and NKT cells is NK1.1 (CD161), which is recognized by the monoclonal antibody PK136. The expression of this antigen is restricted to certain mouse strains such as C57BL/6, SJL, and NZB. In all strains, CD49b (mAb DX5) may be used to identify NK cells with the caveat that it is not expressed by all NK cells. As shown in Fig. 4.1 when one examines NK1.1 vs. DX5 staining in an NK1.1 + strain (C57BL/6 in this case), there are some NK1. 1 + cells that do not express DX5.

To differentiate between NK and NKT cells during flow cytometric analysis, one should stain with both NKl.l and CD3 mAbs so that these populations can be gated upon for subsequent evaluation of the expression of other markers of interest. A host of other molecules are associated with NK cell function, including the Ly-49 and NKG2D receptors. Like other lymphocytes, NK and NKT cells constitutively express a variety of adhesion molecules (CD11a, GD11b and, in a subset, CD11c) and cytokine receptors (such as the beta and gamma chains of the IL-2 receptor, CD122, and CD132, respectively). NK cells also contain intracellular molecules which are essential for their cytotoxic function (perform and granzymes). Both perform and granzymes can be identified by cytoplasmic staining coupled with surface marker staining and flow cytometry analysis.

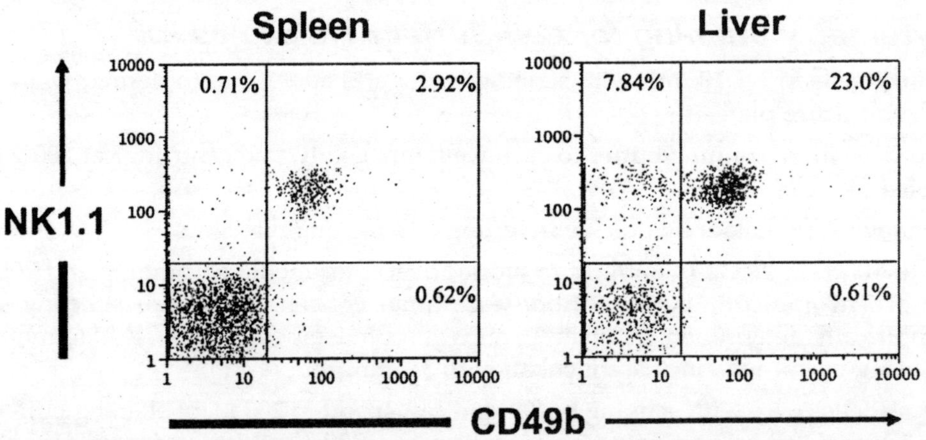

Fig. 4.1. NK1.1 Vs DX5 expression on NK cells. Splenocytes and hepatic mononuclear cells were isolated from C57BL/6 mice and stained with mAbs specific for CD3, NKl.l, and CD49b (DX5). Lymphocytes were gated on CD3(–) events to exclude NKT cells. The majority of NK cells coexpress NKl.l and CD49b, but there is a significant proportion which is negative for expression of CD49b, particularly in the liver.

By definition, NK cells are CD3-whereas NKT cells are CD3+ (Fig. 4.2). Also shown in Fig. 4.2 is that the percentage of NK and NKT cells is much greater in the liver than the spleen, although the absolute numbers may not necessarily be so. NK cells do not express T-cell receptors whereas NKT cells have a restricted expression of certain TCR haplotypes. NKT cells are not homogeneous and may be broadly classified into four groups based upon phenotypic and functional criteria.

Classical (or Type I) NKT cells are the predominant subset found in murine liver and express an invariant TCR alpha chain (Vα14-Jα18). They are restricted by the CD Id glycolipid antigen presenting molecule and are highly reactive with the lipid antigen α-galactosylceramide (α-GalCer), which is derived from the marine sponge *Agelas mauritanius.* CD 1d tetramer reagents complexed with this molecule (obtained from the NIH Tetramer Facility-Emory University Vaccine Center, Atlanta, GA) are particularly useful for the identification of this subset. Staining with this tetramer is shown in Fig. 4.3. Of note in Fig. 4.3 is subset that binds the tetramer but does not express NK1.1 on its surface. The exact nature of this subset is not known but may reflect an NKT cell that either down-modulated or never expressed NK1.1.

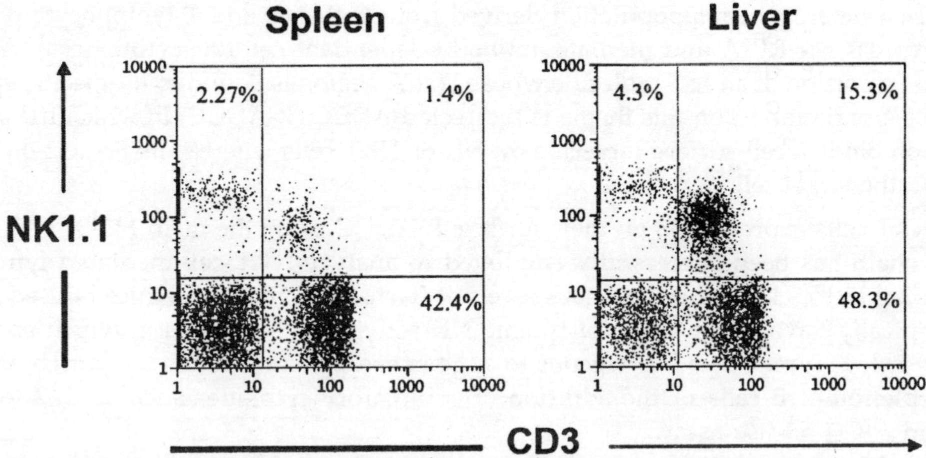

Fig. 4.2. NK1.1 Vs CD3 Expression on spleen and liver mononuclear cells. Splenocytes and hepatic mononuclear cells were isolated from C57BL/6 mice and stained with mAbs specific for CD3 and NK1.1. Double positive NKT cells are detectable in both organs, but are greatly enriched in the liver.

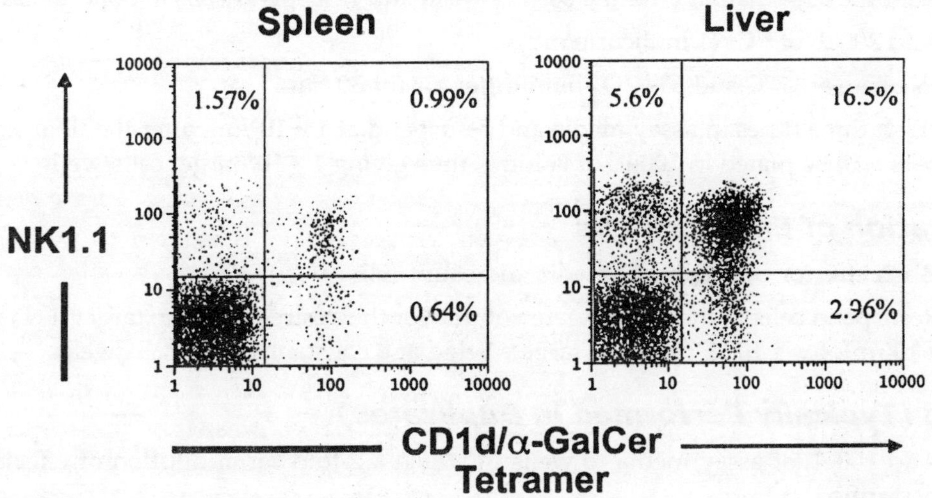

Fig. 4.3. NKT cells bind CDld/α-GalCer tetramers. Splenocytes and hepatic mononuclear cells were isolated from C57BL/6 mice and stained with mAbs specific for CD3, NK 1.1 and the CD 1d/α-GalCer tetramer. Lymphocytes were gated on CD3(+) cells. Note the significant number of tetramer positive NK1.1 negative cells suggesting that some NKT cells may down-modulate or not express NK1.1 on their surface.

Functional Assessment of NK and NKT Cells

NK cells are so named for their ability to lyse susceptible target cells with no need for stimulation. The prototypical target for murine NK cells is the YAC-1 lymphoma. While NK cells can kill directly ex vivo, they may also be stimulated to higher lytic activity through short term in vitro culture with a variety of stimuli including cytokines and certain CpG oligodeoxynucleotides (ODN). Long-term culture of murine splenocytes (> 3 – 5 days) with Interleukin-2 results in the induction of lymphokine-activated killer (LAK) cells, which demonstrate high lytic activity against NK-sensitive targets and also acquire the ability to lyse NK-resistant targets such as EL4 lymphoma and P815 mastocytoma.

LAK are a heterogeneous population derived from both NK and T lymphocyte precursors. NK cells express Fc-γRIIIA and mediate antibody-dependant cellular cytotoxicity (ADCC) by binding the Fc portion of an IgG molecule whose $F(ab)'_2$ region is bound to its specific epitope on a target cell. A variation upon this theme is redirected-ADCC (R-ADCC) in which the antibody's $F(ab)'_2$ region binds a cell-surface molecule on NK or NKT cells whereas its Fc portion binds FC receptors on the target cell.

Since NKT cells express CD3 on their surface, R-ADCC using the mAb 145-2C11 specific for the CD3 ε chain has been successfully employed to analyze NKT cell-mediated lytic activity. Environmental factors impact endogenous levels of fresh NK activity and mice housed in barrier facilities typically have lower levels of splenic NK activity. In situations in which endogenous NK activity is low, it may be advantageous to augment responses through the enrichment of NK cells by depletion of B cells or the addition of stimulatory cytokines such as IL-2 to the 4 hr ^{51}Chromium (^{51}Cr)-release assay.

^{51}Cr Labeling of Target Cells

1. Wash target cells twice in assay media centrifuging for 5 min at 200 g after each wash.
2. Aspirate supernatant after the second wash and resuspend cells in 50μL of assay media.
3. Add 200μL of ^{51}Cr (1 millicurie/mL).
4. Incubate at 37°C and 5% CO_2-humidifier-air for 30 min.
5. Wash three times in assay media and resuspend at 1×10^5/mL after the final wash. Target cells will be plated in 100μL of volume, thus giving 1×10^4 target cells/well.

Preparation of Effector Cells

1. Wash effector cells in assay media and count cells.
2. Resuspend cells in the appropriate volume for the desired effector: target (E:T) ratio (*e.g.* 1×10^7/mL for a 100:1 ratio; the targets being at a constant 1×10^4 cells/well).

Plating (Typically Performed in Duplicates)

1. Add 100μL of assay media to wells in which a 2-fold serial dilution of effectors will be performed.
2. Add 200μL of effectors to the wells which will contain the highest E:T ratio and perform serial dilution leaving 100L per well.

3. Add 100μL of target cells per well for a final volume of 200μL per well. Determination of maximal and spontaneous release is performed by adding 100μL of target cells to 2% acetic acid or assay media, respectively.
4. Centrifuge at 300 g for 30 s to facilitate cell-cell contact.
5. Incubate 4–6h at 37°C in 5% CO_2-humidified-air atmosphere.
6. After incubation, centrifuge plate at 300 g. Carefully harvest 100μL of supernatant and transfer to borosilicate centrifuge, tubes and measure radioactivity (expressed as counts-per-minute, CPM).

Data Analysis

Percent specific lysis is calculated using the following formula :

Percent Specific Lysis = [(experimental CPM-spontaneous CPM)]/[(maximum : release CPM-spontaneous CPM)] x 100

The calculation of lytic units (LUs) is a convenient method to quantify lytic activity and is useful for comparison of the lytic activity among various populations. The method for calculation of LU involves plotting the percentage of specific lysis against the $\log_{10}$ of the various E:T ratios. The linear portion of the $\log_{10}$ [E:T] versus specific lysis curve is selected and the data entered into a computer for. calculation by a standard linear regression program. The E:T ratio required to effect a given percent specific lysis (commonly 30%) of the targets is obtained from the regression line.

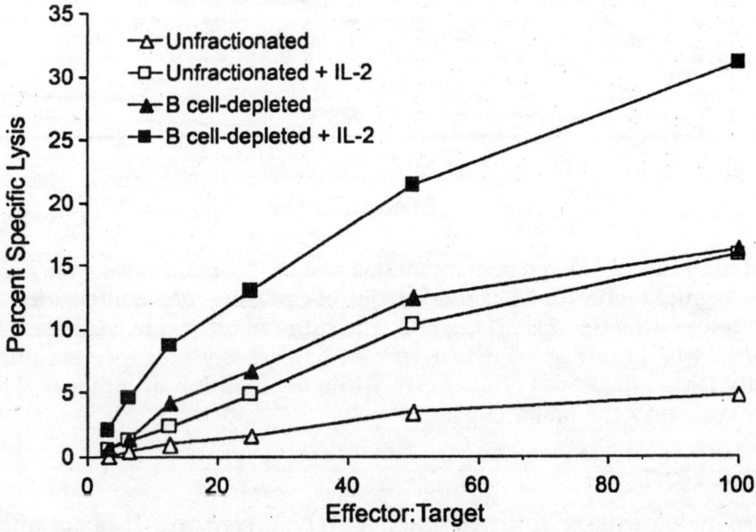

Fig. 4.4. Cytotoxic activity of freshly obtained unfractionated and B cell-depleted splenocytes. Splenocytes were isolated from naive C57BL/6 mice and either used unfractionated or depleted of B cells using magnetic beads coated with goat-anti-mouse IgM and goat-anti-mouse IgG. A standard ^{51}Cr-release assay was performed with rIL-2 [1000U/mL] being added, where indicated, immediately prior to the addition of radiolabeled YAC-T target cells.

Because the number of targets is a constant, one can calculate from this ratio the number of effectors needed to give 30% specific lysis; this is defined as one LU. It is conventional to present the data as the number of LU per million effector cells. An example of a ^{51}Cr-erelease assay is shown in Fig. 4.4. Unfractionated splenocytes have a detectable but low cytotoxic activity; this activity is increased upon depletion of B cells. The activity can be further amplified by the addition of IL-2 directly to the wells of the microtiter plate. IL-2 in this short-term stimulation enhances the killing per NK cell but does not expand the spectrum of susceptible target cells. Another approach useful in the assessment of NK cell function and responsiveness is to culture B cell-depleted splenocytes overnight either with IL-2 or stimulatory CpG ODN. NK cells lose their cytotoxic activity in overnight culture unless stimulated. IL-2 stimulates NK cells directly while CpG ODN stimulated DC which then secrete cytokines that stimulate NK cell activity.

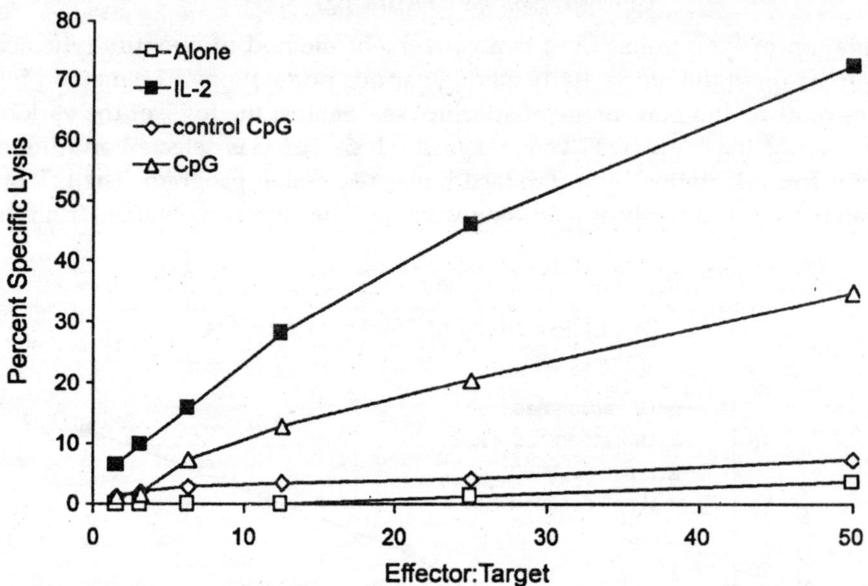

Fig. 4.5. Short-term stimulation of NK cell activity by IL-2 and CpG. Splenocytes were isolated from C57BL/6 mice and cultured overnight with rlL-2 (200u/mL), control CpG oligodeoxynucleotide 2243 (10μg/mL) or activating CpG oligodeoxynucleotide 2336 (10μg/mL). The cultured cells were harvested, counted, and used as effectors in a standard 4-h ^{51}Cr-release assay to determine lytic activity. Splenocytes cultured in the absence of stimulation rapidly lose endogenous NK activity, while the addition of either IL-2 or activating CpG enhance lytic activity against YAC-1 tumor targets.

Cytokine Production

The production of various cytokines, particularly IFN-γ and IL-4, is another important function of NK and NKT cells. Various methods can be employed to analyze cytokine secretion including intracellular cytokine staining and the enzyme-linked immunosorbent assay (ELISA). Intracellular cytokine staining is a flow cytometry-based assay and allows the investigator to assess the production of cytokines by particular cells as defined by simultaneous staining of the cell surface for particular markers (such as NK1.1 and CD3). The analysis of intracellular cytokines by flow cytometry relies upon the blockade of protein transport by inhibitors of the Golgi apparatus such as monensin and brefeldin A. These inhibitors are included in short-term cultures with various NK and NKT agonists such as PMA/Ionomycin, anti-CD3 mAbs and α-GalCer: After 4 – 6 h of culture, cells are stained for cell surface markers then fixed and permeabilized for staining of intracellular markers.

Flow cytometry is then performed to analyze expression of the cytokine of interest by particular cell types. Careful consideration should be given to the selection of fluorochromes for intracellular cytokine staining. Investigators will often select the brightest fluorochromes' such as phycoerythrin (PE) and allophycocyanin (APC) for the anticytokine mAb because the intensity may be low and the cytokine-producing cells infrequent. Additionally, the fixation/ permeabilization process may degrade the intensity of the surface staining hence the use of bright mAb conjugates is essential. An example of surface and cytoplasmic staining of NK cells is shown in Fig. 4.6 where one can clearly see that PMA/Ionomycin stimulation induces the cytoplasmic expression of IFN-γ, which is restricted to NK1.1+ cells. Another approach to examine cytokine secretion by NK and NKT cells is to measure cytokine content in the culture supernatants of stimulated cells.

ELISA yields quantitative information about the amount of cytokine present in culture supernatants as well as in sera and other biological fluids. ELISA is a standard assay in many immunological studies and commercial kits with detailed instructions are readily available. It relies upon the availability of a pair of mAbs directed against distinct epitopes on the analyte of interest. The capture mAb is bound to the surface of a microtiter well, the sample to be analyzed is added to the wells and the presence of the analyte is detected with the second (detection) mAb which is usually labeled with a dye that can be assessed objectively. A recent modification in this methodology is the multiplex suspension array technology in which capture mAbs are linked to microparticles that are complexed with varying amounts of fluorescent dyes. After incubation with the sample and a cocktail of detection antibodies, quantitation of numerous analytes can be performed in a manner analogous to flow cytometry. This method is particularly useful when limited quantities of the sample are available.

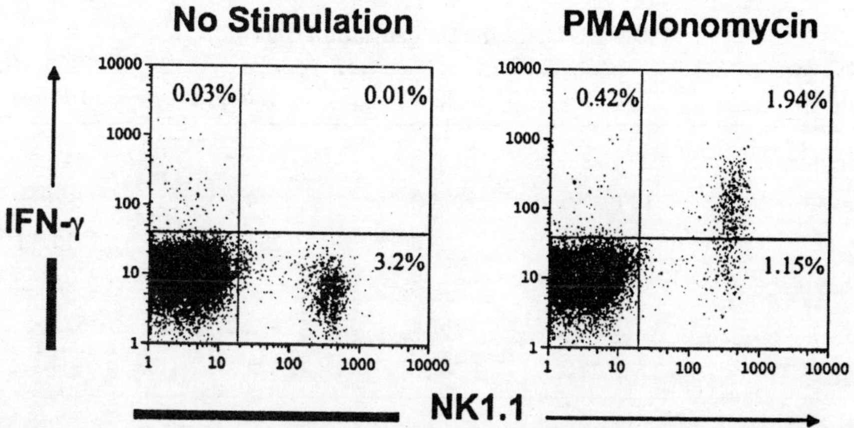

Fig. 4.6. Cytoplasmic IFN-γ expression in NK cells. Splenocytes from C57BL/6 mice were cultured the presence of monensin (2μ*M*) either without stimulation or with PMA (10ng/mL) and ionomycin (200ng/mL) for 5 h. They were then harvested and surface-stained for CD3 and NK1.1. Following fixation and permeabilization the cells were stained for intracellular IFN-γ. Lymphocytes were gated on CD3(-) events to exclude NKT cells.

Stimulation of NKT Cells by α-GalCer

Induction of hepatic NKT cells in vivo can be achieved through injection of α-GalCer (Axxora cat. no. ALX-306-027) either intravenously or intraperitoneally. One can then perform intracellular cytokine staining, within 2 – 3 h, for IFN-γ to identify activated hepatic NKT cells. A host of cytokines are also readily detectable in the serum of treated animals. The effects of NKT activation include secondary stimulation (by virtue of IFN-γ secretion) of NK lytic activity which is detectable within 12 – 18h following α-GalCer injection. α-GalCer-reactive NKT cells increase as a function of age and systemic treatment of older mice with this compound can result in significant liver pathology and death.

Within 24 h after the administration of α-GalCer, NKT cells may no longer be detectable in the liver, a phenomenon originally attributed to activation-induced cell death. It has been suggested, however, that the inability to detect these cells is the result of internalization of the TCR/CD3 complex as well as down-modulation of NK1.1. Additionally, it has been suggested that subsequent to a single dose of α-GalCer, NKT cells proliferate rapidly then enter a state of long-term hyporesponsiveness with many of the hall-marks of anergy. Stimulation of NKT cells by α-GalCer can also be done in vitro. Such stimulation enhances NKT lytic activity and induces them to secrete several cytokines.

As shows in Table 4.1, stimulation of splenic and hepatic mononuclear cells by overnight culture with α-GalCer induces the secretion of significant amounts of IFN-γ, MIP-lα, and IP-10. Interestingly, hepatic, but not splenic, NKT cells also secrete significant amounts of IL-4 and IL-6. In addition to cytokine secretion, α-GalCer enhances the cytotoxic activity of NKT cells. NKT cells do not kill YAC-1 (the classical NK-susceptible target cell) or P815 (an NK-resistant mastocytoma). However, NKT can mediate R-ADCC by using anti-CD3 mAb (which binds to CD3 on the surface of NKT cells) and an Fc-γ-expressing target cells (P815 works well for this assay). As shown in Fig. 4.7, overnight stimulation of hepatic mononuclear cells with α-GalCer markedly enhances the R-ADCC activity of NK cells.

Table 4.1. Cytokine concentration (pg/mL).

Culture conditions	*IL-4*	*IL-6*	*IFN-γ*	*MIP-lα*	*IP-10*
Splenocytes					
Vehicle	0.6	18.0	9.2	188.3	28.2
α–GalCer	35.2	18.0	19,284.0	5,271.6	412.8
Hepatocytes					
Vehicle	3.3	157.4	172.2	37.1	113.5
α–GalCer	3,818.2	1,560.6	22,978.6	9,940.6	616.6

Splenocytes and hepatic mononuclear cells were isolated from naive C57BL/6 mice and cultured with vehicle (DMSO) or α-GalCer (100ng/mL) overnight. Supernatants were harvested and analyzed by multiplex cytokine bead array (Linco Research, St. Charles, MO). Data are expressed as pg/mL.

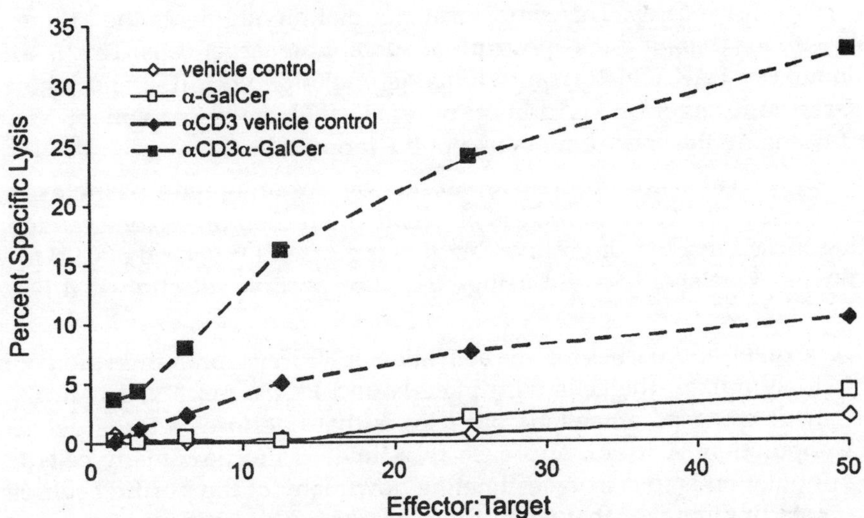

Fig. 4.7. R-ADCC activity of NKT cells stimulated, in vitro, with α-GalCer. Hepatic mononuclear cells were isolated from naive C57BL/6 mice and cultured with vehicle (DMSO) or α-GalCer [100ng/mL] overnight. The cultured cells were harvested, counted and used as effectors in a 4 – h ^{51}Cr-release assay. The targets were the NK-resistant P815 cells. To determine R-ADCC activity, some wells received 2μg/well of the α-CD3 mAb 145-2C11 (represented by the dashed lines). There is little killing of P815 targets without the addition of α-CD3. In vitro culture with α-GalCer greatly enhances R-ADCC activity.

NOTES

1. In general, the methodology for the examination of splenic NK and NKT cell is straightforward. Hepatic mononuclear cells, on other hand, may pose a challenge as to the technique and recovery. It is imperative that the liver circulatory system be flushed in order to get rid of circulating lymphocytes that might otherwise dilute hepatic lymphocytes and NKT cells.

   We found that flushing through the portal vein works best, although it takes some practice. We found it helpful, when teaching the technique, to search for images of the murine hepatic circulation on the Internet so that one can have a schematic as to where the portal vein is. However, even with the best technique, recovery of hepatic lymphocytes is meager. If deemed acceptable by an experimental protocol, one may have to pool the lymphocytes from several livers for adequate numbers.

2. NK cells are part of the innate immune system and are dependent on several cytokines, with IL-15 playing a central role, for their maturation and proliferation. However, mice reared in gnotobiotic environments may have NK cells as determined by surface markers but these cells have little or no cytotoxicity. Upon moving mice to a less-sterile environment, NK cells assert their presence rather quickly. Indeed, a very high spontaneous NK-cell cytotoxic activity should prompt one to worry about a possible infection in the mouse colony.

   Although this creates a problem when one wants to examine the cytotoxic activity of freshly obtained NK cells, one can work around it by depleting B cells (either by plate adherence or using paramagnetic particles coupled with an anti-mouse immunoglobulin reagent). If that is not enough, one can add IL-2 in the wells of the microtiter plates used for the ^{51}Cr-release assay. This short-term stimulation augments the killing activity per NK cell without changing the spectrum of susceptible target cells. This is different from the induction of LAK which require longer stimulation with IL-2 and which are able to kill NK-resistant target cells. Addition of type 1 IFN can also augment NK cell activity without changing the spectrum of susceptible target cells.

3. Enrichment of various lymphocyte subsets by paramagnetic particles is a fast and convenient method. We favor the use of negative selection unless one can identify a non-activating surface marker for positive selection (*e.g.*, use CD5 instead of CD3 for enrichment of NKT cells). One should avoid using NK1.1 for positive selection as it is an activating marker.

   Indeed, one of the approaches for the activation of NK cells for examination of intracellular IFN-γ, is to stimulate the cells with plate-bound PK136 (anti-NKI.1) mAb. A word of caution about using paramagnetic particles is that one loses a large percentage of the starting population. A useful approach to estimate as to how many cells to use in the starting population is to start by estimating how many of the purified subsets one needs and use a starting number that is ten times as much (this ratio applies for NK and NKT not necessarily for T or B cells which exist in much larger percentages).

   Another important issue, which was alluded to already, arises when one uses a mouse strain whose NK cells do not express NK1.1. Although NK from all mouse strains express DX5 (CD49b), when one examines NK1.1 vs DX5 in an NK1.1 + strain, one is struck by a small but consistent subset which is NK1.1 + DX5-. Moreover, there is another rare subset that is NK1.1-DX5 +. The exact characteristics and functions of these two minor subsets are not well delineated.

4. Coupling surface and cytoplasmic staining is a powerful and convenient tool for clearly identifying which cell is the major source of a given cytokine. Although the methodology is relatively straightforward, credible results may be difficult to obtain unless one selects the brightest antibodies possible. The brightness of most surface markers dims considerably after fixation and permeabilization. A useful approach is to first titrate available antibodies using surface staining only. Then, titrate all available antibodies for the cytoplasmic marker using cytoplasmic staining alone. One can then select the brightest antibody for each marker and do the combined surface/cytoplasmic staining.

   However, even with this approach, certain cytokines are very difficult to identify. In general, commercially available antibodies for IFN-γ give bright staining. Most of the available antibodies for IL-4, on the other hand, are extremely dim and one cannot easily make any conclusions with the staining obtained. Nevertheless., this approach is much more informative than measuring cytokines in the supernatants unless one starts with a population that is >99% pure. ELISPOT circumvents the purity problem but is semi-quantitative at best. This is especially important if one is examining an NKT cell population that has some NK cells. Stimulation of NKT cells will induce them to secrete IFN-γ, which will then activate NK cells to induce them to secrete more IFN-γ as well as other cytokines.

CHAPTER 5 Chemistry of Locomotory Organelle

The swimming of bacteria, the flowing motion of the ameba, the rapid contraction of voluntary muscles, and the slower movements of organelles and cytoplasm within cells all depend upon transduction of chemical energy into mechanical work.

A. MOTILITY OF BACTERIA

The smallest organs of propulsion are the bacterial flagella, and we have been able to unravel some of the mystery of movement by looking at them. When a cell of *E. coli* or *Salmonella* swims smoothly, each flagellum forms a left-handed superhelix with an ~2.3 µm pitch. Rotation of these "propellers" at rates of 100 – 200 revolutions/s (100 – 200 Hz) or more in a counterclockwise direction, as viewed from the distal end of the flagellum, drives the bacterium forward in a straight line. Several flagella rotate side-by-side as a bundle. The observed velocities of 20 – 60 µm/s are remarkably high in comparison with the dimensions of the bacteria. Also remarkable is the fact that a cell may travel straight for a few seconds, but then tumble aimlessly for about 0.1 s before again moving in a straight line in a different direction. The tumbling occurs when the flagellum reverses its direction of rotation and also changes from a left-handed to a right-handed superhelix, which has just half the previous pitch.

Such behavior raised many questions. What causes reversal of direction of the propeller? Why do the bacteria tumble? How does a bacterium "decide" when to tumble? How is the flagellum changed from a left-handed to a right-handed superhelix? How does this behavior help the bacterium to find food? Most intriguing of all, what kind of motor powers the flagella? The answers are complex, more than 50 genes being needed to specify the proteins required for assembly and operation of the motility system of *E. coli* or *Salmonella typhimurium*.

1. The Structure and Properties of Bacterial Flagella

Twenty or more structural proteins are present from the base to the tip of a complete bacterial flagellum. However, over most of their length the long thin shafts are composed of subunits of single proteins called *flagellina*. Flagellin molecules have a high content of hydrophobic amino acids and, in *Salmonella,* contain one residue of the unusual N^{ϵ}-methyllysine. The subunits are arranged in a helix of outside diameter ~20 nm in which they also form 11 nearly longitudinal rows or *protofilaments*. Each subunit gives rise to one of the projections. The flagella usually appear under the electron microscope to be super-coiled with a long "wavelength" (pitch) of ~2.5 µm. The supercoiled structure is essential for function, and mutant bacteria with straight flagella are nonmotile. Under some conditions and with some mutant flagellins, straight flagella are formed.

There is a central hole which is surrounded by what appears to be inner and outer tubes with interconnecting "spokes." However, all of the 494-residue flagellin subunits presumably have identical conformations, and each subunit contributes to both inner and outer tubes as well as to the outer projections. *Basal bodies* anchor the flagella to the cell wall and plasma membrane and contain the protic motors that drive the flagella.

Quasiequivalence

There are two distinct types of straight flagella: one (R) in which the protofilaments have a right-handed twist and the other (L) in which the protofilaments have a left-handed twist. These arise from two different conformations of the subunit proteins. Native supercoiled flagella contain a mixture of flagellins in the R-and L-states with all subunits in a given protofilament being in the same state. The supercoiling of the filament cannot be explained by stacking of identical subunits but is thought to arise because of an asymmetric distribution of protofilaments in a given state around the filament. Here, as with the icosahedral viruses quasiequivalence permits formation of a structure that would be impossible with full equivalence of subunits. The corkscrew shape of the flagellum is essential to the conversion of the motor's torque into a forward thrust. Certain mutants of *Salmonella* have "curly" flagella with a superhelix of one-half the normal pitch. The presence of *p*-fluorophenylalanine in the growth medium also produces curly flagella, and normal flagella can be transformed to curly ones by a suitable change of pH. More important for biological function, the transformation from normal to curly also appears to take place during the tumbling of bacteria associated with chemotaxis.

Growth of Flagella

Iino added *p*-fluorophenyl-alanine to a suspension of bacteria, whose flagella had been broken off at various distances from the body. Curly ends appeared as the flagella grew out. Unlike the growth of hairs on our bodies, the flagella grew from the outer ends. Because no free flagellin was found in the surrounding medium, it was concluded that the flagellin monomers are synthesized within the bacterium, then pass out, perhaps in a partially unfolded form, through the 2- to 3-nm diameter hole in the flagella, and bind at the ends. Flagella of *Salmonella* grow at the rate of 1 μm in 2 – 3 min initially, then more slowly until they attain a length of ~15 μm. More recent studies have provided details. The hook region grows first to a length of ~55 nm by addition to the basal-body rod of ~140 subunits of protein *FlgE*. During growth a *hook cap* formed from subunits *FlgD* prevents the *FlgE* subunits from passing out into the medium. Hook subunits are added beneath the cap, moving the cap outward. Hook growth is terminated by protein *FlgK* (also called hook-associated protein Hap1). This protein displaces the hook cap and initiates growth of the main filament.[24] The first 10 – 20 subunits added are those of the FlgK. These are followed by 10 – 20 subunits of *FlgL* (Hap3), a modified flagellin whose mechanical properties can accomodate the stress induced in the flagella by their rotation. FlgJ is also needed for rod formation.

Growth of the flagellum to a length of up to 20 μm continues with subunits of *FliC* that are added at the tip, which is now covered by a dodecamer of the *cap protein FliD* (HAP2). Its 5-fold rotational symmetry means that this "star-cap" does not form a perfect plug for the 11-fold screw-symmetry of the flagellum, a fact that may be important in allowing new flagellin subunits to add at the growing tip. If the cap protein is missing, as in some *FliD* mutants, a large amount of flagellin leaks into the medium.

Still unclear is how the protein synthesis that is taking place on the ribosomes in the bacterial cytoplasm is controlled and linked to "export machinery" at the base of the flagellin. The genetically identified proteins FlhA, FliH, and FliI are involved in the process that sends the correct flagellin subunits through the growing flagellum at the appropriate time. FliI contains an ATPase domain. FliS protein may be an export chaperone.

2. Rotation of Flagella

A variety of experiments showed that the flagellum is a rigid propeller that is rotated by a "motor" at the base. For example, a bacterium, artificially linked by means of antibodies to a short stub of a flagellum of another bacterium, can be rotated by the second bacterium. Rotation of cells tethered to a cover slip has also been observed. Although it is impossible to see individual flagella on live bacteria directly, bundles of flagella and even single filaments can be viewed by dark-field light microscopy. Normal flagella appear to have a left-handed helical form, but curly *Salmonella* flagella, which have a superhelix of one-half the normal pitch, form a right-handed helix. Normal bacteria swim in straight lines but periodically *"tumble"* before swimming in a new random direction. This behavior is part of the system of *chemotaxis* by which the organism moves toward a food supply. Curly mutants tumble continuously. When bacteria tumble the flagella change from normal to curly. The pitch is reversed and shortened. A proposed mechanism for the change of pitch involves propagation of cooperative conformational changes down additional rows of flagellin subunits.

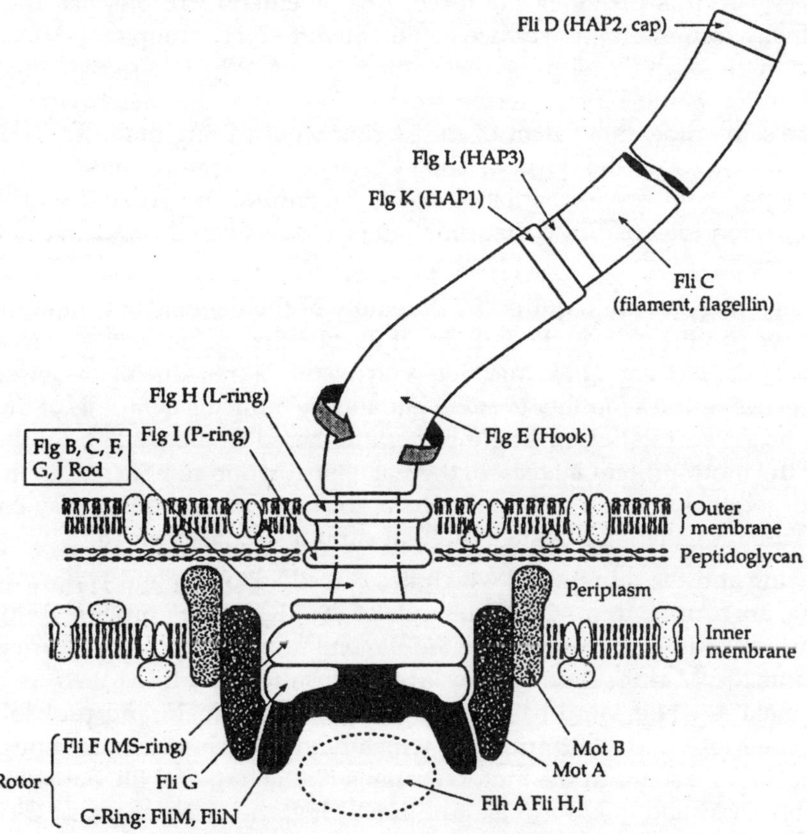

Fig. 5.1. Schematic drawing of bacterial flagellar motor.

There are no muscle-type proteins in the flagella. By incubating flagellated bacteria with penicillin and then lysing them osmotically, Eisenbach and Adler obtained cell envelopes whose flagella would rotate in a counterclockwise fashion if a suitable artificial electron donor was added. This and other evidence showed that ATP is not needed. Rather, the torque developed is proportional to the *protonmotive force* and, under some circumstances, to ΔpH alone. It is the flow of protons from the external medium into the cytoplasm that drives the flagella. Movement of *E. coli* cells in a capillary tube can also be powered by an external voltage. In alkalophilic strains of *Bacillus* and some *Vibrio* species a sodium ion gradient will substitute. Several hundred protons or Na^+ ions must pass through the motor per revolution. Some estimates, based on energy balance, are over 1000. However, Na^+-dependent rotation at velocities of up to 1700 Hz has been reported for the polar flagellum of *Vibrio alginolyticus*. It is difficult to understand how the bacterium could support the flow of 1000 Na^+ per revolution to drive the flagellum.

What kind of protic motor can be imagined for bacterial flagella? Electron microscopy reveals that the flagellar hook is attached to a rod that passes through the cell wall and is, in turn, attached to a thin disc, the *M-ring* (or MS-ring), which is embedded in the cytoplasmic membrane both for gram-positive and gram-negative bacteria. Two additional rings are present above the M-ring of flagella from gram-negative bacteria. The P-ring interacts with the peptidoglycan layer, and the L-ring contacts the outer membrane lipopolysaccharide. A logical possibility is that the M-ring, which lies within the plasma membrane, is the rotor, and a ring of surrounding protein subunits is the stator for the motor. Glagolev and Skulachev suggested in 1978 that attraction between $–COO^-$ and $–NH_3^+$ groups provides the force for movement. Protons passing down an H^+-conducting pathway from the outer surface could convert $–NH_2$ groups to $–NH_3^+$, which would then be attracted to the $–COO^-$ groups on the stator subunits. When these two oppositely charged groups meet, a proton could be transferred from $–NH_3^+$ to $–COO^-$ destroying the electrostatic attraction. At the same time, movement of the M-ring would bring the next $–NH_2$ group to the H^+-conducting pathway from the outside. The –COOH of the stator would now lose its proton through a conducting pathway to the inside of the bacterium, the proximity of the new $–NH_3^+$ assisting in this proton release. Since that time, other models based on electrostatic interactions have been advanced.

Approximately 40 genes are required for assembly of the flagella, but mutations in only five motility genes have produced bacteria with intact flagella that do not rotate. Among these genes are *mot A, Mot B, FliG, FliM,* and *FliN*. Infection with a lambda transducing bacteriophage carrying functional *motB* genes restores motility to *mot B* mutants by inducing synthesis of the motB protein. Block and Berg observed rotation of single bacteria tethered to a coverslip by their flagella. As the synthesis of the motB protein increased, the flagellar rotation rate increased in as many as 16 steps. This suggested that as many as 16 subunits of the *motB* gene product may contribute to the operation of the motor. Later studies suggest eight subunits rather than 16.

Both the M-ring and the thin S-ring, which lies directly above it and is now usually referred to as the MS-ring, are formed from –20-25 subunits of the 61-kDa *FliF* protein. Both the MotA and MotB proteins are embedded in the inner bacterial membrane and appear to form a circular array of "studs" around the M-ring. MotA has a large cytosolic domain as well as four predicted transmembrane helices while MotB has a large periplasmic domain and probably binds to the peptidoglycan. The MotA and MotB proteins, which bind to each other, are thought to form the ~8 functional units in the stator of the motor. Proteins FliG, FliM, and FliN are evidently parts of the rotor assembly. FliM and FliN form an additional ring, the cytoplasmic or *C-ring,* which had been difficult to see in early electron microscopy. As many as 40 of each of these subunits may be present in the ring. A ring of FliG subunits joins the C-ring to the MS-ring. FliE is also a part of the basal body.

From study of mutants it has been concluded that three charged residues of FliG, R279, D286, and D287 are directly involved in generation of torque by the motor. Side chains of these residues may interact with the cytoplasmic domains of MotA and MotB. Residues R90 and E98 of MotA may be involved in controlling proton flow through the motor units. The two prolines P173 and P122 are also essential for torque generation.

There are obvious similarities between the flagellar motors and the protic turbines of ATP synthases, but there are also substantial differences. It apparently takes about 12 protons for one revolution of the ATP synthase but about 1000, or ~125 per motor unit, for rotation of a bacterial flagellum. Elston and Oster propose an ion turbine more complex than that of ATP synthase. They suggest that the rotor might contain about 60 slanted rows of positively charged groups spaced. The motor is reversible, *i.e.,* it can rotate in either direction. One possibility is that the subunits alter their conformations cooperatively in such a way that the slant of the rows of charged groups is reversed. Other possibilities for altering the constellation of charges via conformational changes can be imagined. See also Thomas *et. al.*

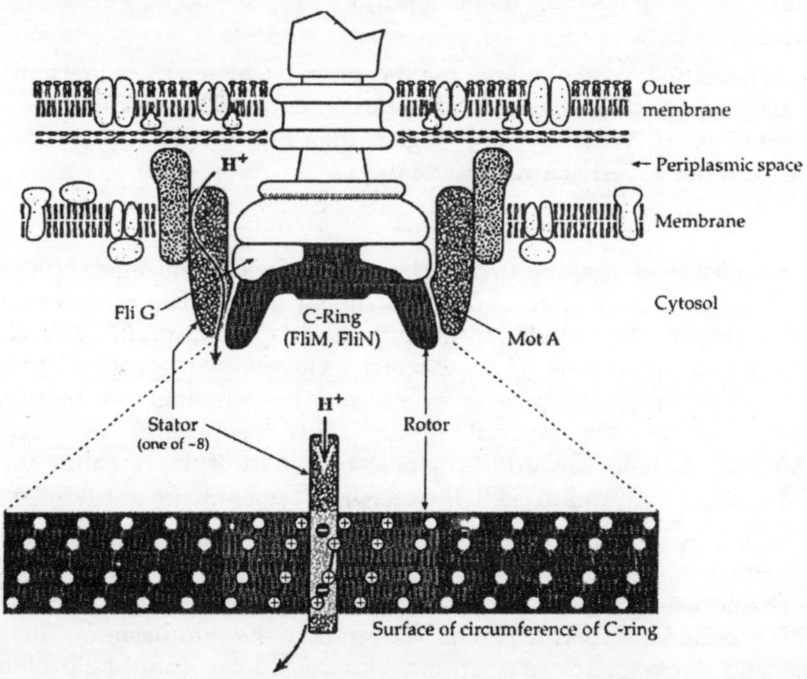

Fig. 5.2. Schematic drawing of a hypothetical configuration of rotor and one stator unit in a flagellar motor as proposed by Elston and Oster.[1] The rotor can hold up to 60 positive charges provided by protons flowing from the periplasm through the stator motor units that surround the C-ring and hopping from one site to the next along the slanted lines. The rotor is composed of 15 repeating units, each able to accommodate four protons. Negative charges on the stator units are 0.5 nm from the rotor charges at their closest approach. For details see the original paper.

3. Chemotaxis

The flagellar motor is reversible, and in response to some signal from the bacterium it will turn in the opposite direction. At the same time, the flagellin sub-units and those of the hook undergo conformational changes that change the superhelical twist. Perhaps synchronous conformational changes in the M-ring also are associated with the change in direction of rotation and are induced by interaction with a *switch complex* that lies below the M-ring. This consists of proteins FliG, FliM, and FliN. Mutations in any one of these proteins lead to the following four phenotypes: absence of flagella, paralyzed flagella, or flagella with the switch biased toward clockwise or toward counterclockwise rotation.

What signals a change in direction of rotation? The answer lies in the attraction of bacteria to compounds that they can metabolize. Bacteria will swim toward such compounds but away from repellent substances, a response known as *chemotaxis.* Cells of *E. coli* swim toward higher concentrations of L-serine (but not of D-serine), of L-aspartate, or of D-ribose. Phenol and Ni^{2+} ions are repellent. By what mechanism can a minuscule prokaryotic cell sense a concentration gradient? It is known that the plasma membrane contains receptor proteins, whose response is linked to control of the flagella. Since the dimensions of a bacterium are so small, it would probably be impossible for them to sense the difference in concentration between one end and the other end of the cell. The chemotatic response apparently results from the fact that a bacterium swims for a relatively long time without tumbling when it senses that the concentration of the attractant is increasing with *time.* When it swims in the opposite direction and the concentration of attractant decreases, it tumbles sooner.

Koshland proposed that as the membrane receptors become increasingly occupied with the attractant molecule, the rate of formation v_f of some compound X, within the membrane or within the bacterium, is increased. When [X] rises higher than a threshold level, tumbling is induced. At the same time, X is destroyed at a velocity of v_d.

$$\xrightarrow{v_f} [X] \xrightarrow{v_d} \quad \text{...(5.1)}$$

Subsequently, a readjustment of v_f and v_d occurs such that the concentration of X falls to its normal steady state level. X would act directly on the flagellar motor. The receptors for L-serine and L-aspartate are 60-kDa proteins encoded by genes *tsr* and *tar* in *Salmonella* or *E. coli.* These proteins span the inner plasma membrane of the bacteria. However, there is still much that is not understood. The symmetric head, whose structure is known has two binding sites, but the aspartate receptor binds only one aspartate tightly. There is substantial evidence that suggests a piston-type sliding of one helix toward the cytoplasm as part of the signaling mechanism. While the flagella are distributed around the cell, the receptors appear to be clustered at the cell poles.

Proteins encoded by genes *che A, cheW, cheY* and *cheZ, cheB,* and *ChR* are all involved in controlling chemotaxis. All of the corresponding protein products have been isolated and purified, and the whole chemotaxis system has been reconstituted in phospholipid vesicles. Gene *CheA* encodes a 73-kDa protein kinase, which binds as a dimer to the cytoplasmic domains of the related aspartate, serine, and ribose/galactose receptors with the aid of a coupling protein, chew. A great deal of effort has been expended in trying to understand how binding of an attractant molecule to the periplasmic domain of the receptor can affect the activity of the CheA kinase, but the explanation is unclear. There is a consensus that a small but distinct conformational alteration is transmitted through the receptor. An apparently α-helical region containing methylation sites (Fig. 5.5) appears to be critically involved in the signaling, responding not only to occupancy of the receptor site but also to intracellular pH and temperature and to methylation. Mutation of the buried Gly 278 found in this region to branched hydrophobic amino acids, such as Val or leucine, locks the receptor in state with a superactivated CheA kinase, while *substitution* of Gly 278 with aspartate leaves the kinase *inactive.* Occupancy of the normal receptor site with ligand (aspartate, serine, etc.,) dramatically decreases the kinase activity.

The CheA protein is an autokinase which, upon activation by the receptor, becomes phosphorylated on N$^{\varepsilon}$ of the imidazole ring of His 48. It then transfers this phospho group from His 48 to the carboxylate of Asp 57 of the 654-residue protein *CheY,* which is known as the *response regulator.* The unregulated flagellum rotates counterclockwise (CCW). *Phospho-CheY carries the message to the flagellar motor to turn clockwise (CW).* This is apparently accomplished through the binding of Che*Y-P* to the N-terminal portion of protein FliM. This presumably induces a conformational change, which is propagated to FliG and to all of the proteins of the rotor and flagellar rod, hook, and filament. The flagella fly apart, and the bacterium tumbles and heads randomly in a new direction.

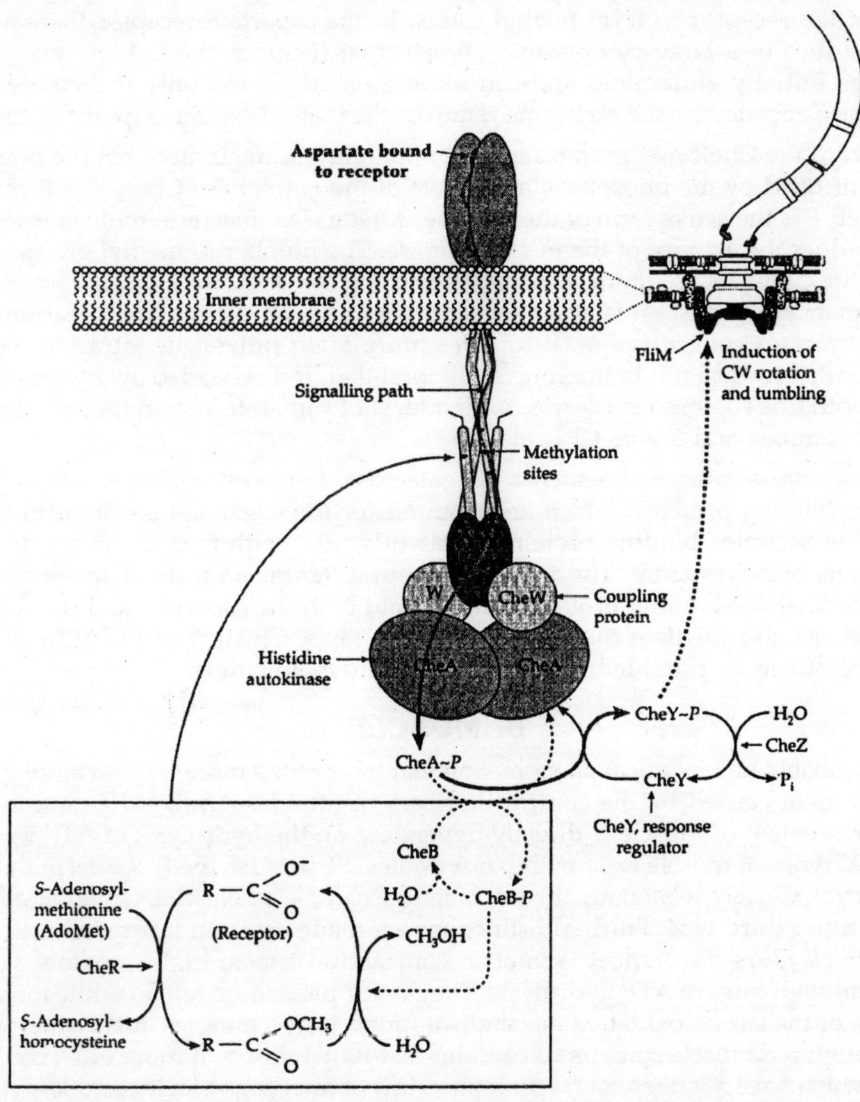

Fig. 5.3. Schematic representation of an important chemotactic system of *E. coli, S. typhimurium,* and other bacteria. The trans-membrane receptor activates the autokinase CheA, which transfers its phospho group to proteins CheY and CheB to form CheY-P andCheB-P. CheY-P regulates the direction of rotation of the flagella, which are distributed over the bacterial surface. CheR is a methyltransferase which methylates glutamate carboxyl groups in the receptor and modulates the CheA activity. CheZ is a phosphatase and CheB-P a methylesterase.

Tumbling occurs most often when receptors are unoccupied, and the bacteria change directions often, as if lost. However, if a receptor is occupied by an attractant, the activity of CheY is decreased and less CheY-P will be made. The carboxyl phosphate linkage in this compound is labile and readily hydrolyzed, a process hastened by the phosphatase *CheZ*. Consequently, in the presence of a high enough attractant concentration the tumbling frequency is decreased, CCW flagellar rotation occurs, and the bacterium swims smoothly for a relatively long time.

There are still other important factors. Occupancy of the receptor by a ligand makes the receptor protein itself a substrate for the chemotaxis-specific methyl-transferase encoded by the *cheR* gene. This enzyme transfers methyl groups from S-adenosyl-methionine to specific glutamate side chains of the receptor to form methyl esters. In the aspartate receptor there are four such glutamate residues in a large cytoplasmic domain that includes the C terminus. Two of these glutamates are initially glutamines and can undergo methylation only if they are deaminated first. An esterase encoded by the *cheB* gene removes the methyl ester groupings as methanol.

The action of the CheR methyltransferase is apparently unregulated, but the esterase activity of CheB is controlled by the phosphorylation state of the autokinase CheA. CheB competes with CheY and CheB-*P* is the active form of the esterase. After a chemotactic stimulus the level of CheA-*P* falls and so does the activity of the methylesterase. The number of methyl groups per receptor rises making the CheA kinase more active and opposing the decrease in kinase activity caused by receptor occupancy. The system is now less sensitive to the attractant; the bacterium has *adapted* to a higher attractant concentration. It tumbles more often unless the attractant concentration rises; if it is headed toward food tumbling is still inhibited. If it is headed away from the attractant the levels of both CheY-*P* and ChB-*P* rise. A high level of fumarate within the cell also acts on the switch-motor complex and favors CW rotation.

For some bacterial attractants such as D-galactose, D-ribose, maltose, and dipeptides the corresponding binding proteins, which are required for the sugar uptake are also necessary for chemotaxis. The occupied binding proteins apparently react with membrane-bound receptors to trigger the chemotactic response. The aspartate receptor (*tar* gene product) appears also to be the receptor for the maltose-binding protein complex, and both the aspartate and the serine receptor (*tsr* gene product) also mediate thermotaxis and pH taxis. Clusters of identical receptors may function cooperatively to provide high sensitivity and dynamic range.

B. MUSCLE

There is probably no biological phenomenon that has excited more interest among biochemists than the movement caused by the contractile fibers of muscles. Unlike the motion of bacterial flagella, the movement of muscle is directly dependent on the hydrolysis of ATP as its source of energy. Several types of muscle exist within our bodies. *Striated* (striped) *skeletal muscles* act under voluntary control. Closely related are the *involuntary striated heart muscles,* while *smooth involuntary muscles* constitute a third type. Further distinctions are made between fast-twitch and slow-twitch *fibers. Fast-twitch fibers* have short isometric contraction times, high maximal velocities for shortening, and high rates of ATP hydrolysis. They occur predominately in white muscle. Because of the absence of the strong oxidative metabolism found in red muscles, fast-twitch fibers fatigue rapidly. Although red muscle sometimes contains fast-twitch fibers, it more often consists of *slow-twitch fibers,* which have a longer contraction time, low shortening velocity, and low ATPase. They are more resistant to fatigue than fast-twitch fibers. Embryonic muscle contains fast-twitch fibers as well as embryonic forms which contract slowly. Some organisms contain specialized types of muscle. For example, the asynchronous flight muscles of certain insects cause the wings to beat at rates of 100 – 1000 Hz. In these muscles nerve impulses are used only to start and to stop the action; otherwise the cycle of contraction and relaxation continues automatically. The adductor muscles, which close the shells of oysters and clams, can sustain large tensions for long periods of time with little expenditure of energy. This is accomplished by a *catch mechanism.*

1. The Structural Organization of Striated Muscle

Skeletal muscles consist of bundles of long *muscle fibers,* which are *single cells* of diameter 10 – 10 μm formed by the fusion of many embryonic cells. The lengths are typically 2 – 3 cm in mammals but may sometimes be as great as 50 cm. Each fiber contains up to 100 – 200 nuclei. Typical cell organelles are present but are often given special names. Thus, the plasma membrane (plasmalemma) of muscle fibers is called the *sarcolemma.* The cytoplasm is *sarcoplasm,* and mitochondria may be called *sarcosomes.* The major characteristic of muscle is the presence of the contractile *myofibrils,* organized bundles of proteins 1 – 2 μm in diameter and not separated by membranes from the cytoplasm. Since they occupy most of the cytoplasm, a substantial number of myofibrils are present in each muscle fiber.

In the light microscope cross striations with a repeating distance of ~2.5 μm can be seen in the myofibrils. The space between two of the dense *Z-discs* (Z lines) defines the *sarcomere,* the basic contractile unit. In the center of the sarcomere is a dense *A-band* (anisotropic band). The name refers to the intense birefringence of the band when viewed with plane polarized light. Straddling the Z-discs are less dense *I-bands* (the abbreviation stands for isotropic, a misnomer, for although the bands lack bire-fringence, they are not isotropic). Weakly staining *M-lines* (usually visible only with an electron microscope) mark the centers of the A-bands and of the sarcomeres.

The fine structure of the sarcomere was a mystery until 1953, when H. E. Huxley, examining thin sections of skeletal muscle with the electron microscope, discovered a remarkably regular array of interdigitated protein filaments. *Thick filaments,* 12 – 16 nm in diameter and ~1.6 μm are packed in a hexagonal array on 40- to 50-nm centers throughout the A-bands. Between these thick filaments are *thin filaments* only 8 nm in diameter and extending from the Z-line for a length of ~1.0 μm. When contracted muscle was examined, it was found that the I-bands had become so thin that they had nearly disappeared and that the amount of overlap between the thick and the thin filaments had increased. This indicated that contraction had consisted of the sliding movement of the thick and thin filaments with respect to each other. In skeletal muscle the sarcomere shortens to a length of ~2 μm, but in insect flight muscle a much smaller shortening occurs repetitively at a very high rate.

2. Muscle Proteins and their Structures

The myofibrillar proteins make up 50 – 60% of the total protein of muscle cells. Insoluble at low ionic strengths, these proteins dissolve when the ionic strength exceeds ~0.3 and can be extracted with salt solutions. Analysis of isolated mammalian myofibrils shows that nine proteins account for 96% or more of the protein; *myosin* which constitutes the bulk of the thick filaments, accounts for 43% and *actin,* the principal component of the thin filaments, 22%.

Actin and the thin Filaments

There are at least six forms of actin in adult mammalian tissues: α-cardiac, α-skeletal muscle, α- and γ-smooth muscle, β- and γ-cytoplasmic. All of them are closely homologous, *e.g.,* the 42-kDa α-skeletal muscle actin differs in only 4 of 375 residues from the α-cardiac form and only in 6 residues from the γ-smooth form. In almost all organisms actins contain one residue of N^{δ}-methylhistidine at position 73. Actin is an unusual protein in that it can exist in both a filamentous and a soluble state. The interconversion between them is of great physiological importance. Actin filaments dissolve in a low ionic strength medium containing ATP to give the soluble, monomeric *G-actin.* Each G-actin monomer usually contains one molecule of bound ATP and a calcium ion.

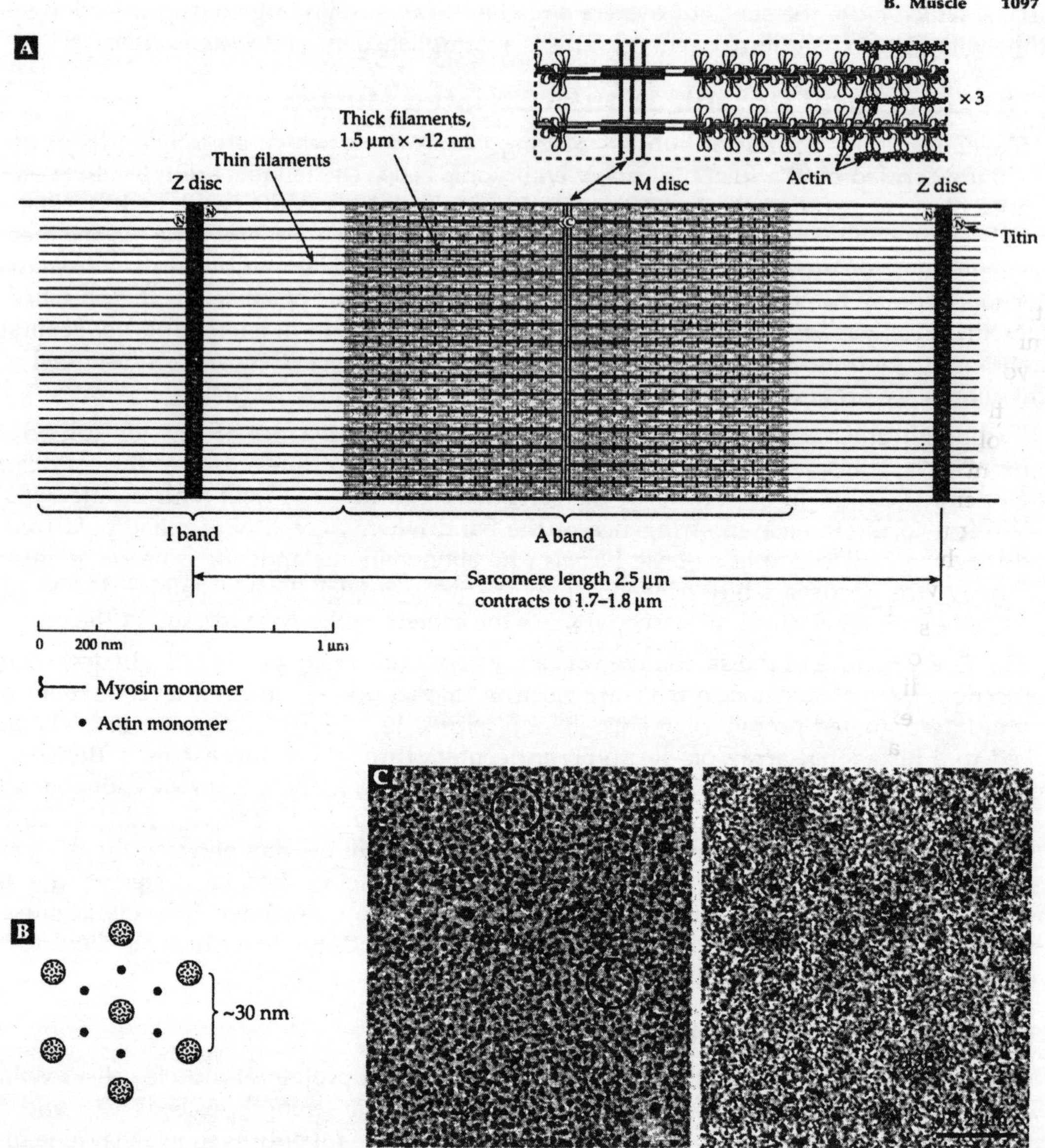

Fig. 5.4 (A) The structure of a typical sarcomere of skeletal muscle. The longitudinal section depicted corresponds to that of the electron micrograph, Fig. 5.7A. The titin molecules in their probable positions are colored green. The heads of only a fraction of the myosin molecules are shown protruding toward the thin actin filaments with which they interact. A magnified section at the top is after Spudich. It shows the interactions of the myosin heads with the thin filaments at the right-hand edge. (B) A sketch showing the arrangement of thick and thin filaments as seen in a transverse section of a striated muscle fiber. (C) Left: electron micrograph of a transverse section of a glycerated rabbit psoas muscle. The hexagonal arrangement of six thin filaments around one thick filament can be seen in the center of the circle. Six other thick filaments form a larger concentric circle as in (B). Right: transverse section of a smooth muscle fiber. Notice the irregular arrangement of thick and thin filaments. Filaments of intermediate diameter are also present, as are dense bodies (DB). The latter are characteristically present in smooth muscle.

Because of its tendency to polymerize, G-actin has been difficult to crystallize. However, it forms crystalline complexes with several other proteins, *e.g.*, deoxyribonuclease I, a fragment of gelsolin, and profilin, which block polymerization and it has recently been crystallized as the free ADP complex. The three-dimensional structure of the actin is nearly the same in all cases. The molecule folds into four domains, the ATP binding site being buried in a deep cleft. The atomic structure resembles that of hexokinase, of glycerol kinase, and of an ATP-binding domain of a chaperonin of the Hsp 70 family. As with the kinases, actin can exist in a closed and more open conformations, one of which is seen in the profilin complex. Addition of 1 mM Mg^{2+} or 0.1 M KCl to a solution of G-actin leads to spontaneous transformation into filaments of *F-actin* each containing 340 – 380 actin monomers and resembling the thin filaments of muscle. The ATP bound in the F-actin filament is hydrolyzed within ~100 s to ADP and P_i. However, the hydrolysis is not as rapid as polymerization so that a *"cap"* of ATP-containing monomers may be found at each end of the filament. There is a striking similarity to the binding of nucleotides to microtubule subunits and in the contractile tail sheath of bacteriophage.

The two ends of the F-actin filaments have different surfaces of the monomer exposed and grow at different rates. This has been demonstrated by allowing the myosin fragment called heavy meromyosin to bind to (or "decorate") an actin filament. The myosin heads bind at an angle, all pointed in one direction. This gives a "pointed" appearance to one end and a "barbed" appearance at the other. When monomeric actin is added to such an HMM-decorated F-actin filament the barbed ends grow much faster than the pointed ends. In the intact sarcomere the ends that become pointed when decorated are free, while the opposite barbed ends of the filaments are attached at the Z-line. The existence in the cytoplasm of proteins that "cap" the fast-growing end of actin filaments thereby preventing further growth suggests that cap proteins may be present at the ends of the thin filaments of the myofibrils.

Titin and Nebulin

The third most abundant protein (10%), titin (also called connectin), is one of the largest of known proteins. Titin cDNA from human cardiac muscle encodes a 26, 926-residue chain. Several tissue-specific isoforms of the protein are created by alternative mRNA splicing. A single titin molecule stretches ~1200 nm from the Z-disc, where the N terminus is bound, to the M-line, where the C-terminal domain is attached. Thoughout much of the A-band titin binds to the thick filament and appears to be part of a scaffold for maintenance of the sarcomere structure. The I-band portion of titin has elastic properties that allow it to lengthen greatly or to shorten as the sarcomere changes length.

Under the electron microscope titin appears as a flexible beaded string ~4 nm in diameter. Most of the molecule is made up of repetitive domains of two types. In human cardiac titin there are 132 folded domains that resemble type III fibronectin repeats and 112 immunoglobulin-like domains. In a "PEVK region," between residues 163 and 2174, 70% of the residues are Pro, Glu, Val, or Lys. The titin molecule may be organized as polyproline helices in this elastic region. At the C terminus of titin 800 residues, including a Ser/Thr protein kinase domain, are found within the M-line.

Another very large protein, *nebulin* (3% of the total protein), appears to be stretched alongside the thin filaments. In the electron microscope it appears as a flexible, beaded string ~4 nm in diameter. Ninety-seven percent of the 6669-residue human nebulin is organized as 185 copies of an ~35-residue module. Nebulin has a proline residue at about every 35th position, possibly corresponding in length to the pitch of the actin helix. At the C terminus is an SH3 domain which is preceded by a 120-residue segment rich in potential phosphorylation sites. This part of the peptide chain is anchored in the Z-discs. The three extreme N-terminal modules of nebulin bind to tropomodulin, which caps the pointed ends of thin filments. Avian cardiac muscle contains a much shorter 100-kDą protein called *nebuletfte* which resembles the C-terminal parts of nebulin. Nebulin has been described as encoding a blueprint for thin filament architecture.

Proteins of the Mi-line and Z-disc

The M-line region contains the structural protein *myomesin,* which binds to both titin and myosin and holds the two together. Fast skeletal and cardiac fibers also contain another *M-protein,* which may bridge between myosin filaments. Both the C-terminal region of nebulin and the N termini of pairs of titin molecules meet in the Z-disc, where they are bound into a lattice containing *α-actinin* and other proteins. The dimeric α-actinin, a member of the spectrin family, has a subunit mass of ~97 kDa. Found primarily in the Z-discs, it is also present in nonmuscle cells in stress fibers and at other locations in the cytoskeleton. It may anchor actin filaments to various structures outside of the sarcomere. In the dense Z-disc of insect flight muscle a regular hexagonal lattice of α-actinin and a large (500 – 700 kDa) modular protein called *kettin* bind the thin filaments of opposite polarity together.

The *C-protein* (thick filaments), myomesin (M-line protein), and α-actinin (Z-line protein) each provide 2% of the protein in the myofibril. Less than 1% each of 11 or more other proteins may also be present within the sarcomere. Several of these, including the cytoskeletal proteins *desmin* and *vimentin,* and *synemin* surround the Z-discs.

The Regulatory Proteins Troponin and Tropomypsin

These two proteins are also associated with the filaments, each one contributing ~5% to the total protein of myofibrils. Tropomyosin is an elongated α-helical coiled-coil molecule, each molecule of which associates with seven actin subunits of an actin filament. Troponin consists of three subunits known as troponins C, T, and I. The elongated troponin T binds to tropomyosin. Troponin I is an inhibitor of the interaction of myosin and actin necessary for muscle contraction. Troponin C, a member of the calmodulin family binds Ca^{2+} and induces conforma-tional changes that relieve the inhibition and allow contraction to occur. Nebulin is also thought to bind to tropomyosin. A possible arrangement of one of the tropomyosin-troponin-nebulin complexes that lie along the length of the thin filaments is shown schematically and as a three-dimensional model in Fig. 5.8C and D. Fig. 5.9 shows a model of the thin filaments with tropomyosin coiled-coil molecules on each side. The troponin subunits are not shown.

Myosins

There are 15 distinct families of proteins within the myosin superfamily. They vary greatly in size, but all of them bind and hydrolyze ATP, and all bind to actin. Most have C-terminal tails. At their N termini are one or two globular heads, which contain the catalytic centers in which ATP hydrolysis occurs. Sizes vary from 93 kDa for a myosin with a very short tail from *Toxoplasma to* over 300 kDa. Myosins I, found in ameboid organisms and also in our own bodies (for example in the microvilli of the brush border of intestinal epithelial cells), are small single-headed molecules. Myosins II are the "conventional" myosins of myofibrils and are often referred to simply as myosin. However, each of the three muscle types (skeletal, cardiac, and smooth) has its own kind

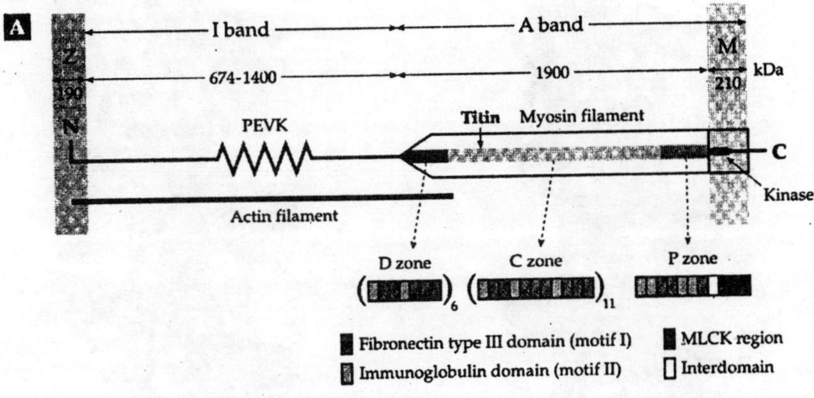

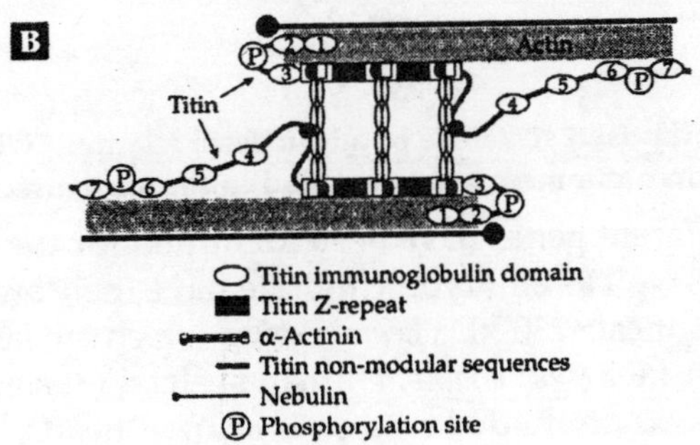

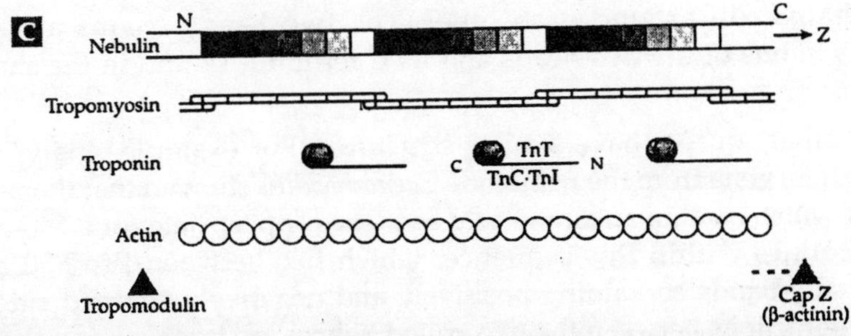

Fig. 5.6. (A) Schematic drawing showing one molecule of titin (connectin) in a half sarcomere and its relationship to thick myosin filaments and thin actin filaments. The complex repeat patterns of fibronectin type III, immunoglobulin, in the three zones D, C, and P are also indicated. (B) Schematic drawing of the molecular structure of the sarcomere Z-disc. Titin, which is thought to parallel the thin filaments through the I-band, consists of various modules that are numbered from the N termini. In the Z-disc titin binds to α-actin, shown here as three vertical rods, and also to actin or actin-binding proteins. The SH3 domain (shown as a sphere) of nebulin and the N terminus of titin may interact. Regulatory phosphorylation sites are marked P. (C) Hypothetical model of a composite regulatory complex containing nebulin, tropomyosin, and troponin on the thin filaments of a skeletal muscle sarcomere. Each seven-module nebulin super-repeat (squares with graded shading) binds one tropomyosin, possible through the seven charge clusters along the length of each tropomyosin, and one troponin complex (shaded spheres with a tail). This complex consists of TnT, TnI, and TnC in orientations indicated by the N and C termini. Each nebulin super-repeat binds to seven actin monomers (open circles) along the thin filament. *Tropomodulin* caps the pointed ends of actin filaments and Cap Z, the *"barbed ends."*

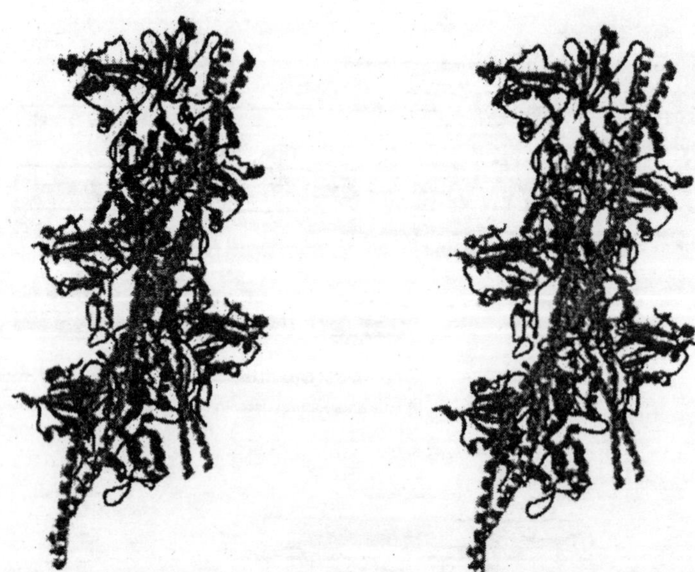

Fig. 5.7. Stereoscopic ribbon drawing of the proposed structure of a thin actin filament with tropomyosin coiled-coils bound on opposing sides. Five actin nomomers are assembled in the structure.

of myosin II. Likewise, at least six different genes have been identified for the light chains of the myosin heads. Fast and slow muscle as well as embryonic muscle have their own light chains. Each myosin II molecule consists of two identical ~230-kDa *heavy chains,* which are largely α-helical, together with two pairs of smaller 16- to 21-kDa *light chains.* Human skeletal muscle heavy chains contain 1938 residues of which the first ~850 are folded into pear-shaped heads, which contain the catalytic sites involved in harnessing ATP cleavage to movement. Following proline 850 nearly all of the remaining 1088 residues form an α-helical coiled-coil rod of dimensions ~160 × 2 nm in which the two chains coil around each other. The two heavy chains are parallel, each having its N terminus in one of the two heads and its C terminus bound in the shaft of the thick filament.

Myosins II from other sources have similar structures. For example, analysis of the DNA sequence for a heavy chain gene from the nematode *Caenorhabditis* showed that the protein contains 1966 residues, 1095 of which contain an amino acid sequence appropriate for a 160-nm long coiled coil. There are no prolines within this sequence, which lies between Pro 850 and Pro 1944. Although there are many bands containing positively and negatively charged side chains along the myosin rod, the interactions between the two coiled helices are largely nonpolar. In *Drosophila* 15 different heavy-chain isoforms are created by splicing of a single mRNA.

While the C-terminal portions of the two parallel myosin heavy chains form a rod, the N-terminal portions fold into two separate heads. Each head also contains two smaller 16- to 21-kDa peptide chains which belong to the calmodulin family. One of these, the *essential light chain,* is tightly bound to the heavy chain. The second, the *regulatory light chain,* is able to bind Ca^{2+} and is less tightly bound to the rest of the head. A short treatment with trypsin or papain cuts the myosin molecule into two pieces. The tail end gives rise to *light meromyosin* (*LMM*), a molecule ~90 nm in length. The remainder of the molecule including the heads is designated *heavy meromysin (HMM).* A longer trypsin treatment leads to cleavage of HMM into one ~62-kDa S2 fragment 40 nm long, and two ~130-kDa S1 fragments, each of the latter representing one of the two heads.

The junction of the head and tail portions of myosin appears rigid. However, there must be considerable conformational flexibility and perhaps some uncoiling of the helices to allow the two heads to interact with a single thin filament as is observed by electron microscopy. There also appears to be a hinge between the S2 and LMM segments.

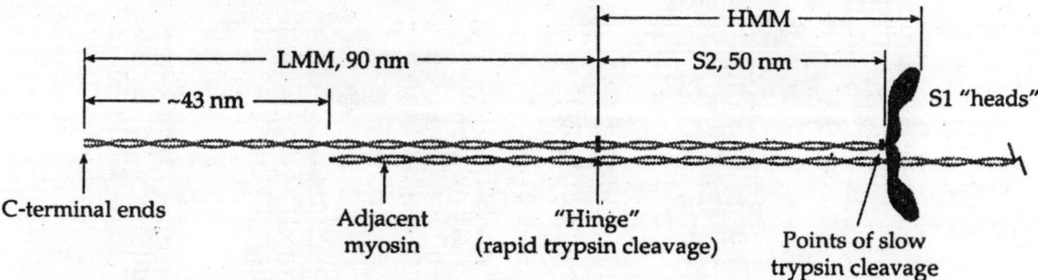

Fig. 5.8. (A) An approximate scale drawing of the myosin molecule. The "hinge" is a region that is rapidly attacked by trypsin to yield the light and heavy meromyosins (LMM and HMM). Total lenght ~160 nm, molecular mass, 470 kDa; two ~200-kDa heavy chains; two pairs of 16- to 21-kDa light chains; heads; ~15 × 4 × 3 nm.

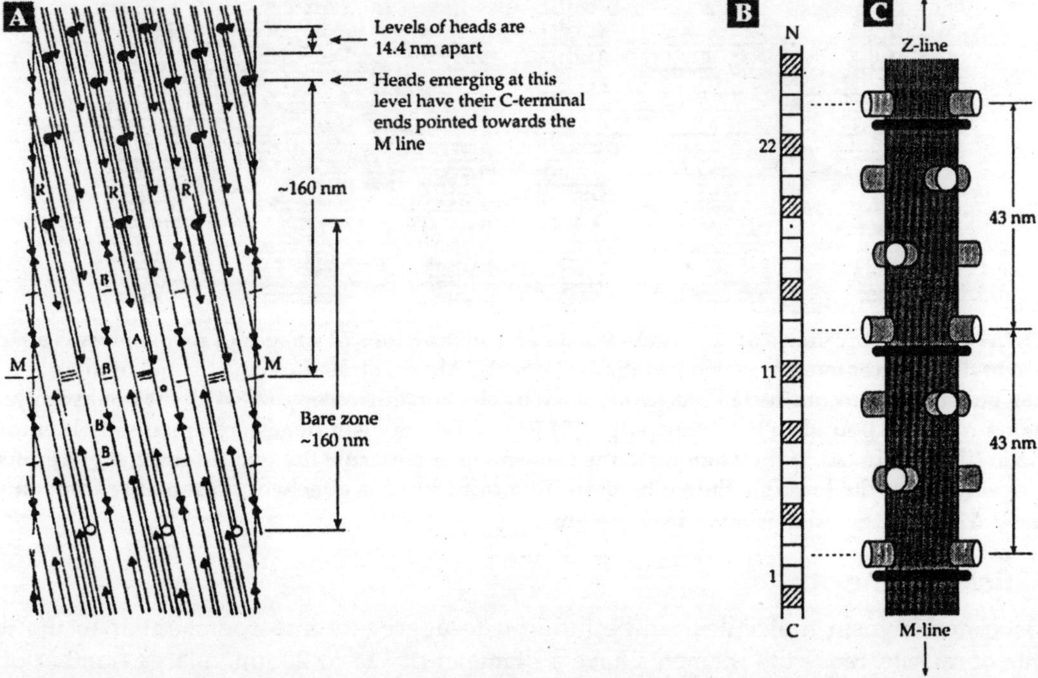

Fig. 5.9. (A) Radial projection illustrating packing of myosin rods as suggested by Squire for thick filaments of vertebrate skeletal muscle. The region of the bare zone at the M-line is shown. The filled circles represent the head ends of the myosin molecules and the arrowheads represent the other end of the rod, *i.e.*, the end of the LMM portion. Antiparallel molecules interacting with overlaps of 43 and 130 nm are shown joined by single and triple cross-lines, respectively. Positions where two arrowheads meet are positions of end-to-end butting. O is an "up" molecule (thin lines) and A a "down" molecule (thick lines). The molecules move from the core at the C-terminal end to the filament surface at the head end. The levels marked B may be the levels of attachment of M-bridge material to the myosin filament. The level M-M is the center of the M-line and of the whole filament. The lateral scale is exaggerated more than threefold. (B) A segment of titin showing the 43-nm 11-domain super-repeat. (C) Model of a segment of a thick filament showing the 43-nm repeat, the C-protein, also bound at 43 nm intervals. (B) and (C).

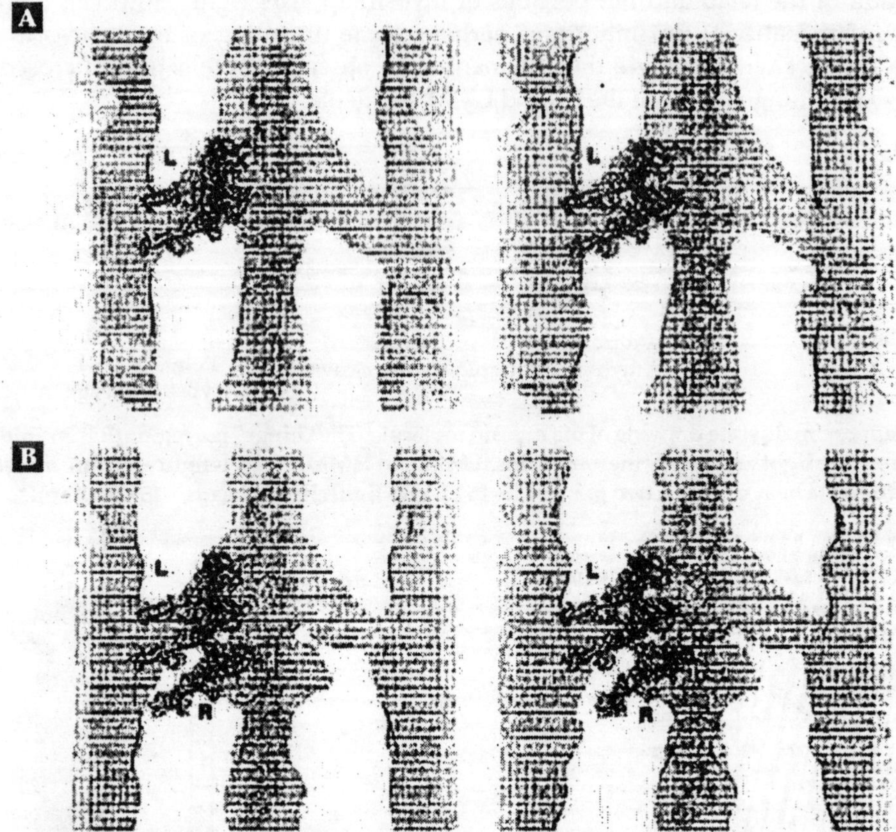

Fig. 5.10. (A) Stereoscopic views of computer-assisted reconstructions of images of myosin heads attached to an F-actin filament centered between two thick filaments. Atomic structures of actin and of myosin heads have been built into the reconstructed images obtained by electron microscopy. (A) With the nonhydrolyzable ATP analog ATPPNP bound in the active sites. (B) Rigor. Two myosin heads are apparently bound to a single actin filament in (A). If they belong to the same myosin molecule the two C-terminal ends must be pulled together from the location shown here. In (B) a third head is attached, presumably from another myosin rod. This configuration is often seen in rigor.

The Thick Filaments

Dissociated myosin molecules can be induced to aggregate into rods similar to the thick filaments of muscle. Since the filaments have a diameter of ~14 to 20 nm, a large number of the thin 2-nm myosin molecules must be packed together. Electron microscopy reveals the presence of the heads projecting from the thick filaments at intervals of ~43.5 nm. However, there is a bare zone centered on the M-line, a fact that suggests tail-to-tail aggregation of the myosin monomers at the M-line in the centers of the thick filaments. A helical packing arrangement involving about 300 myosin molecules (up to 30 rods in a single cross section) in close packing with a small central open core has been proposed for skeletal muscle myosin. There are approximately three heads per turn of the helix, each group of three heads spaced 14.3 nm from the preceding one along the thick filament. It is apparently the zones of positive and negative charge, which are especially prevalent in the LMM segment toward the C termini, that lock the successive myosin molecules into this 14.3-nm spacing. Titin also binds to the LMM segment of the myosin rod, and its

11-domain super repeat of IgG-like and fibronectin-like modules are also 43 nm in length. There are typically 47 – 49 of these super repeats in titin, and if each fits to a turn of the helix, there would be 147 myosin molecules in one-half of the thick filament.

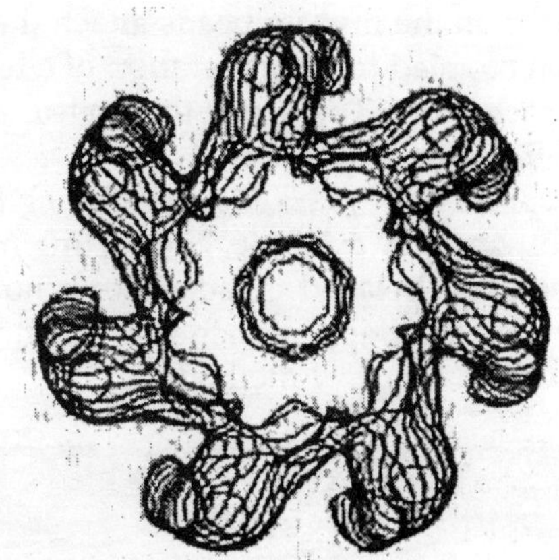

Fig. 5.11. Superimposed sections for the 14 nm thickness of a computer-assisted reconstruction of the myosin filaments of the scallop adductor muscle.

Not all muscles have the thick filament structure of Fig. 5.11. In the tarantula muscle, which has a particularly well-ordered structure there are four myosins per turn. Figure 5.13 shows a reconstruction of scallop myosin which has a 7-fold rotational symmetry. The thick filaments often contain other proteins in addition to myosin. Thus, skeletal muscle contains the C-protein in a series of helical bands along the thick rod. In nematodes, molluscs, and insects the thick filament has a cylindrical core of *paramyosin,* another protein with a structure resembling that of the myosin rod. A minor component of *Drosophila* myosin, the *myosin rod protein,* lacks heads but is transcribed from the myosin heavy chain gene.

3. Actomyosin and Muscle Contraction

That actin and myosin are jointly responsible for contraction was demonstrated long before the fine structure of the myofibril became known. In about 1929, ATP was recognized as the energy source for muscle contraction, but it was not until 10 years later that Engelhardt and Ljubimowa showed that isolated myosin preparations catalyzed the hydrolysis of ATP. Szent-Györgi showed that a combination of the two proteins actin and myosin was required for Mg^{2+}-stimulated ATP hydrolysis (ATPase activity). He called this combination *actomyosin.*

Under the electron microscope the myosin heads can sometimes be seen to be attached to the nearby thin actin filaments as *crossbridges.* When skeletal muscle is relaxed (not activated by a nerve impulse), the crossbridges are not attached, and the muscle can be stretched readily. The thin filaments are free to move past the thick filaments, and the muscle has some of the properties of a weak rubber band. However, when the muscle is activated and under tension, the crossbridges form more frequently. When ATP is exhausted (*e.g.,* after death) muscle enters the state of *rigor* in which the crossbridges can be seen by electron microscopy to be almost all attached to thin filaments, accounting for the complete immobility of muscle in rigor.

In rigor the crossbridges are almost all firmly attached to the thin actin filaments, making an approximately 45° angle to the actin filaments. However, the addition of ATP causes their instantaneous release and the relaxation of the muscle fiber. In contrast, activation by a nerve impulse, with associated release of calcium ions, causes the thin filaments to slide between the thick filaments with shortening of the muscle. An activated muscle shortens if a low tension is applied to the muscle, but at a higher tension it maintains a constant length. Because the maximum tension developed is proportional to the length of overlap between the thick and thin filaments, it was natural to identify the individual crossbridges as the active centers for generation of the force needed for contraction.

The Rowing Hypothesis

H. E. Huxley and A. F. Huxley and R. M. Simmons independently proposed that during contraction the myosin heads attach themslves to the thin actin filaments. The hydrolysis of ATP is then coupled to the generation of a tension that causes the thick and thin filaments to be pulled past each other. The heads then release themselves and become attached at new locations on the actin filament. Repetition of this process leads to the sliding motion of the filaments. The evidence in favor of this *"rowing"* or *"swinging bridge"* hypothesis was initially based largely on electron microscopy. For example, contracting muscle was frozen rapidly and fixed for microscopy in the frozen state. Relaxed muscle shows no attached crossbridges, but contracting muscle has many. However, their appearance was distinct from that seen in rigor. The model was also supported by indirect physical methods.

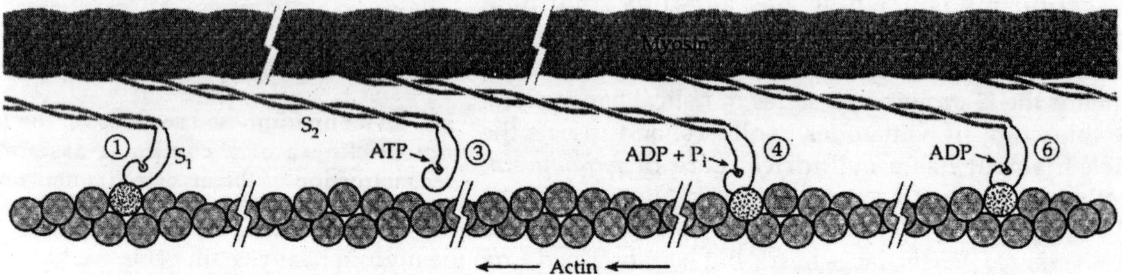

Fig. 5.14. A model for the coupling of ATP hydrolysis to force production in muscle based on proposals of H. E. Huxley, and A.F. Huxley and Simmons. The power stroke is depicted here as a rotation of the crossbridge from a 90° to a 45° configuration. Four representative stages are shown: (1) the rigor complex, (3) the dissociated myosin ATP complex, (4) the actomyosin ADP pre-power stroke state in which the actin-myosin band has reformed but with a different actin subunit, which may be distant from that in (1), and (6) the actomyosin ADP post-power stroke state. Force production and contraction result from crossbridges passing cyclically through the steps depicted from left to right.

An impressive demonstration that myosin heads do move along the actin filaments was provided by Sheetz and Spudich, who found that myosin-coated fluorescent beads ~0.7 μm in diameter will move along actin filaments from cells of the alga *Nitella* in an ATP-dependent fashion at velocities similar to those required in muscle. The myosin heads literally glide along the thick cables of parallel actin filaments present in these algae.

Why Two Heads?

The actin filament is a two-start helix, and it is natural to ask whether the two myosin heads bind to just one or simultaneously to both of the actin strands. Most evidence supports a 1:1 interaction of a single head with just one strand of actin. However, the other actin strand may associate with heads from a different thick filament. Another question concerns the role of the pairs of myosin heads. Could the two heads bind sequentially to the actin and exert their pull in a fixed sequence? In the reconstruction of the actomyosin complex in rigor two different images are seen for the crossbridges. This suggests the existence of two different conformations for the attached myosin heads. Similar images for smooth muscle heavy meromyosin in its inactive (resting) dephosphorylated state show the two heads in very different orientations with one binding to the other of the pair and blocking its movement. Perhaps one head is tightly bound at the end of the power stroke while the other is at a different stage of the catalytic cycle. Nevertheless, single-headed myosin from *Acanthamoeba* will propell organelles along actin filaments and actin filaments will slide across a surface coated with single-headed myosin formed by controlled proteolysis. The additional interactions seen in rigor may be peculiar to that state.

Structure of the Myosin Heads

Myosin and myosin fragments can be isolated in large quantities, but they have been difficult to crystallize. However, Rayment and coworkers purified S1 heads cleaved from chicken myosin by papain and subjected them to reductive methylation (using a dimethylamine-borane complex. With most of the lysine side chain amino groups converted to dimethylamine groups, high-quality crystals were obtained, and a structure was determined by X-ray diffraction. Since that time various forms of both modified and unmodified myosin heads from several species have been studied by X-ray crystallography. Especially clear results were obtained with unmodified myosin from the ameba *Dictyostelium discoideum.* The head structure, includes a 95-kDa piece of the heavy chain and both light chains. A clearer picture of the neck region containing the light chains was provided by the structure of the "regulatory domain" of scallop myosin. Unlike mammalian or avian myosins, molluscan myosins are regulated by binding of Ca^{2+} to a site in the essential light chain.

Cleavage of the ~850-residue S1 heads with trypsin yields mainly three large fragments that correspond to structural domains of the intact protein. They are known as the 25-kDa (N-terminal), 50-kDa, and 20-kDa fragments, and for myosin from *D. discoideum* correspond to residues 1 to 204, 216 to 626, and 647 to 843, respectively. The ATP-binding site is in a deep cleft between the 20-kDa and 50-kDa regions. Fig. 5.16 illustrates the binding of an ATP analog, the beryllirum fluoride complex of MgADP, in the active site. As can be seen, the ATP binds to loops at the C termini of the β strands of the 8-stranded β sheet from the 25-kDa domain. The conserved P-loop, which lies between T178 and E187, curls around the α and β phospho groups, and has the sequence G(179)ESGAGKT. A second conserved loop N(233)SNSSR-G(240) from the 50-kDa domain contributes to the binding of ATP.

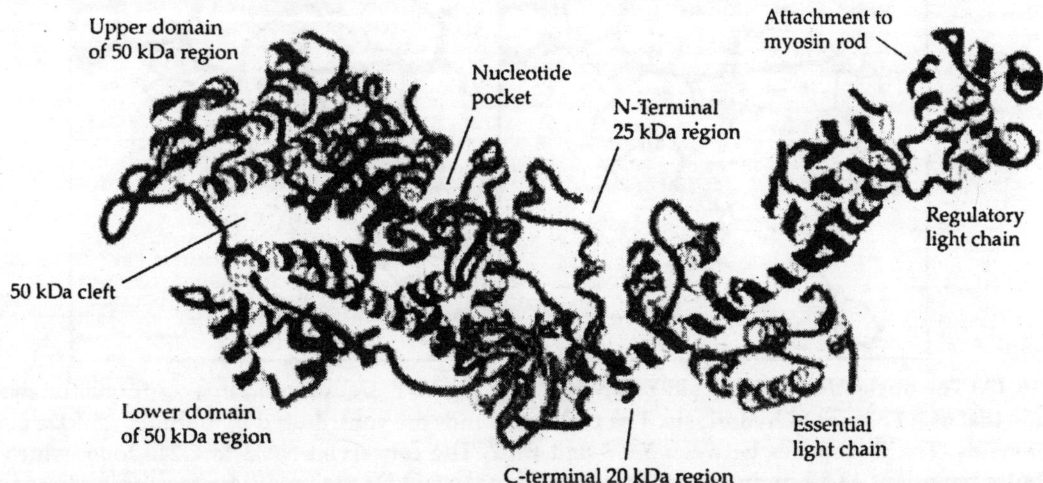

Fig. 5.15. Ribbon representation of chicken-skeletal myosin subfragment-1 showing the major domains and tryptic fragments.

The actin-binding region of the myosin head is formed largely by the 50-kDa segment, which is split by a deep cleft into two separate domains both of which are thought to participate in binding to actin. A surface loop (loop 1) near the ATP-binding site at the junction of the 25- and 50-kDa regions affects the kinetic properties of myosin, probably by influencing product release.

A second loop (loop 2, residues 626-647) at the junction of the 50- and 20-kDa regions interact with actin. Loop 2 contains a GKK sequence whose positive charges may interact with negative charges in the N-terminal part of actin.

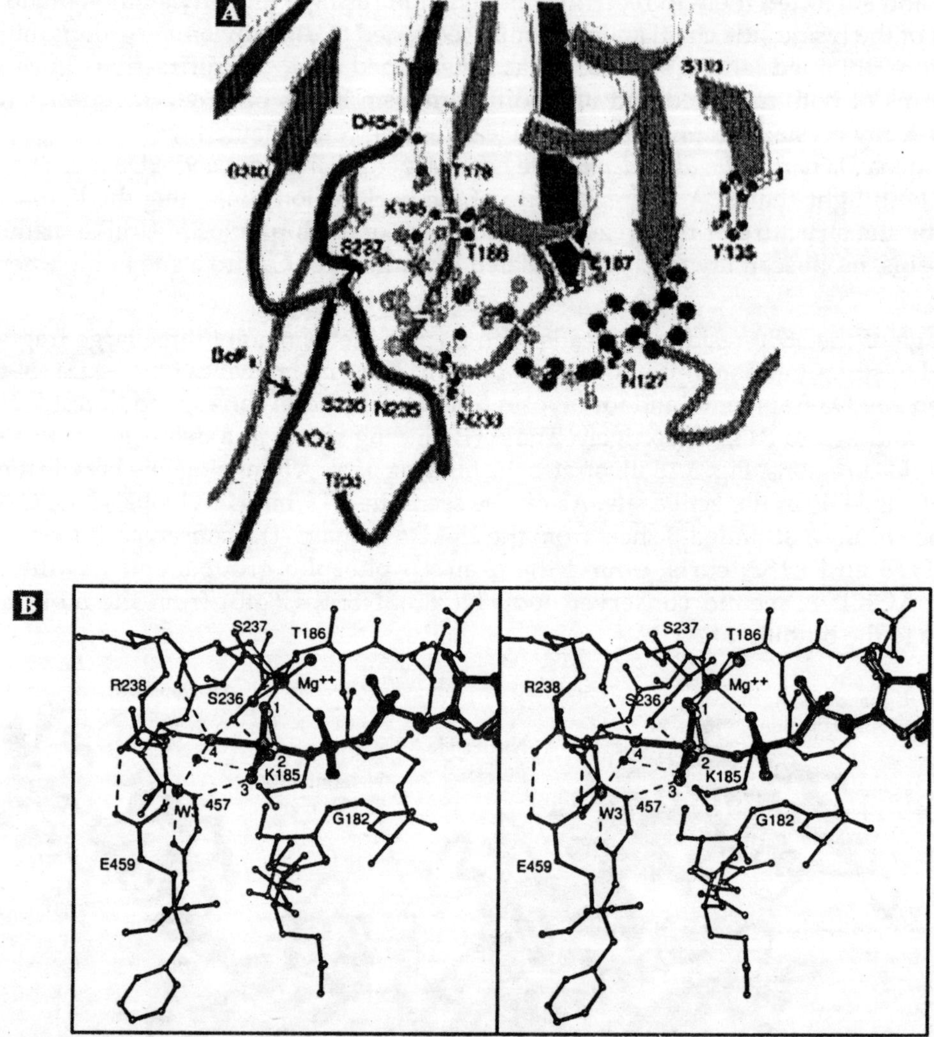

Fig. 5.16. (A) The nucleotide binding site of myosin with MgADP•BeFx bound in a conformation thought to mimic that of ATP prior to hydrolysis. The β-sheet strands are contributed by both the 25-kDa and 50-kDa domains. The P-loop lies between T178 and E187. The conserved N233 to G240 loop, which also contributes important ATP-binding residues, comes from the 50-kDa region. (B) Stereoscopic view of the γ-phospho group binding pocket with the bound MgADP•VO_4 (vanadate) complex. The coordinated Mg^{2+} and associated water molecules are seen clearly.

The C-terminal fragment of myosin contains a globular domain that interacts with both the 20-kDa and 50-kDa regions and contains an α-helical neck that connects to the helix of the coiled-coil myosin rod. This helical region is surrounded by the two myosin light chains. A pair of reactive thiol groups (from C697 and C707) in the globular domain are near the active site. Crosslinking of these cysteines by an –S–S– bridge has been utilized to trap nucleotide analogs in the active site.

How Does Actin Bind?

The actin monomer consists of four subdomains, 1, 2, 3, and 4 numbered from the N terminus. The negatively charged N-terminal region of actin contains the sequence.

$$\text{D–E–D–}\overset{4}{\text{E}}\text{ ---- }\overset{24}{\text{D}}\text{ – }\overset{25}{\text{D}}$$

It may interact with loop 1 of myosin, which contains five lysines. However, to form a strong interaction with the myosin head a conformational change must occur in the myosin. A change may also occur in actin. Modeling suggests that a large nonpolar contact region involves actin residues A144, I341, I345, L349, and F352 and myosin residues P529, M530, I535, M541, F542, and P543. A conformational change in actin, which might involve largely the highly conserved actin subdomain 2, may also be required for tight interaction.

Kinesins and Other Molecular Motors

Before considering further how the myosin motor may work, we should look briefly at the *kinesins,* a different group of motor molecules, which transport various cellular materials along microtubule "rails." They also participate in organization of the mitotic spindle and other microtubule-dependent activities. More than 90 members of the family have been identified. Kinesin heads have much shorter necks than do the myosin heads. A myosin head is made up of ~850 residues, but the motor domain of a kinesin contains only ~345. Like myosin, the 950- to 980-residue kinesins have a long coiled-coil C-terminal region that forms a "neck" of ~50 residues, a "stalk" of ~190 and ~330 residue segments with a Pro/Gly-rich hinge between them, and an ~45 residue "tail."

Crystal structures are known for motor domains of human kinesin and of a kinesin from rat brain. The structures of one of six yeast kinesins, a protein called *Kar3,* and also of a *Drosophila* motor molecule designated *Ncd* have also been determined. The last was identified through study of a *Drosophila* mutant called non-claret disjunctional (Ned). The motor domains of various members of the kinesin family show ~40% sequence identity and very close structural identity. Although the sequences are different from those of the myosin heads or of G proteins, the folding pattern in the core structures is similar in all cases. An 8-stranded β sheet is flanked by three α helices on each side and a P-loop crosses over the ATP-binding site. Further similarity is found in the active site structures, which, for a monomeric kinesin KIFlA, have been determined both with bound ADP and with a nonhydrolyzable analog of ATP. Although there is little similarity in amino acid sequences the structures in the catalytic core are clearly related to each other, to those of dimeric kinesins, to those of myosins, and to those of the GTP-hydrolyzing G proteins.

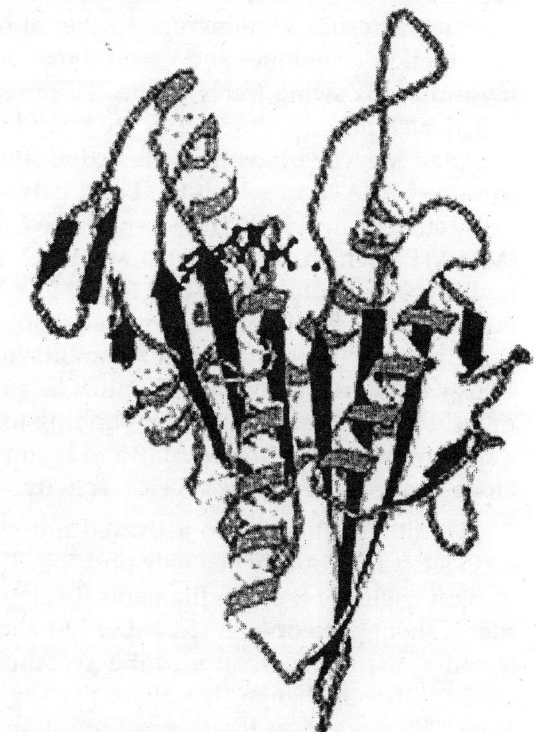

Fig. 5.17. Ribbon drawing of human kinesin with bound Mg · ADP.

A puzzling discovery was that the motor domain of kinesin, which binds primarily to the β subunits of tubulin and moves toward the fast growing *plus* end of the microtubule, is located at the N terminus of the kinesin molecule, just as is myosin. However, the Ned and *Kar3* motor domains are at the C-terminal ends of their peptide chains and move their "cargos" toward the *minus* ends of microtubules. Nevertheless, the structures of all the kinesin heads are conserved as are the basic chemical mechanisms. The differences in directional preference are determined by a short length of peptide chain between the motor domain and the neck, which allows quite different geometric arrangements when bound to micro tubules. Like Ned, myosin VI motor domains also move "backwards" toward the pointed (minus) ends of actin filaments.

Other major differences between kinesins and myosin II heads involve kinetics and processivity.[173] Dimeric kinesin is a *processive* molecule. It moves rapidly along microtubules in 8-nm steps but remains attached. Myosins V and VI are also processive but myosin II is not. It binds, pulls on actin, and then releases it. The many myosin heads interacting with each actin filament accomplish muscle contraction with a high velocity in spite of the short time of attachment. Ned and Kar3 are also nonprocessive and slower than the *plus* end-oriented kinesins.

The ATPase Cycles of Actomyosin and of the Kinesins

The properties of the protein assemblies found in muscle have been described in elegant details, but the most important question has not been fully answered. How can the muscle machinery use the Gibbs energy of hydrolysis of ATP to do mechanical work? Some insight has been obtained by studying the ATPase activity of isolated myosin heads (SI) alone or together with actin. Results of numerous studies of ATP binding, hydrolysis, and release of products using fast reaction techniques and cryoenzymology are summarized in Fig. 5.18. In resting muscle the myosin heads swing freely in the ~20-nm space between the thick and thin filaments. However, in activated muscle some heads are bound tightly to actin as if in rigor complex A•M. When ATP is added MgATP binds into the active site of the myosin inducing a conformational change to form A•M*•ATP in which the bond between actin and myosin is weakened greatly, while that between myosin and ATP is strengthened. The complex dissociates (step *b)* to give free actin and (M*•ATP), which accumulates at –15°C. However, at higher temperatures the bound ATP is hydrolyzed rapidly (step *c)* to a form M**•ADP•P_i in which the ATP has been cleaved to ADP + P_i but in which the split products remain bound at the active site. All of these reactions are reversible. That is, the split products can recombine to form ATP. This fact suggests that most of the Gibbs energy of hydrolysis of the ATP must be stored, possibly through a conformational change in the myosin head or through tighter bonding to ATP. As long as calcium ions are absent, there is only a slow release of the bound ADP and P_i and replacement with fresh ATP takes place. Thus, myosin alone shows a very weak ATPase activity.

On the other hand, in activated muscle the head with the split ATP products will bind to actin (step *d)*, probably at a new position. The crossbridges that form appear to be attached almost at right angles to the thin filaments. In step *e*, P_i is released following a conformational alteration that is thought to open a *"back door"* to allow escape of the phosphate ion. In the final two steps (*f* and *g)* the stored energy in the myosin head (or in the actin) is used to bring about another conformational change that alters the angle of attachment of myosin head to the thin filament from ~90° to ~45°. At least some indication of such a change can be observed directly by electron microscopy. Such a change in angle is sufficient to cause the actin filament to move ~10 nm with respect to the thick filaments to complete the movement cycle, if the head is hinged at the correct place. However, the existence of at least four different conformational states suggests a more complex sequence. Examination of the three-dimensional structures available also suggest a complex sequence of alterations in structure and geometry. X-ray crystallographic structures of myosin heads, in states thought to correspond to states 1 and 3 are also in agreement on an ~10 (5 – 12) nm movement of the lever arm. Six states of the actomyosin complex are depicted in Fig. 5.18, but a complete kinetic analysis requires at least eight and possibly 12 states.

Observing Single Molecules

A major advance in the study of molecular motors has been the development of ways to observe and study single macromolecules. The methods make use of *optical traps* (optical "tweezers") that can hold a very small (~1 μm diameter) polystyrene or silica bead near the waist of a laser beam focused through a microscope objective. In one experimental design an F-actin filament is stretched between two beads, held in a pair of optical traps. The filament is pulled taut and lowered onto a stationary silica bead to which a few myosin HMM fragments have been attached. If ATP is present, short transient movements along the filament are detected by observation of displacements of one of the beads when the actin filament contacts HMM heads. An average lateral displacement of 11 nm was observed. Each HMM head exerted a force of 3 – 4 pN, a value consistent with expectations for the swinging bridge model. From the duration of a single displacement (≤7ms) and an estimated K_{cat} for ATP hydrolysis of 10 s^{-1}, the fraction of time that the head was attached during one catalytic cycle of the head was therefore only 0.07. This ratio, which is called the *duty ratio,* is low for actomyosin. However, many myosin heads bind to each actin filament in a muscle. Each head exerts its pull for a short time, but the actin is never totally unattached. Similar measurements with smooth muscle revealed similar displacements but with a 10-fold slower sliding velocity and a 4-fold increase in the duty ratio. This may perhaps account for an observed 3-fold increase in force as compared with skeletal muscle.

Other single molecule techniques involve direct observation of motor molecules or of SI myosin fragments tagged with highly fluorescent labels. All measurements of single molecule movement are subject to many errors. Brownian motion of the beads makes measurements difficult. Not all results are in agreement, and some are difficult to understand. Most investigators agree that there is a step size of ~4 – 10 nm. Kitamura *et. al.,* found 5.3 nm as the average. However, they also reported the puzzling observation that some single S1 molecules moved 11 – 30 nm in two to five rapid successive steps during the time of hydrolysis of a single molecule at ATP. They suggested that some of the energy of ATP hydrolysis may be stored in S1 or in the actin filament and be released in multiple steps. Veigel *et. al.,* observed that a brush border myosin I from chicks produced ~6 nm movements, each of which was followed by an additional ~5.5 nm step within ~5 ms. They attribute these steps to two stages in the power stroke, *e.g.,* to steps *f* and *g*. A value of ~10 nm was reported recently by Piazzesi *et. al.* Myosin V moves along actin filaments with very large 36-nm steps.

Motion of kinesin heads has been observed by movement of microtubules over biotinylated kinesin fixed to a steptavidin-coated surface, by direct observation of fluorescent kinesin moving along microtubules, and by optical trap interferometry. Kinesin heads move by 8-nm steps, evidently the exact length of an αβ tubulin dimer in the microtubule structure. One ATP is apparently hydrolyzed for each 8-nm step. However, shorter steps of ~5 nm have sometimes been seen.

The movement is processive, kinesin motors typically taking 100 steps before dissociating from the microtubule. Kinesin is bound to the microtubule continuously. Its duty ratio is nearly 1.0 (the same is true for the bacterial flagellar motor; Fig. 5.4). However, single kinesin heads,

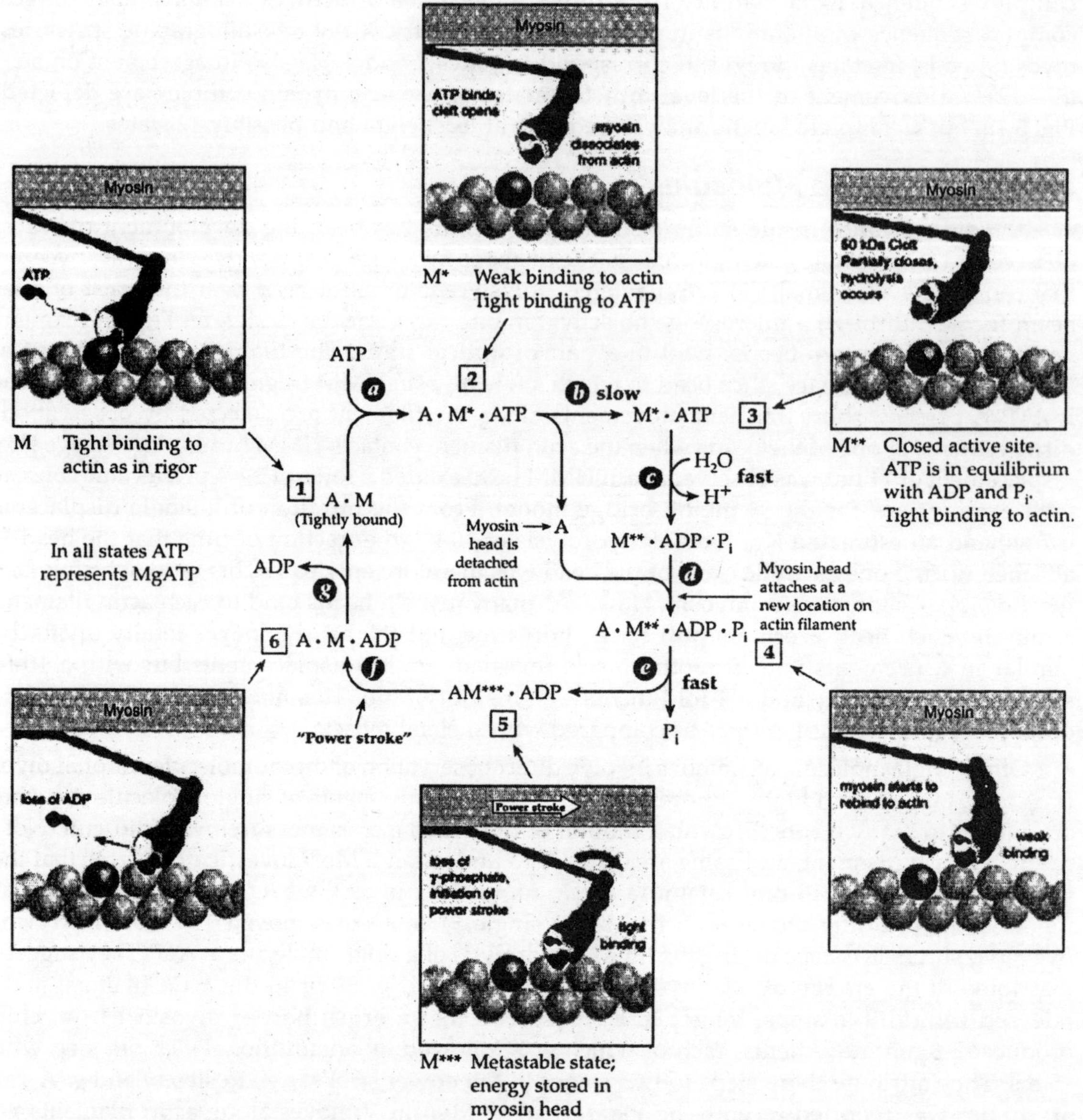

Fig. 5.18. Simplified view of the ATP hydrolysis cycle for actomyosin. A similar cycle can be written for kinesins and dyneins. Here A stands for fibrous actin and M, M*, M, and M*** for four different conformations of the myosin heads.**

which lack the coiled-coil neck region, have a duty ratio of <0.45. The movement is nonprocessive. The Ncd motor is also nonprocessive. As mentioned previously, the Ncd and kinesin motor domains are at opposite ends of the peptide chain, and the motors move in opposite directions along microtubules. The critical difference between the two motor molecules was found in the neck domains, which gave rise to differing symmetrics in the two heads. The latter are shown in Fig. 5.20, in which they have been docked onto the tubulin protofilament structure. One head, both of ncd and of kinesin, occupies a similar position on the microtubule, but the other head points toward the microtubular plus end for kinesin but toward the minus end for Ncd. Cryoelectron microscopy also supports this interpretation.

Still not fully understood is the processive action of kinesin. It is often assumed that this protein moves in a hand-over-hand fashion with the two heads binding alternatively to the microtubule. Some substantial reorganization of the peptide chain in the hinge region at the end of the neck is presumably involved. An alternative "inchworm" mechanism has been suggested.

Thinking about Chemical Mechanisms

We have now examined the active sites of kinases that cleave ATP, ATPases that pump ions by cleaving ATP, ATP synthases that form ATP from ADP and P_i and GTP hydrolyzing enzymes that cause movement and shape changes that control metabolic processes. It is striking that the active site regions where the ATP or GTP bind have such a highly conserved structure. This suggests that the secret of movement can be found in the very strong interactions of the nucleotides and their split products with the proteins. In every case there is at least one tight binding or closed conformation in which a large number of hydrogen bonds and ionic interactions bind the nucleotide. During the actomyosin reaction several conformational changes must occur. Not only does the affinity for the bound nucleotide vary, but also the binding of actin to myosin can be strong, as in the nucleotide-free state or in the presence of bound ADP. Binding is weak when ATP or the split product ADP + P_i are in the active site.

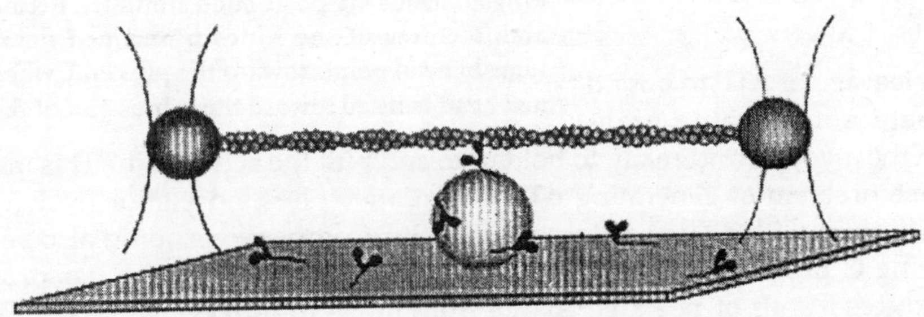

Fig. 5.19. Schematic drawing (not to scale) illustrating the use of two optical traps that are focused on beads attached to a single actin filament. The filament is lowered onto a stationary silica bead sparsely coated with HMM fragments of myosin. In the presence of ATP the myosin heads bind transiently for a few milliseconds to the actin, moving it in one direction and displacing the beads from their positions in the optical traps. An image of one of the beads is projected onto photodiode detectors capable of measuring small displacements. The displacing force can also be recorded. For details of the experiments and of the optical traps and measuring devices.

To understand these differences we should look at the structure of ATP itself. The triphosphate group has many negative charges repelling each other. What must happen to allow the binding of ATP to myosin to break the actin-myosin bond? The electrostatic attraction of these phospho groups for active site groups is doubtless one cause of the observed conformational changes. Could it be that electrostatic repulsion, via a proton shuttle mechanism, is also induced at the right point in the actin-myosin interface? Many studies with analogs of ATP have contributed to our understanding. Neither the purine nor the ribose ring of ATP is absolutely essential. The compound 2-[(4-azido-2-nitrophenyl) amino] diphosphate (NANDP) and related nonnucleotide analogs support muscle contraction and relaxation in the same way as does ATP. An analog with a rigid five-membered ring, 2',3'-dideoxydidehydro-ATP, is also active. A comparison of kinetic data and X-ray structural data supports the proposal that the ATP must be bound in the conformation shown for MgADP•BeF_x in Fig. 5.16A. When the two SH groups of C697 and C707 of the myosin head are crosslinked by various reagents, this NANDP analog can be trapped at the active site. Because of the presence of its azide ($-N_3$) group the trapped compound can serve as a photo-affinity label, attaching itself to a tryptophan side chain upon activation with visible light.

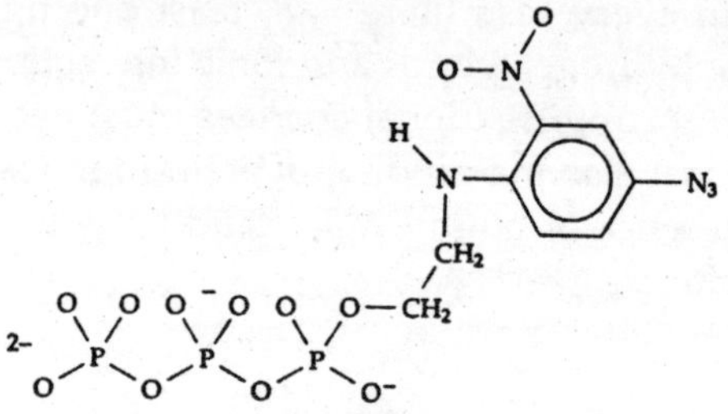

NANTP

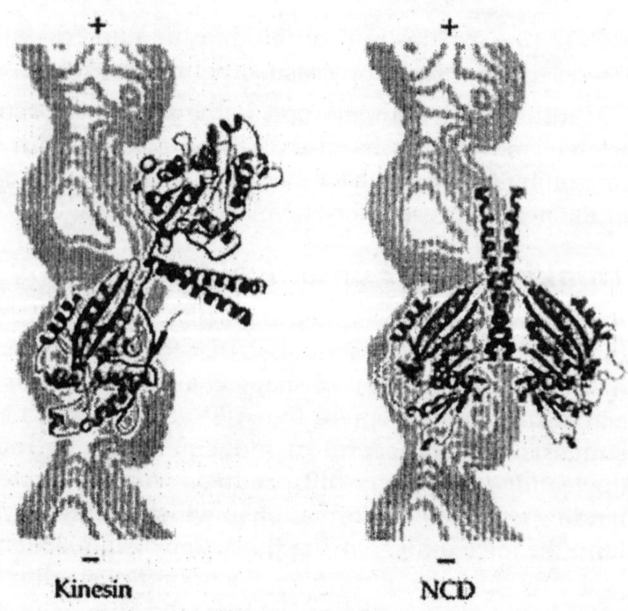

Fig. 5.20. Model showing the ncd and kinesin dimer structures docked onto a tubulin protofilament. The bound ncd and kinesin heads are positioned similarly. Because of the distinct architectures of the kinesin and ncd necks, the unbound kinesin head points toward the plus end, whereas the unbound ncd head is tilted toward the minus end of the protofilament.

How can cleavage of ATP to bound ADP + P_i create a metastable high-energy state of the myosin head ready to hold onto and pull the actin chain? This may be compared with the inverse problem of generating ATP in oxidative phosphorylation, in which ADP and P_i coexist in equilibrium with ATP in a closed active site. Comparison should also be made with the GTP-hydrolyzing G proteins (Fig. 5.7). During hydrolysis of GTP by the Ras protein binding to the protein induces a shift of negative charge from the γ- to β-oxygens of GTP facilitating bond cleavage as in Eq. 5.2. G proteins also couple substrate hydrolysis to mechanical motion. We should also think about the fact that when ATP is cleaved within myosin there will necessarily be a flow of electrical charge from the water to the ADP. This will be followed by some accomodation of the protein to the new charge constellation. As we have seen previously, movement of protons is often the key to conformational changes. In this case, the initial change must be to create a high-energy state of myosin which, following loss of the orthophosphate ion, can cause the major conformational change that swings the lever arm of the myosin. The conformational changes may occur in several steps in which the packing of groups within the myosin head is always tight in some places and rather loose in others. Movement within the head is being observed not only by X-ray crystallography but by *fluorescence resonance energy transfer* and by the newer *luminescence resonance energy transfer* (LRET). For example, a terbium chelate of azide-ATP was photochemically bound in the active site, and a fluorescent dye was attached to Cys 108 in the regulatory light chain. The terbium ion was irradiated, and fluorescence of the dye was observed. Distance changes, measured in the absence and presence of ATP, were consistent with the swinging arm model. Dyes have been attached to –SH groups engineered into various locations in the myosin molecule to permit other distance measurements. In another elegant application of the FRET technique the *green fluorescent protein* of *Aequoria* was fused to the C terminus of the motor domain of myosin giving a fluorescent lever arm. Energy transfer to blue fluorescent protein fused to the N terminus of the S1 head was measured. The distance between these was estimated by the FRET technique and was also consistent with expectations for the *"rowing model."*

The "rowing model" is generally accepted, but other quite different processes have been proposed to account for the elementary cycle of muscle contraction. Muscle contracts nearly *isovolumetrically;* thus, anything that expands the sarcomere will cause a contraction. It has been suggested that the hydrolysis of ATP deposits negatively charged phospho groups on the actin filaments, and that the electrostatic repulsion is responsible for lateral expansion of the sarcomere. Still other ideas have been advanced.

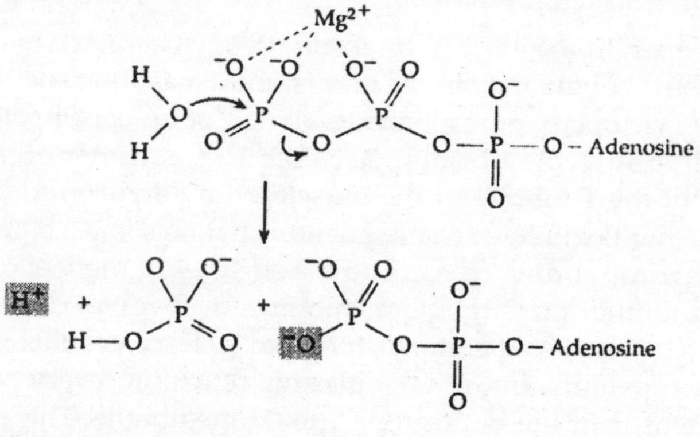

4. Control of Muscle Contraction

Skeletal muscle must be able to rest without excessive cleavage of ATP but able to act rapidly with a high expenditure of energy upon nervous excitation. Even a simple physical activity requires that a person's muscles individually contract and relax in rapid response to nerve impulses from the brain. To allow for this control the endoplasmic reticulum *(sarcoplasmic reticulum, SR*) of striated muscle fibers is organized in a striking regular manner. Interconnecting tubules run longitudinally through the fibers among the bundles of contractile elements. At regular intervals they come in close contact with infoldings of the outer cell membrane (the *T system* of membranes. A nerve impulse enters the muscle fiber through the neuromuscular junctions and travels along the sarcolemma and into the T tubules. At the points of close contact the signal is transmitted to the longitudinal tubules of the sarcoplasmic reticulum, which contain a high concentration of calcium ions.

Calcium Ions in Muscle

A nerve signal arriving at a muscle causes a sudden release of the calcium ions into the cytoplasm from cisternae of the sarcoplasmic reticulum (SR) that are located adjacent to the T-tubules. Diffusion of the Ca^{2+} into the myofibrils initiates contraction. In smooth muscle the signals do not come directly from the nervous system but involve hormonal regulation. Again, calcium ions play a major role. Muscle contains a large store of readily available Ca^{2+} in lateral cisternae of the SR. The free intracellular Ca^{2+} concentration is kept low by a very active ATP-dependent calcium ion pump which is embedded in the membranes of the SR. Within the vesicles Ca^{2+} is held loosely by the ~63-kDa protein *calsequestrin,* which binds as many as 50 calcium ions per molecule. When the cytoplasmic concentration of free Ca^{2+} falls below ~10^{-6} M, contraction ceases. In fast-contracting skeletal muscles the Ca^{2+}-binding protein *parvalbumin* may assist in rapid removal of free Ca^{2+} from the cytoplasm. Contraction is activated when Ca^{2+} is released from the SR through the *calcium release channels,* which are often called *ryanodine receptors.* The name arises from their sensitivity to the insecticidal plant alkaloid ryanodine, which at low concentration (≤60 μM) causes the channels to open, but a high concentration closes the channels. These calcium release channels consist of tetramers of ~5000-residue proteins. The bulk of the 565-kDa polypeptides are on the cytosolic surface of the SR membranes, where they form a complex "foot" structure that spans the ~12 ran gap between the SR vesicles and the T-tubule membrane. Ryanodine receptor function is modulated by NO, which apparently binds to –SH groups within the Ca^{2+} channel. Some ryanodine receptors are activated by cyclic ADP-ribose (cADPR,). Some have an oxidoreductase-like structural domain.

The release channels open in response to an incompletely characterized linkage to the *voltage sensor* that is present in the T-tubule membrane and is known as the *dihydropyridine receptor.* This too, is a Ca^{2+} channel, which opens in response to arrival of an *action potential* (nerve impulse; that move along the T-tubule membrane. Because the action potential arrives almost simultaneously throughout the T-tubules of the muscles, the dihydropyridine receptors all open together. It isn't clear whether the linkage to the calcium channels is via stimulation from released Ca^{2+} passing from the dihydropyridine receptor to the surfaces of the feet of the release channels, or is a result of depolarization of the T-tubule membrane, or involves direct mechanically linked conformational changes. The close cooperation of the Ca^{2+} release channel and the voltage sensor is reflected in their close proximity. In the sarcoplasmic reticulum every second release channel is adjacent to a voltage sensor in the opposing T-tubule membrane. The essential nature of the voltage sensor is revealed by a lethal mutation *(muscular disgenesis)* in mice. Animals with this autosomal recessive trait generate normal action potentials in the sarcolemma but Ca^{2+} is not released and no muscular contraction occurs. They lack a 170-kDa dihydropyridine-binding subunit of the sensor.

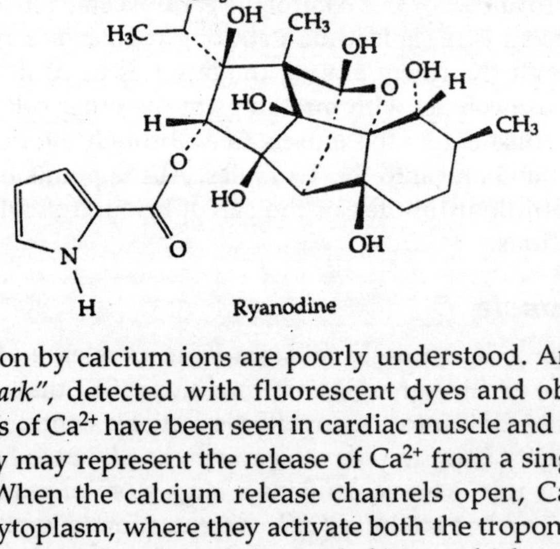

Ryanodine

Some aspects of regulation by calcium ions are poorly understood. Another phenomenom is the observation of Ca^{2+} *"spark"*, detected with fluorescent dyes and observation by confocal microscopy. These small puffs of Ca^{2+} have been seen in cardiac muscle and in a somewhat different form in smooth muscle. They may represent the release of Ca^{2+} from a single release channel or a small cluster of channels. When the calcium release channels open, Ca^{2+} ions flow from the cisternae of the SR into the cytoplasm, where they activate both the troponin-tropomyosin system and also the Ca^{2+}-calmodulin-dependent *light chain protein kinase,* which acts on the light chains of the myosin head. These light chains resemble calmodulin in their Ca^{2+}-binding properties. The function of light chain phosphorylation of skeletal muscle myosin is uncertain but it is very important in smooth muscle.

The regulatory complex of tropomyosin and troponin is attached to the actin filaments. The latter shows a model at near atomic resolution but without side chains on the tropomyosin and without the troponin components. When the regulatory proteins are completely removed from the fibrils, contraction occurs until the ATP is exhausted. However, in the presence of the regulatory proteins and in the absence of calcium, both contraction and hydrolysis of ATP are blocked. Tropomyosin (Tm) is a helical coiled-coil dimer, a 40-nm rigid rod, in which the two 284-residue 33-kDa monomers have a parallel orientation as in the myosin tail. However, an 8– to 9– residue overlap at the ends may permit end-to-end association of the Tm molecules bound to the actin filament. As with other muscle proteins there are several isoforms, whose distribution differs in skeletal and smooth muscle and in platelets. The elongated Tm rods appear to fit into the grooves between the two strands of actin monomers in the actin filament. In resting muscle the Tm is thought to bind to actin near the site at which the S1 portion of the myosin binds. As a consequence, the Tm rod may block the attachment of the myosin heads to actin and prevent actin-stimulated hydrolysis of ATP. The 40-nm Tm rod can contact about seven actin subunits at once. Thus, one Tm-troponin complex controls seven actin subunits synchronously.

Troponin (Tn) consists of three polypeptides (TnC, TnI, TnT) that range in mass from 18 to 37 kDa. The complex binds both to Tm and to actin. Peptide TnT binds tightly to Tm and is thought to link the TnI•TnC complex to Tin. TmI interacts with actin and inhibits ATPase activity in the absence of Ca^{2+}. It may work with the other two peptides to keep the Tm in the proper position to inhibit ATP hydrolysis. TnC binds calcium ions. This ~160-residue protein has a folding pattern almost identical to that of calmodulin with four Ca^{2+}-binding domains arranged in two pairs at the ends of a long 9-turn helix. When Ca^{2+} binds to TnC, a conformational change occurs. This induces changes in the troponin•tropomyosin•actin complex, releases the inhibition of actomyosin ATPase, and allows contraction to occur. In the heart additional effects are exerted by β-adrenergic stimulation, which induces phosphorylation of two sites on TnI by the action of the cAMP-dependent protein kinase PKA. Dephosphorylation by protein phosphatase 2A completes a regulatory cycle in which the doubly phosphorylated TnI has a decreased sensitivity to calcium ions. Cardiac muscle also contains a specialized protein called *phospholamban.* An oligomer of 52-residue subunits, it controls the calcium ion pump in response to β-adrenergic stimulation. Unphosphorylated phospholamban inhibits the Ca ATPase, keeping $[Ca^{2+}]$ high in the cytoplasm. Phosphorylation of phospholamban by cAMP and/or calmodulin-dependent protein kinase activates the Ca^{2+} pump, removing Ca^{2+} and ending contraction.

X-ray diffraction and electron microscopy in the 1970s suggested that when calcium binds to troponin the tropomyosin moves through an angle of ~20° away from S1, uncovering the active site for the myosin-ATP-actin interactions. Tropomyosin could be envisioned as rolling along the surface of the actin, uncovering sites on seven actin molecules at once. Side-chain knobs protruding from the tropomyosin like teeth on a submicroscopic gear might engage complementary holes in the actin. At the same time a set of magnesium ion bridges between zones of negative charge on tropomyosin and actin could hold the two proteins together. This proposal has been difficult to test. Although the older image reconstruction is regarded as unreliable, recent work still supports this *steric blocking* model. Image reconstruction from electron micrographs of thin filaments shows that, in the presence of Ca^{2+}, the tropomyosin does move 25° away from the position in low $[Ca^{2+}]$. However, instead of two states of the thin filament, *"on"* and *"off,"* there may be at least three, which have been called *"blocked," "closed"* and *"open."* The closed state may be attained in rigor. In addition, the possibility that changes in the conformation of actin as well as of myosin occur during the contraction cycle must be considered.

Smooth Muscle

The primary regulation of smooth muscle contraction occurs via phosphorylation of the Ser 19 –OH group in the 20-kDa regulatory light chains of each myosin head. The phosphorylated form is active, participating in the contraction process. Removal of Ca^{2+} by the calcium pump and dephosphorylation of the light chains by a protein phosphatase restores the muscle to a resting state. The N-terminal part of the myosin light chain kinase binds to actin, while the catalytic domain is in the center of the protein. The C-terminal part binds to myosin, and this binding also has an activating effect.

Another protein, *caldesmon,* binds to smooth muscle actin and blocks actomyosin ATPase. It is present in smooth muscle in a ratio of actin:tropomyosin:caldesmon of ~14:2:1. Inhibition can be reversed by Ca^{2+}, but there is no agreement on the function of caldesmon. It is an elongated ~756-residue protein with N-terminal domain, which binds to myosin, and a C-terminal domain, which binds to actin, separated by a long helix. Caldesmon may be a substitute for troponin in a tropomyosin-type regulatory system, or it may promote actomyosin assembly. Another possibility is that it functions in a *latch state,* an energy-economic state of smooth muscle at low levels of ATP hydrolysis. In molluscan muscles Ca^{2+} binds to a myosin light chain and activates contraction directly. Some molluscan smooth muscles *(catch muscles)* also have a latch state, which enables these animals to maintain muscular tension for long periods of time, *e.g.,* holding their shells tightly closed, with little expenditure of energy. Catch muscles contain myosin plus a second protein, *catchin* which is formed as a result of alternative mRNA splicing. Catchin contains an N-terminal sequence that may undergo phosphorylation as part of a regulatory mechanism. However, recent experiments indicate that twitchin (see next paragraph), rather than catchin, is essential to the catch state and is regulated by phosphorylation. Regulation of the large groups of unconventional myosins is poorly understood. Phosphorylation of groups on the myosin heavy chains is involved in ameba myosins and others.

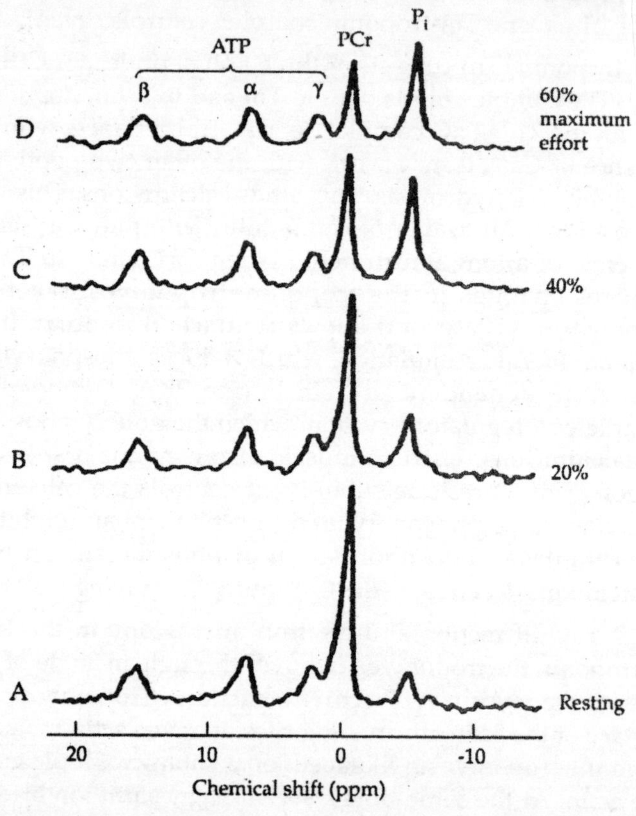

Fig. 5.22. Phosphorus-31 magnetic resonance spectra of wrist flexor muscles of the forearm of a trained long-distance runner at rest and during contraction at three different levels of exercise. Ergometer measurements indicating the percent of initial maximum strength (% max) were recorded over each 6-min period. Spectra were obtained during the last 3 min of each period. Times of spectral data collection: A, resting; B, 4 – 6 min; C, 10 – 12 min; and D, 16–18 min. The pH ranged from 6.9 to 7.0.

An unexpected aspect of regulation was discovered from study of the 40 or more genes of *Caenorhabditis elegans* needed for assembly and function of muscle. The mutants designated *unc*-22 showed a constant twitch arising from the muscles in the nematode's body. The gene was cloned using transposon tagging and was found to encode a mammoth 753-kDa 6839-residue protein which has been named *twitchin.* Twitchin resembles titin and like titin has a protein kinase domain, which is normally inhibited by the end of its peptide chain, which folds over the active site of the kinase. Perhaps the protein kinase activities of twitchin, titin, and related proteins are required in assembly of the sarcomere.

5. The Phosphagens

ATP provides the immediate source of energy for muscles but its concentration is only ~5 mM. *Phosphagens,* such as *creatine phosphate,* are also present and may

$$\text{Creatine-}P + \text{ADP} \longrightarrow \text{Creatine} + \text{ATP}$$

attain a concentration of 20 mM in mammalian muscle. This provides a reserve of high-energy phospho groups and keeps the adenylate system buffered at a high phosphorylation state ratio. The concentration of both ATP and creatine phosphate as well as their rates of interconversion can be monitored by ^{31}P NMR within living muscles. Phospho groups were observed to be transferred from creatine phosphate to ADP to form ATP with a flux of 13 mmol kg^{-1} s^{-1} in rat legs. The reverse reaction must occur at about the same rate because little cleavage of ATP to P_i was observed in the anaesthetized rats. Use of surface coils has permitted direct observation of the operation of this shuttle system in human muscle as well as in animal hearts. Only a fraction of the total creatine present within cells participates in the shuttle, however.

C. MOTION IN NONMUSCLC CELLS

Table 5.1. Some Actin-Binding Proteins

Function	*Name*
Bind and stabilize monomeric actin	Profilin[a, b, c, d, e, f, g, h]
	ADF/Cofilin[i, j, h, k]
	Thymosin[f, l]
Cap actin filament ends	CapZ[m, n]
barbed end	Fragmin[o, p]
pointed end	β-Actin[q]
	Tropomodulin[r]
	Arp2/3, a complex of seven polypeptides
Sever or dissociate actin filament	Gelsolin[s, t, u, v, w]
	Depactin
	Profilin[d, e]
	ADF/cofilin[h, k, x, y]
Crosslink actin filaments or monomers	
Tight bundles	Villin[c, z, aa]
Loose bundles	α-Actinin[bb]
Spectrin[bb]	
	Fascin[cc]
	MARCKS[a]
Network	Filamin[bb, c]
	Gelactins
Bind actin filaments to membrane	Talin[dd]
	"ERM" proteins[ee, ff]

At one time actin and myosin were thought to be present only in muscles, but we know now that both actin proteins of the myosin family are present in all eukaryotic cells. Ameboid movement, the motion of cilia and flagella, and movement of materials along microtubules within cells also depend upon proteins of this group.

1. Actin-based Motility

Ameboid movements of protozoa and of cells from higher organisms, the ruffling movements of cell membranes, phagocystosis, and the cytoplasmic streaming characteristic of many plant cells have all been traced to actin filaments or actin cables rather than to microtubules. Actin is one of the most abundant proteins in all eukaryotes. Its network of filaments is especially dense in the lamellipodia of cell edges, in microvilli, and in the specialized stereocilia and acrosomal processes. Actin filaments and cables are often formed rapidly and dissolve quickly. When actin filaments grow, the monomeric subunits with bound ATP are added most rapidly at the "barbed end" and dissociate from the filament at the "pointed end". The rate of growth may be ~20-200 nm/s, which requires the addition of 10 – 100 subunits/s. Various actin-binding proteins control the growth and stability of the filaments. The actin-related proteins *Arp2* and *Arp3,* as a complex Arp2/3, together with recently recognized *formins,* provide nuclei for rapid growth of new actin filaments as branches near the barbed ends.

Growth of the barbed ends of actin filaments is stimulated by phosphoinositides and by members of the Rho family of G proteins through interaction with proteins of the *WASp* group. The name WAS comes from the immune deficiency disorder Wiskott-Aldrich Syndrome. Yet another family, the *Ena/VASP* proteins, are also implicated in actin dynamics. They tend to localize at focal adhesions and edges of lamellipodia. Profilin stabilizes a pool of monomeric actin when the barbed ends of actin filaments are capped. However, it catalyzes both the addition of actin monomers to uncapped barbed ends and rapid dissociation of subunits from pointed ends, leading to increased *treadmilling.* Actin-severing proteins such as *actin depolsmerizing factor ADF* or *cofilin,* promote break-down of the filaments. Treadmilling in the actin filaments of the lamellipods of crawling cells or pseudopods of amebas provides a motive force for many cells ranging from those of *Dictyostelium* to human leukocytes. A series of proteins known as the *ezrin, radixin, moesin (ERM)* group attach actin to integral membrane proteins and may interact directly with membrane lipids. Bound ATP in the actin subunits is essential for polymerization, and excess ATP together with crosslinking proteins stabilize the filaments. However, when the bound ATP near the pointed ends is hydrolyzed to ADP the filaments become unstable and treadmilling is enhanced. Thus, as in skeletal muscle, ATP provides the energy for movement.

Bacteria also contain filamentous proteins that resemble F-actin and which may be utilized for cell-shape determination. Actin-based motility is used by some bacteria and other pathogens during invasion of host cells. It is employed by sperm cells of *Ascaris* and of C. *elegans,* which crawl by an ameboid movement that utilizes treadmilling of filaments formed from a motile sperm protein, which does not resemble actin. Cells are propelled on a glass surface at rates up to ~1 μm/s.

Various nonmuscle forms of myosin also interact with actin without formation of the myofibrils of muscle. ln most higher organisms nonmuscle myosins often consist of two ~200-kDa subunits plus two pairs of light chains of ~17 and 24 kDa each. These may form bipolar aggregates, which may bind to pairs of actin filaments to cause relative movement of two parts of a cell. Movement depending upon the cytoskeleton is complicated by the presence of a bewildering array of actin-binding proteins.

2. Transport along Microtubules by Kinesin and Dynein

Many materials are carried out from the cell bodies of neurons along microtubules in the axons, which in the human body may be as long as 1 m. The rates of this *fast axonal transport* in neurons may be as high as 5 μm/s or 0.43 m/day. The system depends upon ATP and kinesin and permits small vesicles to be moved along single microtubules. Movement is from the minus end toward the plus end of the microtubule. Slower *retrograde axonal transport* carries vesicles from the synapses at the ends of the axons back toward the cell body. This retrograde transport depends upon the complex motor molecule *cytoplasmic dynein* which moves materials from the plus end of the microtubule toward the minus end. In addition to these movements, microtubules often grow in length rapidly or dissociate into their tubulin subunits. Growth occurs at one end by addition of tubulin subunits with their bound GTP. The fast growing *plus-ends* of the microtubules are usually oriented toward the cell periphery, while the minus-ends are embedded in the *centrosome* or *microtubule-organizing center.* Just as with actin, in which bound ATP is hydrolyzed to ADP, the bound GTP in the β-tubulin subunits of microtubules is hydrolyzed to GDP decreasing the stability of the microtubules, a phenomenon described as *dynamic instability.* Various *microtubule-associated proteins* (MAPs) have strong effects on this phenomenon. The MAPs are often regulated by phosphorylation-dephosphorylation cycles involving serine/threonine kinases. Microtubules also undergo posttranslational alterations not seen in other proteins. These include addition or removal of tyrosine at the C terminus. Polyglycyl groups containing 3 – 34 glycine residues may be bound covalently to γ-carboxyl groups of glutamate side chains in both α- and β-tubulins. This stabilizes the microtubules and is important to the long-lived microtubules of the axonemes of flagella and cilia. Polyglutamyl groups of 6 – 7 glutamates are also often added.

Both dynein and several kinesins act as motors for formation of the spindle and for movement of chromosomes toward the minus ends of spindle microtubules during mitosis and meiosis. In the genome of *Saccharomyces cerevisiae* there is only one dynein gene, but genes for six different kinesin-type motor molecules are present. In higher organisms there may be even more genes for kinesins but there is apparently only one dynein in most species. *Axonemal dyneins* drive the motion of eukaryotic flagella and cilia. As with the cytoplasmic dyneins a complete molecule consists of two or three heavy chains with molecular mass -520 kDa, some localized in the dynein tail, and several lighter chains. Like myosin dynein is an ATPase.

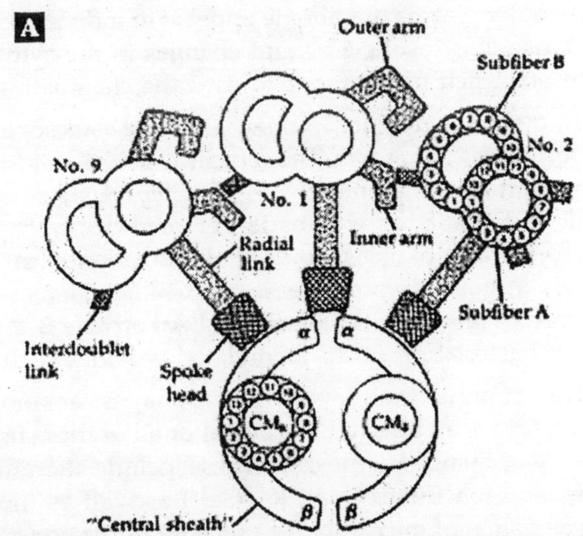

Fig. 5.23. (A) Diagram of a cross-sectional view of the outer portion of a lamellibranch gill cilium. This has the 9 + 2 axoneme structure (B). The viewing direction is from base to tip. (B, C) Thin-section electron micrographs of transverse (B) and longitudinal (C) sections of wild-type *Chlamydomonas* axonemes. In transverse section labels A and B mark A and B subtubules of microtubule doublets; oa, ia, outer and inner dynein arms, respectively; sp, spokes; cpp, central pair projections; bk, beaks.

3. Eukaryotic Cilia and Flagella

The motion of eukaryotic flagella

involves a sliding of the microtubular filaments somewhat analogous to the sliding of muscle filaments. Sliding between the outer doublet microtubules via their inner and outer arms (dynein compounds) is thought to provide the characteristic bending waves. The movement is powered by dynein and ATP hydrolysis. Force and displacement measurements made by optical trapping nanometry suggest that the characteristic rhythmic beating of flagella results from an oscillatory property of the dynein. The extremely complex structure of flagella is illustrated in Fig. 5.23. About 250 individual axonemal proteins have been detected in flagella of the alga *Chlamydomonas* and a large number of mutants with various defects in their flagella have been isolated. The radial spokes alone contain 17 different proteins. These spokes protrude at ~29-nm intervals while the dynein molecules lie between pairs of the outer microtubule doublets at ~24-nm intervals. The dynein "arms" protrude from the "A" microtubule of each outer doublet and make contact with the incomplete "B" microtubule of the next doublet. Although the shapes of the molecules are quite different, the basic chemistry of the ATPase activity of the dynein-microtubule system resembles that of actomyosin. However, the complexity of the dynein arms which exist in two types, inner and outer, suggests a complex contraction cycle.

4. Chemotaxis

As described already, the ameboid cells of the slime mold *Dictyostelium discoideum* are attracted to nutrients such as folic acid during their growth stage. Later, as the cells undergo developmental changes they become attracted by pulses of cyclic AMP. Occupancy of 7-helix receptors for cAMP on the outer plasma membrane appears to induce methylation of both proteins and phospholipids and a rise in cytosolic Ca^{2+} and changes in the cytoskeleton that result in preferential extension of the actin-rich pseudopods toward the chemoattractant.

In a similar manner, human ameboid leukocytes are attracted to sites of inflammation by various *chemotactic factors*. These include the 74-residue cleavage product C5a formed from the fifth component of complement, various *lymphokines* secreted by lymphocytes, and peptides such as VGSE and AGSE, as well as larger peptides released by mast cells, basophils, or stimulated monocytes and oxoicosenoids. Polymorphonuclear leukocytes, upon engulfing sodium urate crystals in gouty joints, release an 8.4-kDa chemotactic protein which may cause a damaging response in this arthritic condition. Leukotriene B is a potent chemotactic agent as are a series of specific bacterial products, formylated peptides such as N-formyl-MLF.

Neutrophils, monocytes, macrophages, eosinophils, basophils, and polymorphonuclear leukocytes are all affected by several or all of these factors. Binding to specific receptors results in a variety of changes in the cells. These include alterations in membrane potential, cyclic nucleotide levels, and ion fluxes (Na^+, K^+, Ca^{2+}) as well as increased methylation of specific proteins. A reorganization of microtubules and actin fibrils occurs, probably in response to an altered gradient of Ca^{2+}. The morphology of the cells changes, and they begin immediately to crawl toward the chemoattractants. It appears that these ameboid cells detect a gradient of attractant concentration between one end of the cell and the other, even though the anticipated difference may amount to only 0.1% of the total.

5. Other Forms of Movement

Movement is characteristic of life and is caused not only by motor proteins but by various springs and ratchets which may be energized in a number of ways. A striking example, which any one with a microscope and some fresh pond water can observe, is contraction of the stalk of protozoa of the genus *Vorticella*. Apparently first reported in 1676 by Leeuwenhoek the organism's

2 – 3 mm-long stalk contracts into a coiled spring when the animal is disturbed. Application of calcium ions causes contraction within a few milliseconds to ~40% of the original length. The process reverses slowly after a few seconds. Contraction is caused by a spring-like organ the *spasmoneme,* which is a bundle of short 2 nm-diameter fibrils inside the stalk. The fibrils are thought to be weakly cross-linked and held in the extended state by electrostatic repulsion between the negatively charged rods. Addition of Ca^{2+} neutralizes the charges permitting an entropy-driven collapse of the fibers.

Another remarkable example is extension of the acrosomal process from a sperm cell of the horseshoe crab *Limulus polyphemus* at fertilization. A bundle of actin filaments in a crystalline state lies coiled around the base of the nucleus. At fertilization the bundle uncoils and slides through a tunnel in the nucleus to form a 60 μm-long acrosomal process within a few seconds. The uncoiled bundle is also crystalline. The coiled bundle is apparently overtwisted and an actin crosslinking protein *scruin* mediates the conformational alteration that takes place. A somewhat related process may be involved in contraction of bacteriophage tails.

Some bacteria glide with a twitching movement induced by rapid retraction of pili. Another type of movement involves the pinching off of vesicles, *e.g.,* of clathrin-coated pits. This is a GTP-driven process that requires a mechanoenzyme called *dynamin.*

HERDITARY DISEASES OF MUSCLE

Considering the numerous specialized proteins in muscle it is not surprising that many rare hereditary muscle diseases are known. The most frequent and most studied of these is *Duchenne muscular dystrophy.* An X-linked disease of boys, it may not be recognized until two to three years of age, but victims are usually in a wheelchair by age 12 and die around age 20. Individual muscle fibers disintegrate, die, and are replaced by fibrous or fatty tissue. The disease strikes about 1 out of 3500 boys born. The less serious *Becker muscular dystrophy* arises from defects in the same gene but affects only 1 in 30,000 males, some of whom have a normal life span. Because of its frequency and the knowledge that the gene must lie in the X-chromosome, an intensive search for the gene was made. It was found in 1986 after a five-year search. This was the first attempt to find a faulty gene whose protein product was totally unknown. The project, which relied upon finding restriction fragment polymorphisms that could serve as markers in the genome, made use of the DNA from patients with a range of related diseases. The very rare female patients in whom the faulty gene had been translocated from the X-chromosome to an autosome also provided markers. DNA probes obtained from a young man with a large X-chromosome deletion that included genes related to retinitis pigmentosa and several other diseases provided additional markers. The result was a triumph of "reverse genetics" which has since been applied to the location of many other disease genes.

The muscular dystrophy gene may be the largest human gene. It consists of 2.3 million base pairs, which include 79 exons which encode a huge 427 kDa protein named *dystrophin.* The protein consists of four main domains. The N-terminal domain binds to actin and is homologous to β-actinin. The central domain is an elongated rod resembling spectrin. It contains repetitive coiled-coil segments and four hinge regions. The third domain is rich in cysteine and binds Ca^{2+}, while the fourth domain has a structure that is shared by several other proteins of the dystrophin family. Dystrophin is quantitatively a minor protein of muscle. It forms part of the cytoskeleton, lying adjacent to the sarcolemma (cell membrane) along with β-spectrin and vinculin.

While one end of the dystrophin molecule binds to actin filaments, the C-terminal domain associates with several additional proteins to form a *dystrophin glycoprotein complex* (see figure). Dystrophin is linked directly to the membrane spanning protein *β-dystroglycan,* which in the outer membrane surfaces associates with a glycoprotein *α-dystroglycan.* The latter binds to laminin-2, a protein that binds the cell to the basal lamina. Four other membrane-spanning proteins, α-, β-, γ-, and δ-*sarcoglycans,* are among additional members of the complex.

Patients with Duchenne muscular dystrophy are deficient not only in dystrophin but also in the dystroglycan and sarcoglycan proteins. Evidently, dystrophin is needed for formation of the complex which plays an essential role in muscle. In both types of X-linked muscular dystrophy there are individuals with a wide range of point mutations, frame-shift mutations, and deletions in the dystrophin gene. The essential function of dystrophin and associated proteins is uncertain but may be related to the linkage from actin filaments through the membrane to laminin. Individuals with Becker muscular dystrophy also have defects in dystrophin, but the protein is partially functional. Some other muscular dystrophies are caused by defects in the autosomal genes of any of the four sarcoglycan subunits or in that of laminin αl chain. The arrows in the accompany-ing drawing indicate chromosome locations of the sarcoglycan subunits, which are sites for mutations causing *limb girdle muscular dystrophy.*

Dystrophin, shorter isoforms, and related proteins are found in many tissties including the brain. One related protein, *utrophin* (dystrophin-related protein 1), is present in the neuromuscular junction of adult skeletal muscle. One approach to therapy of Duchenne muscular dystrophy is to stimulate a higher level of expression of the utrophin gene. Because the dystrophin gene is so large treatment by gene transfer is not practical, but transfer of parts of the gene may be. Myoblast transfer has not been successful, bxit new approaches will be devised.

Myotonic dystrophy is a generalized adult-onset disorder with muscular spasms, weakness, and many other symptoms. It is one of the *triple-repeat diseases.* The affected gene encodes a protein kinase of unknown function. The corresponding mRNA transcript has ~2400 nucleotides. The gene has a CTG repeat $(CTG)_n$ near the 3′-end with $n < 30$ normally. For persons with the mildest cases of myotonic dystrophy n may be over 50 while in severe cases it may be as high as 2000. As in other triple-repeat diseases the repeat number tends to increase in successive generations of people as does the severity of the disease.

For some individuals, muscular dystrophy causes no obvious damage to skeletal muscle but affects the heart producing a severe *cardiomyopathy,* and persons with Duchenne muscular dystrophy often die from heart failure. Heart failure from other causes, some hereditary, is a major medical problem, especially among older persons. Hereditary forms are often autosomal dominant traits that may cause sudden death in young persons. At least seven genes for cardiac sarcomeric proteins including actin,[z] myosin, both heavy and light chains, three subunits of troponin, tropomyosin, and protein C may all carry mutations that cause cardiomyopathy.

A hereditary disease common in Japan results from a defect in migration of neurons and is associated with brain malformation as well as muscular dystrophy. In *nemaline myopathy* a defect in nebulin leads to progressive weakness and often to death in infancy. A characteristic is the appearance of "nemaline bodies" or thickened Z-discs containing Z-disc proteins. Some hereditary diseases involve nonmuscle myosins. Among these is *Usher syndrome,* the commonest cause of deaf-blindness. The disease, which results from a defect in the myosin VIA gene, typically causes impairment of hearing and retinitis pigmentosa.

MALIGNANT HYPERTHERTHERMIA AND STRESS-PRONE PIGS

Very rarely during surgery the temperature of a patient suddenly starts to rise uncontrollably. Even when heroic measures are taken, sudden death may follow within minutes. This *malignant hyperthermia syndrome* is often associated with administration of halogenated anesthetics such as a widely used mixture of halothane (2-bromo-2-chloro-1, 1, 1-trifluoroethane) and succinylcholine. There is often no warning that the patient is abnormally sensitive to anesthetic. However, development of an antidote together with increased alertness to the problem has greatly decreased the death rate. Nevertheless, severe damage to nerves and kidneys may still occur. Biochemical investigation of the hyperthermia syndrome has been facilitated by the discovery of a similar condition that is prevalent among certain breeds of pigs. Such "stress-prone" pigs are likely to die suddenly of hyperthermia induced by some stress such as shipment to market. The sharp rise in temperature with muscles going into a state of rigor is accompanied by a dramatic lowering of the ATP content of the muscles.

The problem, both in pigs and in humans susceptible to malignant hyperthermia, was found in the Ca^{2+} release channels (ryanodine receptors). Study of inheritance in human families together with genetic studies in pigs led to the finding that the stress-prone pigs have cysteine replacing arginine 615 in the Ca^{2+} channel protein. This modification appears to facilitate opening of the channels but to inhibit their closing. A similar mutation has been found in some human families in which the condition has been recognized. However, there is probably more than one site of mutation in humans. Similar mutations in the nematode *C. elegans* are being investigated with the hope of shedding light both on the problem of hyperthermia and on the functioning of the Ca^{2+} release channels.

BOX-19-C ACTIN-BASED MOTILITY AND BACTERIAL INVASION

Listeria monocytogenes is a dangerous food-borne bacterium that has become a major problem in the United States. This is one of the best understood *intracellular* pathogens. It is able to enter cells, escape from phagocytic vesicles, spread from cell to cell, and cross intestinal, blood-brain, and placental barriers. Within cells these bacteria move using actin-based motility. Actin subunits polymerize at one end of a bacterium leaving *"comet tail"* of crosslinked fibrous actin behind (see micrographs). Actin polymerization occurs directly behind the bacteria with subunits of monomeric actin adding to the fast growing *"barbed end"* of the actin strands. Growth has been described as a "Brownian ratchet." Continual Brownian movement opens up spaces behind the bacteria, spaces that are immediately filled by new actin subunits. This provides a propulsive force adequate to move the bacteria ahead at velocities of about 0.3 μm/s.

Polymerization of actin is induced by interaction of a dimer of a 610-residue bacterial protein *Act A* with proteins of the host cell. Act A is a composite protein with an N-terminal region that protrudes from the bacterial cell, a central region of proline-rich repeats that appear to be essential for recognition by host cells, and a C-terminal hydrophobic membrane anchor. There are also regions of close sequence similarity to the human actin-binding proteins vinculin and and zyxin. The number of host proteins needed in addition to monomeric actin are: the two actin-related proteins, *Arp2* and *Arp3* which stimulate actin polymerization and branching; *ADF/cofilin,* which increases the rates, both of growth at the barbed ends and dissociation from the pointed ends of the filaments; and *CapZ* which caps barbed ends. The need for ADF/cofilin and Cap Z seems paradoxical. Cap Z may cap mostly older and slower growing filaments, restricting rapid filament assembly to the region closest to the bacterium. The need for ADF/cofilin is unclear. Growth rates are also enhanced by the human protein called *VASP* (vasodilator-stimulated phosphoprotein). The proline-rich region of the bacterial ActA may bind to VASP to initietate polymerization. Both profilin and the crosslinking protein α-actinin also stimulate comet tail growth. Myosin does not participate in actin-based motility, but the hydrolysis of ATP drives the process through its link to actin polymerization.

Although *Listeria* has been studied most, actin-based motility is employed by other pathogens as well, *e.g., Shigella flexneri* (the dysentery bacterium), *Rickettsia,* and vaccinia virus. Although enteropathogenic *E. coli* do not use this method of movement, they induce accumulation of actin beneath the bacteria. They also promote formation of actin-rich adherent pseudopods and highly organized cytoskeletal structures that presumably assist the bacteria in entering a cell.

CHAPTER

6

Proteome in Liver Disease

It is widely recognized that a key component in the development of alcohol-induced liver disease is the diminished capacity of liver to generate and maintain sufficient levels of ATP. This resulting decrease in bioenergetic capacity is thought to contribute to decreased hepatocyte viability and ultimately the pathological changes that occur in alcohol-dependent hepatotoxicity. The importance of mitochondrial dysfunction in this disease process has long been appreciated due to chronic alcohol-mediated defects on the oxidative phosphorylation system, which leads to decreased ATP synthesis.

Moreover, evidence indicates that these chronic alcohol-induced alterations in mitochondria structure and function also contribute to increased production of reactive oxygen and nitrogen species in mitochondria. This increase in mitochondrial oxidant production could result in post-translational oxidative modifications and inactivation of mitochondrial macromolecules, particularly proteins, thereby further compromising mitochondria function in the chronic alcohol consumer. Although the mechanisms responsible for alcohol-dependent mitochondrial dys-function have been studied, the impact of chronic alcohol consumption on the overall content of mitochondrial proteins, that is, the global mitochondrial proteome, has only recently been investigated.

Early studies by Cunningham and colleagues demonstrated that chronic alcohol consumption decreases the synthesis of the 13 mitochondrial-encoded proteins that are components of complexes I, III, IV, and V *(6,7)*, as the result of defects in mtDNA and a decrease in functional mitochondrial ribosomes. It is important to note, however, that there are well over 600 proteins that comprise the mitochondrial proteome, and that close to 100 of these are components of the oxidative phosphorylation system, most of which are encoded by the nuclear genome. Because very little information is available on the effects of chronic alcohol consumption on the total mitochondrial proteome, especially effects on nuclear encoded proteins, we have used two complementary proteomics approaches, two-dimensional isoelectric focusing (IEF)/SDS-PAGE and blue native-PAGE (BN-PAGE) to improve our understanding of the impact these changes have on mitochondrial functionality and their contribution to the development of alcohol induced liver pathology.

MATERIALS

Isoelectric Focusing

1. Buffer for rehydration of IEF gel strips: 7*M* urea, 2*M* thiourea, 2% CHAPS, 0.5% *N*-dodecyl-β-D maltoside (*i.e.,* lauryl maltoside), and 0.002% bromophenol blue. Store at –20°C in 1.0-mL aliquots.

2. Equilibration buffer for IEF gel strips: 6M urea, 2.0% sodium dodecyl sulfate (SDS), 0.375*M* Tris-HCl, 20% glycerol, and 0.002% bromophenol blue, pH 8.8 (with HCl). Store at –20°C in 2.0-mL aliquots. Before using, warm to 37°C to redissolve urea into solution.
3. DTT stock solution: 1*M* DTT in water. Store at –20°C in 50-µL aliquots. Do not refreeze and reuse DTT aliquots for future experiments.
4. Ultrapure, low melting-temperature agarose (Aqua*Por*™ LM, National Diagnostics, cat. no. EC-204) for sealing IEF gel strips onto SDS-polyacrylamide gel electrophoresis (PAGE) gels: 1.0% agarose (*w*/*v*) in IX SDS-PAGE running buffer. Make fresh on day of experiment.
5. The Invitrogen ZOOM IPG runner (cat. no. ZM0001) is used to perform IEF in combination with Invitrogen ZOOM strips (cat. no. ZM0011, pH 3 – 10), ZOOM IPG Runner Cassettes (cat. no. ZM0003), and the ZOOM Dual Power Supply (cat. no. ZP10002). Please refer to manufacturer's manual for additional details on setup and use.
6. The ampholyte carriers used for IEF are Ampholines Electrophoresis Reagent (Sigma, cat. no. A5174, pH 3 – 10). Other carrier ampholines or ampholytes can be substituted.

Blue Native-polyacrylamide Gel Electrophoresis

One-dimensional Blue Native-polyacrylamide Gel Electrophoresis

1. Cathode buffers: a) High-blue cathode buffer: 50 mM tricine, 15mM BisTris, and 0.02% Coomassie brilliant blue G-250, pH 7.0. b) Low-blue cathode buffer: 50 mM tricine, 15 mM BisTris, and 0.002% Coomassie brilliant blue G-250, pH 7.0. Store at 4°C.
2. Anode buffer: 50 mM BisTris, pH 7.0. Store at 4°C.
3. 3X gel buffer: 1.5M aminocaproic acid, 150 m*M* BisTris, pH 7.0. Store at room temperature.
4. Extraction buffer: 0.75 M aminocaproic acid, 50 mM BisTris, pH 7.0. Store at 4°C.
5. Coomassie brilliant blue G-250 suspension: 0.5 M aminocaproic acid and 5% Coomassie brilliant blue G-250, pH 7.0. Store at 4°C.
6. Lauryl maltoside solution: 10% (w/v) *N* dodecyl-(β-D-maltoside in water. Store at –20°C.
7. Molecular weight standards for 1-D BN-PAGE: High-molecular weight native marker kit (Amersham Biosciences, cat. no. 17–0445–01). Dissolve contents of 1 vial into 200 µL of extraction buffer, 25µL lauryl maltoside, and 12 µL of Coomassie brilliant blue. Store at –20°C.

Two-dimensional Blue Native-polyacrylamide Gel Electrophoresis

1. Cathode buffer: 100m*M* Tris, 100m*M* tricine, and 0.1% SDS, pH 8.25. Store at room temperature.
2. Anode buffer: 200m*M* Tris, pH 8.9 with HCl. Store at room temperature.
3. 2-D BN-PAGE Gel buffer: 3*M* Tris-HCl and 0.3% SDS, pH 8.45 with HCl. Store at room temperature.
4. SDS/β-Mercaptoethanol solution: 20µL of β-mercaptoethanol, 200µL of 10% SDS into 1780µL of ultrapure H_2O. Make fresh on day of experiment.
5. Agarose solution for sealing 1-D BN-PAGE gel strips onto SDS-PAGE gels: 100 mg of low melting temperature agarose (Aqua Por™ LM, National Diagnostics, cat. no. EC-204), 1 mL 10% SDS, 100µL of β-mercaptoethanol into 9.0 mL of ultrapure H_2O. Make fresh on day of experiment.

Total Protein Staining for Gels

1. Coomassie blue stain: 0.3 g of Coomassie blue R-250, 100mL of glacial acetic acid, 250mL of isopropanol, and 650mL of ultrapure H_2O. Use 100mL per one mini-gel. Stain gel overnight and do not reuse stain. Gels are destained the next day using 10% glacial acetic acid solution. A small piece of a paper towel is added to the gel container to absorb the Coomassie blue as it "leaches" from gel. Change blue-stained paper towels every 2 – 4 h until clear background on gels is achieved.
2. SYPRO® Ruby Protein Gel Stain (Invitrogen, cat. no. S-12000): After electro-phoresis, fix gel in 100 mL of a 40% methanol and 10% glacial acetic acid solution for 1 h. After fixation step, discard fix solution and incubate gel with 60 mL SYPRO Ruby solution overnight. After staining gel, discard SYPRO Ruby solution and destain gel in a 10% methanol/7% glacial acetic acid solution for 1 – 2 h before visualizing gels.

   It is recommended that gels are destained for at least 16 – 24 h using multiple changes in destain solution to minimize background (*i.e.,* decrease speckling) of gels. A number of imaging platforms have been validated for visualizing SYPRO Ruby-stained gels. Please refer to the SYPRO Ruby product information sheet provided by Invitrogen to obtain the excitation sources and emission filters for optimal visualization of protein in gels.

METHODS

It is hypothesized that chronic alcohol-induced alterations to the mitochondrial proteome negatively affects mitochondrial bioenergetics and signaling, which contribute, in part, to the development of liver pathology. To address this question, our laboratories have begun to profile the mitochondrial proteome using both conventional 2-D IEF/SDS-PAGE and BN-PAGE. Recent advancements in the field have now made it possible to identify post-translational modifications to mitochondrial proteins by ROS/RNS, identify defects in the assembly of multi-protein complexes in mitochondria, and separate the more hydrophobic respiratory complex proteins of the inner membrane using newly refined gel electrophoresis methods. A scheme depicting our strategy for detecting alterations to the mitochondrial proteome in models of chronic alcohol-induced liver injury.

Briefly, mitochondria isolated from the liver of control and alcohol-fed animals can be subjected to conventional 2-D IEF/SDS-PAGE and BN-PAGE. The high-resolution protein "maps" generated via these electrophoresis methods can then be analyzed for changes in protein abundance or a variety of well-defined ROS/ RNS-mediated post-translational modifications using immunoblotting techniques. Finally, proteins of interest can then be identified using several mass spectrometry techniques. Proteomic studies from our laboratories are considered significant in that novel changes in several key energy metabolism pathways, including fatty acid oxidation, TCA cycle, and oxidative phosphorylation system were found to be altered in mitochondria isolated from animals exposed to ethanol chronically. These alterations were in response to alcoholic-mediated metabolic stress and correlated to those pathways with a direct impact on the etiology of disease. A detailed discussion of these techniques is provided in the sections to follow.

Preparation of Mitochondrial Samples

1. Mitochondria are prepared from fresh liver tissue by standard differential centrifugation techniques and should exhibit tightly coupled respiration with respiratory control ratios in the range of 4 – 8. Moreover, mitochondrial protein yield should average 25 – 30 mg of mitochondrial protein/g wet weight liver and cytochrome *c* oxidase and citrate synthase

activities should be determined as additional markers of mitochondria purity and yield. It is imperative that mitochondria used for proteomics studies are isolated by methodologies that result in functional and pure mitochondrial preparations.

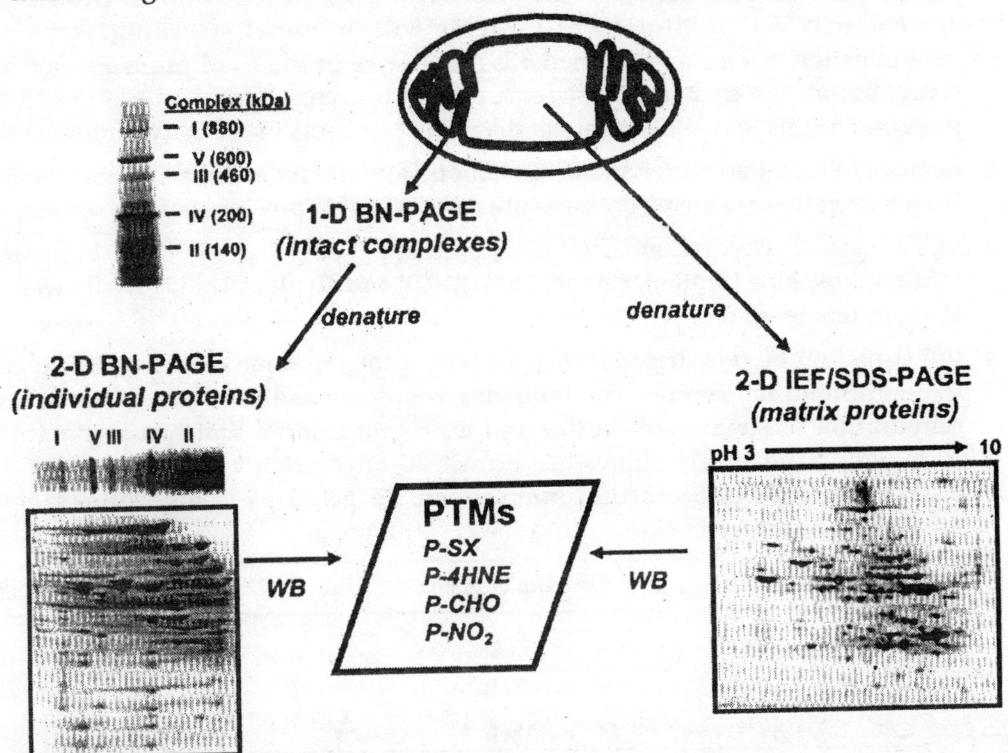

Fig. 6.1. Analysis of the mitochondrial proteome using 2-D IEF/SDS-PAGE and BN-PAGE proteomics. 2-D IEF/SDS-PAGE is used to generate high resolution maps of the soluble matrix proteins (right panel), whereas 1-D and 2-D BN-PAGE is used to define changes to respiratory chain proteins (left panels). The right-side panel illustrates a 2-D IEF/SDS-PAGE separation of proteins from whole liver mitochondria using the Invitrogen IPGRunner® System in combination with pH 3-10 IEF gel strips, whereas the left side panels demonstrate the BN-PAGE separation of inner membrane proteins. Roman numerals are used to identify the five oxidative phosphorylation complexes in the BN-PAGE gels. During nondenaturing 1-D BN-PAGE, the five complexes remain intact (first step) whereas under the denaturing conditions used for 2-D BN-PAGE the individual proteins of each complex are resolved by size (second step). Thus, those proteins that comprise each complex are aligned vertically within the gel. Note that both gel platforms can be subjected to image analysis to look for changes in protein abundances or subjected to western blotting (WB) to reveal post-translational modification (PTMs) by reactive oxygen and nitrogen species, as well as electrophilic lipids, as a number of antibodies have been developed against these modifications. P-SX, protein thiol modifications; P-4HNE, 4-hydroxynonenal protein adduct; P-CHO, protein carbonyls; P-NO_2 3-nitrotyrosine protein modification.

After isolation and determination of the mitochondrial protein concentration, the appropriate volume to achieve 1.0 mg of protein is pippetted into a 1.5-mL microcentrifuge tube and mitochondria are spun at 15,000 × g for 10 min to form a 10 mg protein pellet. The supernatant is removed and pellets are stored and dried at –80°C before used in experiments. Mitochondria samples can then be resuspended in the appropriate buffer for BN-PAGE proteomic experiments. For IEF focusing studies mitochondria are typically stored in suspension at a concentration of at least 10 mg of protein per milliliter.

Running One-dimensional Isoelectric Focusing Gels

1. Mitochondria samples are thawed and kept on ice immediately before use. The optimal protein concentration for a 2-D IEF/SDS-PAGE gel is 50 – 200 mg of protein per gel to achieve maximal resolution of proteins with minimal streaking. Repeat protein determination using standard methods (*i.e.,* Lowry or Bradford protein assay) as protein concentration can change over time. A protein concentration of no less than 5 mg/mL is preferred as this will allow minimal dilution of reagents in IEF rehydration buffer.
2. Remove rehydration buffer and DTT aliquots from freezer and thaw to room temperature. Do not keep these reagents on ice as urea and DTT will precipitate from solution.
3. Into 1.0 mL of rehydration buffer add 10 μL of pH 3 – 10 ampholines, 10 μL of 200 mM tributylphosphine (BioRad, cat. no. 163 – 2101), and 40 μL of 1M DTT. Mix well and keep at room temperature.
4. IEF strips can be rehydrated overnight with a total volume of 150 – 160 μL containing the mitochondrial sample. The following example conditions are given below for the rehydration of strips with buffer and a protein sample that has a concentration of 10 mg/mL. Samples are allowed to extract in rehydration buffer for at least 30 min at room temperature before loading samples onto IEF gel strips. Gently vortex samples every 5 – 10 min during extraction.

Protein amount	*Quantity of sample*	*Quantity of rehydration buffer*
50 μg	5μ L	155μ L
100 μg	10μ L	150μ L
200 μg	20μ L	140μ L

5. Following the procedure described by the specific manufacturer for the instrumentation being used, slowly load 160 μL of sample onto the strip. Samples will load along the entire lane by capillary action. Pipet slowly to minimize the introduction of bubbles into the gel lane as this could result in uneven rehydration of the IEF strips and loss of sample. Load all samples into cassette lanes before removing IEF gel strips from freezer.
6. After loading samples, remove the IEF strip pack from the freezer and keep strips on ice to minimize condensation. For each sample, carefully remove an IEF strip from the plastic backing, holding it with clean forceps by the marked negative (–) end. Holding the IEF strip from the (–) end and with the gel side up, gently slide the IEF gel strip into the cassette well until the strip reaches the (+) end of the cassette.
7. After rehydration of the IEF gel strips is complete, assemble the electrophoresis apparatus according to the manufacturer's instructions.
8. Perform IEF gel electrophoresis using the following conditions: 175 V for 20 min, ramp to 2000 V for 45min, 2000 V for 30 min, ramp down to 500 V for 30 min, and 500 V for 2 h. As IEF proceeds, the bromophenol blue dye front will migrate from the top to the bottom of the IEF strips and a green/blue/yellow band will appear at the anode (+) after IEF.
9. After IEF, seal cassettes in plastic wrap and freeze at –80°C until ready to perform second dimension SDS-PAGE.

Running 2-D IEF/SDS-PAGE Gels

1. For second dimension SDS-PAGE gels a wide range of equipment is commercially available that is compatible with the first dimension IEF. Homogenous (10 or 12%) or gradient (8 – 15%) acrylamide gels can be used for 2-D SDS-PAGE. It is important that spacer plates are selected that provide sufficient width for inserting the IEF gel strips.
2. Remove IEF gel strips from freezer and allow IEF gel strips to thaw for a few minutes before beginning the equilibration step. For equilibration, prepare 1.0-mL equilibration buffer per IEF gel strip. Remove equilibration buffer and DTT aliquots from freezer and thaw at room temperature. To 1.0 mL of equilibration buffer, add 50 μL of 1 *M* DTT.
3. To remove the IEF gel strips from cassette, remove plastic covering gently and using forceps place each strip into a 15-mL conical tube, gel side up, and cover with 1.0 mL of equilibration buffer. Gently rock strips with equilibration buffer for 15 min. During the equilibration step, prepare the agarose solution and keep warm. The aga-rose solution is used to "seal" the IEF gel strips to the top of SDS-PAGE gel.
4. After equilibration, remove strips from conical tubes, cut small plastic end from the anode (-) side, rinse strips in IX SDS-PAGE running buffer, and slide the IEF gel strip between the plates of the SDS-PAGE gel. Gently pipet the warm agarose solution over strip to "seal" the IEF gel strip on top of the SDS-PAGE gel. Do not allow bubbles to form between the IEF gel strip and the top of the SDS-PAGE gel. Slide a "tooth" from a 1.5-mm gel comb between glass plates at the (–) end of the gel strip. This will serve as the "well" for molecular weight markers. Gels are placed upright for 15 – 30 min to allow the agarose to set before beginning electrophoresis.
5. Once the agarose has set, gently remove the gel comb "tooth" and load broad range molecular weight markers to the well. Assemble the gel apparatus per the manufacturer's instructions, fill inner and outer chambers with standard 1 X SDS-PAGE gel electrophoresis running buffer, and run gels for l – 2 h at 100 V or until the dye front reaches the bottom of gel.
6. After electrophoresis, gels can be stained for total protein using conditions described above (Coomassie Blue or SYPRO Ruby) or subjected to western blotting. Typically, we stain the 2-D IEF/SDS-PAGE gels with SYPRO Ruby stain to increase sensitivity of low abundance proteins. An example of a 2-D IEF/SDS-PAGE gel stained with SYPRO Ruby stain is presented in Fig. 6.1.

Running 1-D BN-PAGE Gels

1. BN-PAGE is a specialized method of electrophoresis that facilitates the high-resolution analysis of membrane protein complexes in their native state from biological membranes. A detailed discussion of the development and theory of BN-PAGE for analysis of the oxidative phosphorylation system proteins is presented in the seminal papers of Schagger and von Jagow.
2. For BN-PAGE, the procedure more recently described by Brookes et al. is used with the 1-D BN-PAGE, typically performed with 5 – 12% or 5 – 16.5% gradient gels with 1.5-mm spacer plates. Recipes for gels are as follows

	Light-5%	Resolving Gel solutions Heavy-12%	or Heavy-16.5%	Stacking gel solution 4%
Protogel	0.66mL	1.60mL	1.91mL	0.60mL
Water	1.97mL	0.56mL	–	2.40mL
3XGel Buffer	1.34mL	1.34mL	1.16mL	1.50mL
Glycerol	26.0µ	26.0µL	23.0µL	70.0µL
TEMED	4.0µL	4.0µL	3.50µL	9.0µL

3. On the day of the experiment, remove the 1.0 mg mitochondrial pellets from the –80°C freezer and keep on ice. Add 100 µL of extraction buffer and 12.5 µL of 10% *N*-dodecyl-β-D-maltoside to each 1.0 mg of mitochondrial pellet. Gently resuspend mitochondrial protein by pipetting to extract and dissociate the protein. Place samples on ice for 30 – 60 min with gentle vortexing every 5 – 10 min. After this extraction step, centrifuge samples at 15,000 × g for 5 min at 4°C to pellet any non-dissolved material, remove supernatant to a fresh microcentrifuge, and then determine the protein concentration of the extract. A typical protein concentration of sample prepared by the procedure described here is 5 – 8 mg/mL.
4. After determining the protein concentration in each sample, add 6.3µL of the cold Coomassie brilliant blue G-250 suspension (Solution #5,) to each tube of mitochondrial extract and gently vortex. Load samples immediately onto gel for 1-D BN-PAGE.
5. After assembling the gel apparatus per manufacturer's directions, fill wells of the stacking gel with cold Hi-Blue cathode buffer, and load 75 – 250µg of protein per lane. Fill the inner (*i.e.* upper) buffer chamber with cold Hi-Blue cathode buffer and the outer chamber with cold anode buffer Electrophoresis is performed in the cold room at 40V for 1 h or until samples have migrated into the resolving gels. At this time the cathode buffer is changed to cold Low-Blue cathode buffer and electrophoresis is continued for an additional 3 – 4 h or until dye front reaches the bottom of the gel.
6. At this point, gels can be stained with Coomassie blue to visualize the content of the intact oxidative phosphorylation complexes in samples or gels can be processed for 2-D BN-PAGE by the procedures described below. A representative 1-D BN-PAGE gel is shown in the upper left panel of Fig. 6.1.

Running 2-D BN-PAGE Gels

1. To resolve the individual polypeptides that comprise each oxidative phosphorylation complex, the intact gel lane containing all five complexes is cut from the 1-D gel, rotated 90°, and laid on top of a Tris-Tricine/SDS-PAGE gel. The recipe for preparing one, 1.5-mm thick 2-D BN-PAGE gel is as follows leaving a 1.0-cm gap at the top of the gel.

2-D BN-PAGE Gel Buffer	2.98 mL
Protogel	2.98 mL
H_2O	2.31 mL

Glycerol	0.72 mL
10% AMPS	60.0 µL
TEMED	6.00 µL

2. To apply the 1-D BN-PAGE gel strip on to the top of the SDS-PAGE resolving gel, raise-up the gel plates in the clamps and place the top, that is, a 1– to 2-cm gap on an angled hot plate or heat block. Pour approximately 4 – 5 mL of hot agarose into the gap at the top of the gel sandwich, and using the back plate as a "staging area" and the hot agarose as a "lubricant", gently slide the gel lane down between the plates until the strip lays on top of the SDS-PAGE resolving gel. Once the gel is in place, insert a single "tooth" of a gel comb on the end to serve as a well for molecular weight markers.
3. Remove the gel plate assembly from the hot plate and allow the agarose to set for 20 – 30 min. Excess agarose is then removed from the top of the gel with a scalpel blade and then the gel is overlaid with a thin layer of the SDS/β-mercaptoethanol solution every 5 – 10 min for 30 – 45 min to denature proteins.
4. Fill the inner chamber with Cathode buffer (Solution #1,) and the outer chamber with the anode buffer load molecular weight markers into well, and run gels at 30V for 45 min followed by 110V for 1.5 to 2 h.
5. After electrophoresis, gels can be stained for total protein using the conditions (Coomassie Blue or SYPRO Ruby) or subjected to western blotting. Typically we stain 2D BN Page gels with Coomassie Blue stain.

Image Analysis of Gels

Instructions for image analysis of gels are detailed in the following publications from our laboratories. In brief, scanned TIFF images for 1-D and 2-D BN-PAGE gels can be analyzed using either Scion Image Beta 4.02 (Scion Corp) or Quantity One (BioRad Laboratories) using the directions provided for each software program. For 2-D IEF/SDS-PAGE image analysis, gels are scanned, saved as TIFF files, and analyzed for differences in protein abundance between control and ethanol samples using PDQuest Image Analysis software (BioRad Laboratories).

Mass Spectrometry for Identification of Proteins

Proteins found by image analysis to change in relative abundance in response to a treatment are cut from gels and subjected to matrix-assisted laser desorption ionization time-of-flight (MALDI-TOF) mass spectrometry by standard methodologies as described in to identify proteins. The peptide masses identified using MALDI-TOF mass spectrometry are entered into the MASCOT search engine (see www.matrixscience.com) and the NCBI data base is searched to match the tryptic peptide fingerprint with a parent protein.

ANALYSIS DISEASES

Mitochondrial ribosomes (mitoribosomes) are responsible for the translation of 13 proteins coded for by the mitochondrial genome. These proteins (ND1, ND2, ND3, ND4, ND4L, ND5, ND6, Cyt b, COI, COII, COIII, ATPase6, and ATPase8) are required for the successful assembly of the respiratory complexes NADH dehydrogenase, ubiquinol: cytochrome c oxidoreductase, cytochrome c oxidase, and ATP synthase, respectively. Accurate assembly and subsequent activity of mitoribosomes is therefore essential for the maintenance and correct functioning of the electron transport chain.

The rat liver mitochondrial ribosome has a molecular mass of approximately 3.57MDa, dimensions of 26.2 nm × 23.6 nm, and a sedimentation coefficient of 55S. It has a chemical composition of 75% protein and 25% RNA and comprises a large 39S subunit (LSU) and a small 28S subunit (SSU). Subunits are assembled inside the mitochondrial matrix from 12S (SSU) and 16S (LSU) rRNA, encoded for by the mitochondrial genome, and ribosomal proteins, encoded for by the nucleus and imported into the mitochondrion. One of the earliest observations seen in the liver during chronic ethanol feeding is decreased mitochondrial protein synthesis.

Although a number of potential molecular mechanisms may explain this decrease, one specific mechanism involves the impaired assembly of mitoribosomes. Investigations conducted by the Cunningham lab have revealed that mitoribosomes isolated from young male rats (200 g) fed ethanol (Lieber-DiCarli diet) for 31 d exhibit decreased sedimentation rates through sucrose density gradients when compared with their paired controls. Further, sedimentation velocity experiments reveal a significant decrease in the average sedimentation coefficient for the intact ethanol mitoribosome (52.2, ethanol-fed; 53.7, control) as well as for the small ribosomal subunit (27.0, ethanol-fed; 28.3, control).

In addition, mitoribosomes isolated from ethanol-fed animals exhibit significant decreases in their translational diffusion coefficients ($1.02 \times 10^{-7} cm^2 s^{-1}$, ethanol-fed; 1.10, control) and larger hydrodynamic diameters (42.1 nm, ethanol-fed; 39.1, control). These ethanol-elicited alterations in the physicochemical properties of the hepatic mitoribosome are accompanied by significant amounts of disassociation of the intact 55S monosome (8 – 14%) into its constituent subunits. Further, in vitro studies on the translation capacity of mitoribosomes have revealed that while total translation activity is depressed by 29% in the ethanol-fed animals relative to the controls, no difference is detected between the two treatment groups in the activity of the intact monosomes which suggests that the decrease in translation activity observed in ethanol-fed animals is caused by either increased disassociation or decreased association of functional ribosomes rather than impaired activity of intact monosomes.

Whether this means that ethanol feeding induces increased disassociation or decreased association of mitoribosomes in vivo or whether it causes a more relaxed and loosely assembled monosome which becomes disassociated upon isolation has yet to be determined. The sedimentation properties rnitochondrial ribosomes are affected not only by the size of the ribosomes but also by their shape. Any ethanol-elicited alterations in the ratios of constitutive mitochondrial ribosomal proteins (MRPs) may lead to a change in the shape of the ribosome and consequently a decrease in sedimentation rates. Additionally, altered MRP levels may lead to formation of inaccurately assembled ribosomes that are easily dissociated into subunits or unable to bind to functionally important docking sites on the inner mitochondrial membrane (IMM). Further, it is unclear as to whether any of the MRPs are susceptible to the kinds of protein damage commonly incurred during chronic ethanol feeding, that is oxidative modification, nitrosative damage and formation of adducts such as acetaldehyde (AA), malondialdehye-acetaldehyde (MAA), and the hydroxyethyl radical (HER). Ethanol-elicited modifications of MRPs may result in an alteration in their binding to other MRPs, rRNA or IMM components, as well as alter their rates of degradation. In addition to altering the overall structure of the mitochondrial ribosome, ethanol feeding has been shown to significantly decrease the levels of a specific population of MRPs. The reasons for the ethanol-elicited decreases in a specific population of MRPs as opposed to a global down-regulation are unclear; however, a recent proteomics study by Chevallet et al.

has demonstrated the selective depression of nine MRPs upon depletion of mitochondrial DNA (mtDNA) in 143B osteosarcoma cells. Our laboratory has shown that mtDNA depletion is a common occurrence during chronic ethanol feeding.

In this chapter, we detail methodologies for the isolation of mitochondrial ribosomes from hepatic mitochondria, determination of their activities, extraction of MRPs from the ribosomes and two-dimensional nonequilibrium pH gradient electrophoretic (NEPHGE) analysis of MRPs. Separated proteins can then be excised from the gels and subjected to mass spectrometry or transferred to nitro-cellulose membranes and probed for oxidative and nitrosative modifications by western blotting.

MATERIALS

Isololation of Mitochondrial Ribosomes from Mitochondria

1. Isolation buffer: 0.25 *M* sucrose and 2 *mM* HEPES, pH 7.4 (pH with *6M* KOH). Store at 4°C.
2. RC DC Protein Assay Reagents (BioRad, cat. no. 500 – 0120).
3. Digitonin (solid, free of water-insoluble constituents, Sigma cat. no. D-141).
4. Buffer A: 0.26M sucrose, 40mM KC1, 15mA/ $MgCL_2$, 14mM Tris-HCl, pH 7.5, and 0.8 mM ethylene diamine tetraacetic acid (EDTA); Store at 4°C. Just before use add β-mercaptoethanol to a final concentration of 5 mM.
5. Buffer B: 100m*M* KC1, 20 m*M* $MgCL_2$, and 20 mM triethanolamine pH 7.5. Store at 4°C. Just before use, add β-mercaptoethanol to a final concentration of 5 m*M*.
6. Puromycin dihydrochloride (solid, Sigma, cat. no. P 7255).
7. Magnesium-deficient buffer B: buffer B without the $MgCl_2$ but including 1 m*M* EDTA.

Ribosome Activity Assay

1. Ribosome activity buffer: 50 m*M* KC1, 50 m*M* Tris-HCl, pH 7.8, and 20 m*M* $MgCl_2$. Store at 4°C.
2. Transfer RNA (from baker's yeast, Type, X., Sigma cat. no. R9001).
3. Pyruvate kinase (from rabbit muscle, Type VII, cat. no. P7768).
4. Phosphoenol/pyruvate (Sigma, cat. no. P 7127).
5. ATP (Sigma, cat. no. A 7699) and GTP (Sigma, cat. no. G 5884).
6. L-[4-^{3}H]Phenylalanine (Amersham, cat. no. TRK204-250UC).
7. Econo-Safe (RPI, cat. no. 111175).

Preparation of Soluble Mitochondrial Translation Factors

1. Hypotonic buffer: 20m*M*Tris-HCl, pH 7.5, 20mM $MgCl_2$, and 40mM KCl Store at 4°C. Just before, use add β-mercaptoethanol to a final concentration of 5 mM.
2. 2*M* KCl.
3. Sephadex G-25 (Amersham, cat. no. 17 – 0031 – 02).

Extraction of Ribosomal Proteins from Pelleted Mitochondrial Ribosomes

Both MRP extraction procedures produce similar 2DE profiles.

Lithium Chloride-Urea Procedure

This procedures is used in the lab when purified ribosomes or subunits are first pelleted; starting material usually two livers.

1. Ribosomal protein solubilization solution: 6*M* urea, 3*M* LiCl, 50m*M* KCl, and 5m*M* β-mercaptoethanol, pH 3.5 – 5.5, with 3N HC1. Can be stored at –20°C in 500 μL aliquots.
2. 3N HC1.
3. 20% (w/v) trichloroacetic acid (TCA).
4. Ethanol/ether (1:1 vol/vol).

Acetic Acid-Acetone Procedure

This procedure is used when MRPs are extracted directly from sucrose density gradient fractions without prior sedimentation of purified ribosomes or subunits; starting material can be one liver.

1. Glacial acetic acid.
2. 1*M* magnesium chloride.
3. Acetone.

2D-Electrophoretic Analyses of Mitochondrial Ribosomal Proteins (NEPHGE)

1. 40% Biolyte 3/10 ampholyte (BioRad, cat. no. 163 – 1113) and 40% Biolyte 5/7 ampholyte (BioRad, cat. no. 163 – 1153).
2. Ribosomal protein lysis buffer: 9.5 M urea (ultrapure™ urea, Invitrogen, cat #. 15505 – 035), 2% (w/v) Triton X-100, 2% ampholytes (1% Biolyte 3/10 ampholyte, 1% Biolyte 5/7 ampholyte). Store at –20°C as 500 μL aliquots.
3. 30% acrylamide/bis solution, 29, 1 ratio (BioRad, cat. no. 161 – 0156). Store at 4°C.
4. 10% (w/v) Triton X-100.
5. 10% (w/v) ammonium persulfate (Sigma, cat #. A3678). Make up fresh every 2 wk.
6. TEMED (BioRad, cat. no. 161 – 0801).
7. Equilibration buffer: 10% (w/v) glycerol, 2.3% (w/v) sodium dodecyl sulfate (SDS), 0.002% bromophenol blue, and 62.5 mM Tris-HCl, pH 6.8.
8. 0.5% (w/v) agarose (ultrapure™ agarose, Invitrogen, cat. no. 15110 – 019) in equilibration buffer.
9. Dithiothreitol (DTT), solid (BioRad, cat. no. 161 – 0610).
10. Possible staining procedures: SYPRO® Ruby protein gel stain (Invitrogen, cat no. S-12000), Brilliant Blue R staining solution (Sigma, cat. no. B6529) and SilverQuest™ Silver Staining kit (Invitrogen, cat. no. LC6070).
11. Destaining solution (for Brilliant Blue R staining only): 30% (v/v) ethanol and 10% (v/v) acetic acid.

EQUIPMENT

1. Nitrocellulose filter disks (Millipore).

2. Gel extrusion needles (BioRad, cat. no. 165 – 1944).
3. Appropriate ultracentrifuge, rotors and centrifuge tubes.
4. Microcentrifuge.
5. UV plate reader or spectrophotometer.
6. Scintillation counter.
7. Gradient fractionator (e.g. Labconco, Auto-Densiflow).
8. Liquid nitrogen.
9. Probe sonicator (*e.g.*, Fisher Sonic Dismembrator 300, with Microtip).
10. Lyophilizer.
11. Glass tubes for NEPHGE (*e.g.*, 7 mm OD, 5 mm ID, 15 cm length).
12. Electrophoresis unit (*e.g.*, Hoefer Scientific Instruments, GT1).
13. Image analysis system and suitable software.
14. Microwave (optional).

METHODS

Isolation of Mitochondrial Ribosomes from Mitochondria

This procedure provides sufficient yields of mitochondrial ribosomes for sedimentation profile analyses and, when used in conjunction with silver staining, 2D/SDS-PAGE mini-gel analyses of ribosomal proteins. If 55S ribosomal activity is to be determined or 2D electrophoretic analysis of ribosomal proteins using larger gels is intended we suggest using two rat livers as the starting tissue.

1. Isolate hepatic mitochondria using standard procedures.
2. Perform a protein assay on the isolated mitochondria. We use the Lowry protein assay but any detergent compatible assay will work just as well. Digitonin needs to be added to the mitochondria at a ratio of 0.11 mg of digitonin per milligram of mitochondrial protein. Dissolve a suitable amount of digitonin in mitochondrial isolation buffer such that when the digitonin solution is added to the mitochondria it not only gives the required detergent to protein ratio but that it also dilutes the mitochondrial protein concentration down to 20 mg/mL.
3. Let the digitonin solution cool down before adding it to the mitochondria on ice. Stir the mitochondrial suspension gently for exactly 15 min.
4. Dilute the suspension down 10-fold with ice-cold isolation buffer, allow to continue stirring for a further 10s, then centrifuge immediately at 12,000 g for 10 min.
5. Collect the pelleted mitoplasts and resuspend them in isolation buffer to wash them. Re-pellet the mitoplasts and resuspend them in buffer A. Mitoplasts can be stored overnight at –70°C.
6. Perform a protein assay on the mitoplasts and dilute them down with buffer A to a concentration of 20 mg/mL. Lyse them with Triton X-100 (2% w/v final concentration).
7. Clarify the suspension at 12,000 g for 45 min. Remove the supernatant and layer it onto 15ml cushions of 1M sucrose and 1% (w/v) Triton X-100 dissolved in buffer B. Centrifuge

at 200,000 g) for 4 h (a 50.2Ti is a good choice of rotor). The pellets consist of mitochondrial ribosomes and some mitoplast membrane fragments.

8. Decant the supernatant, resuspend the crude mitochondrial ribosomal pellets in 1 ml of buffer B and layer them on top of a 10 – 30% linear sucrose gradient (30mL) in buffer B and centrifuge at 53,000 g for 14 h (SW27 rotor). Fractionate the gradients and monitor the $A_{260\ nm}$ of the fractions for ribosomal RNA. We detect the $A_{260\ nm}$ of 50-μ.L aliquots of our gradient fractions using a 96-well fluorescent plate reader that is also capable of reading U.V. Alternatively, fractions can be monitored spectrophotometrically. Absorbance peaks of free ribosomal particles corresponding to the small (28S) and large (39S) ribosomal subunits and the intact 55S monosome should be detected (see Fig. 6.1). A small peak is also detected at the bottom of the gradient. This corresponds to a population of ribosomes that may be attached to inner mitochondrial membrane fragments via unique attachments, for example, nascent polypeptide chains being inserted into the inner mitochondrial membrane.
9. If the aim is to isolate mitochondrial ribosomal proteins without any contaminating nascent polypeptides or, alternatively, if it is required to strip from the IMM those ribosomes attached via newly synthesized polypeptides, then incubate the crude ribosomal pellet in 1 ml buffer B containing 1 mM puromycin for 15 min at 37°C before layering on top of the sucrose gradients.
10. Fractions corresponding to the intact ribosomes or the ribosomal subunits can be centrifuged at 230,000 g for 5 h (60Ti rotor) and stored as a pellet at –70°C. Alternatively, the proteins associated with the intact 55S ribosome or the respective subunits can be subjected directly to electrophoresis and detected by silver staining.
11. If it is desired to separate 55S mitochondrial ribosomes into their 28S and 39S, small and large ribosomal subunits, respectively, resuspend the pelleted 55S monosomes or crude ribosomal pellets in magnesium-deficient buffer B and sediment them through 10 – 30% sucrose gradients made up in the same buffer.

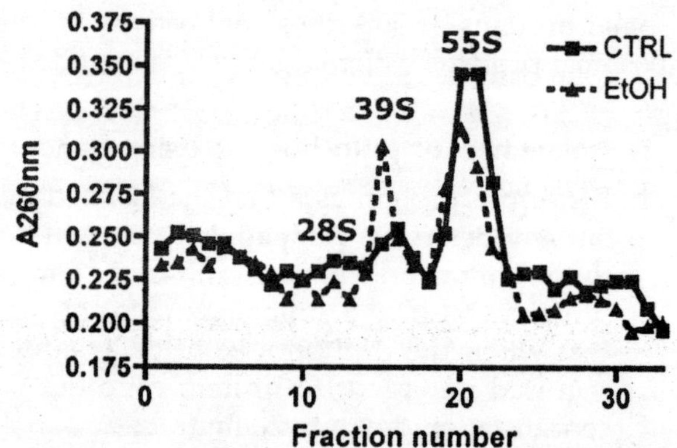

Fig. 6.2. Effects of chronic ethanol feeding (12mo) upon the sucrose density gradient profiles of mitochondrial ribosomes and their subunits isolated from a single control and alcoholic liver. Sucrose cushion purified hepatic mitochondrial ribosomes were separated through 10-30% sucrose gradients into monosomes and subunits. The $A_{260\ nm}$ of the 50-μL fraction aliquots was monitored using a 96-well fluorescent plate-reader (in UV mode).

Ribosome Activity Assay

1. Incubate 2 – 5 pmol of 55S mitochondrial ribosomes in ribosome activity buffer supplemented with (final concentrations) 500 mg/mL tRNA, 100 mg/ml pyruvate kinase, 2 mM phosphoenolpyruvate, 5 mM ATP, 0.3 mM GTP, 0.2 mg/mL translation factors and 10 μCi [^{3}H]phenylalanine in a final volume of 50 μL.
2. Remove 5 μL of the reaction mixture and spot onto a nitrocellulose filter disk. This represents time zero. Start the reaction by adding poly (U) to a final concentration of

1 mg/ml, remove 5 μL aliquots at suitable time points up to 1 h (we find that radiolabel incorporation is linear for this time period) and spot them onto nitrocellulose filter disks (1 cm^2).

3. Incubate the filter disks in ice-cold 10% TCA (w/v, 2mL per 1 cm^2 filter) for 30 min with gentle swirling to stop the reaction and to precipitate the protein onto the filter.
4. Wash the disks in fresh ice-cold 10% TCA for 10 min with gentle swirling.
5. Place the disks in 5% TCA and heat to 90°C for 15 min.
6. Quantify the radioactivity present on the filters by adding 5 μL of a scintillation cocktail of choice (Econo-Safe works well for us). Express the phenylalanine polymerization as pmol [^{3}H] phenylalanine/mg RNA/pmol mitochondrial ribosomes.

Preparation of Soluble Mitochondrial Translation Factors

1. Resuspend mitochondria in hypotonic buffer at a concentration of 5 mg/mL and incubate on ice for 15 min to allow the mitochondria to swell.
2. Subject the mitochondria to three cycles of freeze-thawing using liquid nitrogen then sonicate (we use 4 × 30s pulses at 50%; to disrupt the mitochondria.
3. Clarify the mitochondrial suspension at 16,000 g for 20 min at 4°C. Remove the supernatant and adjust it to 500 mM KCl.
4. Centrifuge at 200,000 g for 4 h at 4°C to pellet mitochondrial ribosomes.
5. Pass the supernatant over a Sephadex G-25 column (to remove endogenous amino acids) and collect the void volume. This contains the mitochondrial translation factors. Lyophilize the translation factors and store at –70°C.

Extraction of MRPs from Pelleted Mitochondrial Ribosomes

Lithium Chloride-Urea Procedure

1. Resuspend 55S mitochondrial ribosomes or mitoribosomal subunits in 150 μL of ribosomal protein solubilization solution in a 1.5 mL microcentrifuge tube.
2. Place a micro stir bar in the centrifuge tube (the stir bar will stand on its end) and stir the ribosomes overnight at 4°C (make sure the sample is stirring and it is not the tube that is spinning).
3. Pellet the RNA at 200,000 g (we use a TLA-100 rotor with 200 μL capacity tubes) for 1 h. Remove the supernatant and reserve. Re-extract the RNA pellet by adding another 150μL of ribosomal protein solubilizing solution. Stir for 6 h and pellet the RNA at 200,000 g for 1 h, Remove the supernatant and combine it with the supernatant from the first extraction.
4. The RNA pellet should contain rRNA and any mRNA that remains associated with the ribosome.
5. Precipitate the ribosomal proteins overnight at 4°C by adding 2 volumes of 20% TCA. Centrifuge at 10,000 g in a microcentrifuge for 15 min. Remove the supernatant and wash the pellet in 1:1 (v/v) ethanolether.
6. Allow to air-dry for 5 – 10min before resuspending the ribosomal proteins in an electrophoresis buffer of choice. We suggest using ribosomal protein lysis buffer supplemented with DTT to a final concentration of 1% (w/v) just before use.

Acetic Acid-Acetone Procedure

We have used this procedure on as little as $0.1A_{260\ nm}$ units of mitochondrial ribosomes.

1. Collect fractions from sucrose density gradients corresponding to intact 55S monosomes or ribosomal subunits.
2. Add magnesium chloride to a final concentration of 100 m*M*.
3. Add 2 volumes of glacial acetic acid. Invert a couple of times. Leave for 45 min on ice.
4. Pellet precipitated RNA at 5000 *g* for 10 min.
5. Add 5 volumes of ice-cold acetone. Invert a couple of times. Leave at –20°C overnight.
6. Pellet precipitated protein at 5000 g for 10 min.
7. Wash pellets with ice-cold acetone. Re–pellet protein at 5000 g for 10 min.
8. Remove supernatant and allow pellet to air-dry (approximately 10min).
9. Resuspend pellet in ribosomal protein lysis buffer supplemented with DTT to a final concentration of 1% (w/v) just before use.

2D-Electrophoretic Analyses of Mitochondrial Ribosomal Proteins (NEPHGE)

Early isoelectric focusing methodologies in which ampholytes were used to form pH gradients suffered from gradient drift. The development in recent years of immobilines (buffering acrylamide derivatives that contain either a free carboxylic acid group or a tertiary amino group) has allowed the formation of stable pH gradients in the range of pH 3 – 12. These immobilized pH gradients (IPGs) are reproducible and insensitive to disturbances from sample components. We have used the ZOOM® IPGRunner™ system from Invitrogen in conjunction with silver staining to analyze the protein content of the large ribosomal subunit from control and ethanol mitochondrial ribosomes. Figure 1 shows a sucrose density gradient profile of mitochondrial ribosomes isolated from control and ethanol-fed animals. Ethanol-fed animals show decreased levels of intact 55S ribosomes but an increase in the level of 39S ribosomal subunits. This is suggestive of an ethanol-elicited increase in ribosome dissociation or a decrease in ribosome association. The 39S peaks from Fig. 6.1 were extracted using the acetic acid-acetone procedure and separated in the first dimension by isoelectric focusing using IPGs (ZOOM strips, Invitrogen) of range pH 6 – 10 followed by second dimension SDS-polyacrylamide gel electrophoresis (PAGE; Fig. 6.2). A significantly greater amount of ribosomal protein was extracted from the 39S peak in the ethanol-fed animals when compared with the paired control. This greater yield of ribosomal protein reflects the increase in 39S subunit seen in Fig. 6.1 and suggests that the increase in 39S fraction as measured by A_{260nm} is the result of increased levels of intact 39S subunits and not due to increased levels of mRNA and rRNA associated with the 39S fraction.

The methodologies involved in using Invitrogen's ZOOM® IPGRunner™ system are described in detail in the chapter "Proteomic Approaches to Identify and Characterize Alterations to the Mitochondrial Proteome in Alcoholic Liver Disease" by Bailey et al. in this volume. We will therefore focus here on an alternative and excellent method of separating ribosomal proteins, that is, NEPHGE. Ribosomal proteins are, for the most part, extremely basic in charge. The 39S ribosomal subunit for example is believed to comprise 48 proteins with pis ranging from 6.6 to 12.3 with an average of 9.6. Even, if these proteins were to be focused using IPGs, it would be difficult to include them all on the same 2D/IEF/SDS-PAGE gel. During NEPHGE proteins are not necessarily focused to their pIs but instead migrate at different rates across the gel due to their charge. At the end of the run, some proteins will have focused to their pIs while the others will still be migrating. The protein pattern produced on .the gel is dependent on the accumulated volt hours so to ensure reproducibility this must remain constant for a specific population of proteins. The method described below is intended for use with the larger 2D system, that is, 15-cm tube gels and 16cm × 16cm SDS-PAGE slab gels which we find to be excellent for resolving the complete population of ribosomal proteins. Mini-systems such as the Mini-PROTEAN® Tube Cell Module from Bio-Rad can be used but we find the larger set-up easier to manipulate.

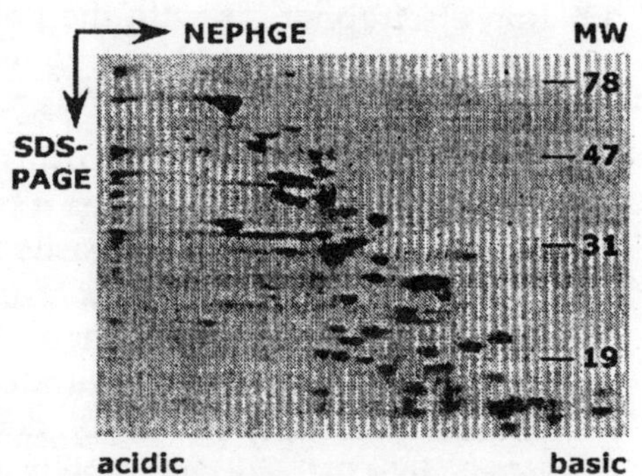

Fig. 6.3. Two-dimensional (NEPHGE/SDS-PAGE) proteomic analysis of mitochondrial ribosomal proteins extracted from intact 55S monosomes (modified from). Proteins were extracted from the 55S monosome using lithium chloride and urea separated in the first dimension by NEPHGE (pH 3-10) and in the second dimension by 12% SDS-PAGE. Proteins were stained with Coomassie Brilliant Blue.

1. Rinse out the glass tubes with water followed by acetone. Allow to air dry.
2. Stopper one end of the tubes with parafilm.
3. To make 10 mL of gel mix, place 5.5 g of urea, 1.33 mL of 30% acrylamide mix, 2 mL of 10% Triton X-100, 1.97mL of dH_2O, 0.25 mL of Biolyte 3/10 ampholyte (40%), and 0.25 mL of Biolyte 5/7 ampholyte (40%) into a 125-mL side-arm flask.
4. Heat slightly to dissolve urea.
5. Add 30 μL of AMPS.
6. De-gas for 1 min.
7. Add 21 μL of TEMED.
8. Pour tube gels using a suitable syringe to approx 1 cm from the top. Overlay with dH_2O. Leave to polymerize for 1 – 2 h. To test whether the gel mix has polymerized successfully we suggest sucking up any remaining gel mix, after the tubes have been poured, into a plastic pipet. Once it has polymerized inside the pipet it is usually safe to assume that it has also polymerized inside the tubes.
9. Fill the lower reservoir of the electrophoresis apparatus with 0.02 *N* NaOH. Remove the parafilm from the bottom of the tubes. Place the tubes inside the tube holder according to apparatus instructions.
10. Load ribosomal protein sample solubilized in ribosomal protein lysis buffer (50 μL is a good volume) onto polymerized gel. For Coomassie blue staining of gels, we suggest 200μg per tube gel, and for silver staining of gels we suggest 50 μg. We suggest using the RC DC Protein Assay (BioRad, cat. no. 500 – 0120) to quantify sample protein levels. This assay is compatible with a number of components of the ribosomal protein lysis buffer, for example, ampholytes and thiol reducing agents that would normally interfere with protein determination.
11. Overlay with ribosomal protein lysis buffer diluted 1 to 4 leaving a small space at the top.
12. Fill the tube to the top with upper reservoir solution, that is, 0.01 M phosphoric acid.

13. Run electrophoresis with the polarity reversed *i.e.* red lead from apparatus into black port on power pack, black into red. Run at 400 V for 1 h and 30 min depending upon the desired separation profile. This corresponds to 600 Vh.
14. Extrude gel from glass tube using a syringe with a long thin gel extrusion needle. Fill the syringe with dH_2O and insert it gentle inside the glass tube down the side of the gel. The needles recommended can easily be slid inside the tubes to just over half way. Pipet water gently while sliding the needle around the outside of the gel. Then withdraw the needle, turn the glass tube around and reinsert the needle into the other side of the tube. Gently pipet water as before. On withdrawing the needle from the tube this time, the gel inside should slide easily out. If nol, determine which end of the gel is sticking and reinsert the needle into that side and pipette. We suggest placing a sheet of parafilm or aluminum foil on the lab bench to catch the tube gel should it slide out unexpectedly during extrusion.
15. Place tube gel into 15mL conical tube and add 5mL of equilibration buffer supplemented with DTT (0.5% w/v) just before use. Incubate with gentle rocking for 30 min.
16. Pour an SDS slab gel (8–20% resolving gel works well, 5% stacking gel) according to standard procedures. Assemble the electrophoresis apparatus according to manufacturer's instructions.
17. Add a small amount of 1% agarose dissolved in equilibration buffer containing 0.5% (w/v) DTT to the top of the gel to serve as a platform for the tube gel to sit on. Once the agarose has set, lay the tube gel on top. Allow some room at one end to add molecular weight markers. We normally embed a small volume of markers in about 20μL of 1% agarose in equilibration buffer, containing 0.5% DTT in the bottom of a 1.5mL centrifuge tube. After the agarose has set pop out the markers plug and sit it next to the tube gel on the agarose platform.
18. Pipet more of the agarose/equilibration buffer solution over the top of the tube gel and markers plug to hold them in place.
19. Run the gel at 25 mA for 7 h or until the Bromophenol blue has migrated to the bottom of the gel.
20. Detect ribosomal proteins using a suitable staining procedure. We have successfully used Coomassie Brilliant Blue R250, SYPRO® Ruby protein gel stain and silver staining procedures. All these staining systems are compatible with the excision of protein spots from the gels, tryptic digestion of the proteins and analysis by mass spectrometry.
21. Alternatively, the proteins can be transferred from the gel to nitrocellulose membranes and analyzed for specific protein modifications by western blotting.

Analysis of Proteins Separated by 2DG

A number of current systems and software products are available for protein analyses in 2D gels. Our laboratory possesses a Kodak ImageStation 440CF with Kodak Digital Science ID & 2D Software. This is convenient for protein staining via Coomassie Brilliant Blue, SYPRO® Ruby and silver.

NOTES

1. Adjust the pH of these solutions at 4°C because gels are run in the cold room (4°C).
2. The 3X gel buffer can be stored at room temperature. However, the pH must be adjusted at 4°C because gels are run in the cold room (4°C).

3. These gel volumes can be proportionally increased to prepare a resolving gel that has an extended length. This is typically done if the 1-D BN-PAGE gels are to be run overnight to improve the resolution (*i.e.,* separation) of the complexes. Similarly, when gels are run overnight the cathode buffer can be switched to a "blue-free" buffer, that is, 50 mM Tricine, 15mM BisTris, pH 7.0, with no added coomassie blue G-250. This helps to minimize the high blue background when 1-D BN-PAGE gels are destained for imaging and densitometry purposes.
4. For the separation of large membrane protein complexes it is recommended that the concentration of the coomassie blue G-25.0 in the sample be one quarter of the lauryl maltoside concentration for electrophoresis. Therefore, based on the given stock solutions and protocol for the extraction of inner membrane proteins this works out to be approximately 6.3μL of a 5% (w/v) coomassie blue G-250 solution. Addition of a molar excess of dye to protein also helps to remedy one problem of native electrophoresis, that is, the tendency of membrane proteins to form aggregates in the presence of detergents. However, the problem of aggregation is minimized by the inclusion of the coomassie blue G-250. This dye binds to the hydrophobic regions on the proteins surface, which induces a negative surface charge thereby reducing protein aggregation. In addition, the anionic nature of coomassie blue G-250 induces the needed "charge shift" on proteins such that they will migrate to the anode at pH 7.5.
5. A high level of magnesium ions (15 – 20 mM) is necessary to ensure that mitochondrial ribosomal subunits remain associated during the isolation procedure.
6. To dissolve digitonin, place it in a suitable amount of buffer in a glass beaker and heat it up in a microwave. Watch for the buffer to just begin to boil then stop the microwave and remove the beaker.
7. Digitonin is a detergent and is being added in a precise amount to the mitochondria to remove the outer mitochondrial membrane and in doing so eliminate the possibility of contamination by cytoplasmic ribosomes. If it is left in contact with the mitochondria for too long it will begin to permeabilize the inner mitochondrial membrane.
8. Solubilization of the mitoplasts occurs immediately and is accompanied by a darkening of the mitoplast suspension as matrix contents are released.
9. Puromycin is an antibiotic compound that binds to the ribosome as an analogue of aminoacyl-tRNA and arrests the formation of nascent polypeptide chains. The polypeptides chains are then liberated from the ribosome. If using puromycin be warned that the compound also has an absorbance at 260 nm and this will mask the presence of the 28S subunit. If the 28S subunit is required we suggest removing the puromycin by pelleting the crude ribosomes through a second sucrose cushion before layering them onto the sucrose gradients.
10. We have found that complete dissociation of rat liver hepatic mitochondrial ribosomes is achieved by incubating them in a buffer containing no Mg^{2+} rather than low levels of Mg^{2+} as is suggested in some publications. The inclusion of 1 mM EDTA into the Mg^{2+}-deficient buffer B ensures a complete absence of Mg^{2+}.
11. On the basis of the molecular weight of the 55S ribosome 1 $A_{260\ nm}$ unit approximates to 40 pmol.

12. It is necessary to remove nucleic acid from protein samples that are to be used in isoelectric focusing or NEPHGE analyses because it can bind to the proteins and cause artifactual migration and streaking.
13. Air bubbles can inhibit gel polymerization.
14. Make sure that no air bubbles are present at the bottom of the tubes as they sit in the lower reservoir solution. If there are, they can easily be removed by blowing reservoir solution across the bottom of the tubes using a glass pipet that has been heated gently and curved at its narrow end.
15. It is very important to pipette the agarose solution gently to avoid the formation of bubbles. These can impede electrophoresis and cause distortions of the protein patterns. If air bubbles form, suck them out with a glass pipette. Also make sure that the marker plugs have no air bubbles present.

CHAPTER 7 Detecting Kupffer Cells

The development of alcoholic liver disease (ALD) is characterized by an increase in hepatic steatosis and inflammation, which in turn can lead to fibrosis and then cirrhosis. Kupffer cells, the resident macrophages of the liver, are critical to the onset of ethanol-induced liver injury. Ablation of Kupffer cells prevents the development of steatosis and inflammation in rats chronically exposed to ethanol via intragastric feeding. After chronic ethanol exposure, Kupffer cells exhibit increased sensitivity to lipopolysaccharide (LPS) and consequently exhibit enhanced LPS-stimulated inflammatory cytokine production. Tumor necrosis factor-α (TNF-α) production, resulting from Kupffer cell activation, has been shown to be a critical component in the development of ALD. Recent studies have shown that increased sensitivity to LPS after chronic ethanol exposure also leads to an increased production of reactive oxygen species (ROS).

Isolation of Kupffer cells from rats fed an ethanol-ontaining diet provides sufficient material for studies using cellular and molecular biology approaches aimed at understanding the effects of ethanol on Kupffer cell function. For example, isolation of mRNA from Kupffer cells from pair- and ethanol-fed rats has shown that LPS-stimulated increases in TNF-α expression can be inhibited by adiponectin. In this chapter, we will focus on methods used for isolating Kupffer cells from animals fed the Lieber-DeCarli ethanol diet and their pair-fed controls, as well as how to isolate mRNA from these primary Kupffer cell cultures.

MATERIALS

Surgical Portion of the Procedure

Materials needed for the surgical portion of the Kupffer cell isolation can be found in Table 7.1. Sources used and catalog numbers associated with them are also included in the table.

Elutriator

Materials needed for the elutriation portion of the Kupffer cell isolation can be found in Table 7.2. Also included are the sources used and the catalog numbers associated with them.

Isolation and Culture of Kupffer Cells

Materials for the isolation and culture of the Kupffer cells can be found in Table 7.3 along with the sources used and catalog numbers.

Table 7.1. Materials of Surgical Portion of Procedure

Material	Vendor	Catalog number	Concentration or amount needed
1. Autoclaved water		6%	
2. Hydrogen peroxide			
3. 8-channel peristaltic pump	Rainin	7103-058	
4. Four-"butterfly needles" (Blood Collection Infusion Set 19 gauge ¾ with 12" tubing)	BD Biosciences	364919	
5. Waterbath at 39°C			
6. Hanks-(Hank's Balanced Salt Solution (w/o Ca^{2+}, Mg^{2+}, $NaHCO_3$)	Mediatech	55–021-PC	250mL/Rat
7. HEPES	Fisher	BP310–500	10mM
7. EGTA	Sigma	E4378	1mM
9. Williams E	Sigma	W4125	250mL/rat
10. Collagenase type IV (from *Clostridium histolyticum*)	Sigma	C5138	50mg/rat
11. 9" Pasteur Pipets			
12. 95% O_2 5%, CO_2 gas mixure	Praxair	mmoxcd5-k	
13. Heparin	Sigma	H3393	1000U/mL
14. Ketamine (100mg/mL)	Henry Schein	995-2949	
15. Xylazine (20mg/mL)	NLS Animal Health	105650	
16. Acepromazine (10mg/mL)	Henry Schein	356-7290	
17. Eight 1-mL syringes	Fisher	14–823–2F	
17. Four 26-gauge 5/8-inch needles	Fisher	14-826-6A	
19. Four 21-gauge needles	Fisher	14-826C	
20. Kimwipes			
21. Buttonhole twist thread (2 colors- ~5")			
22. Two surgical pans (9 × 13 baking pan)			
23. Two racks (cookie coling racks must fit over pan)			
24. Two pairs of sharp/sharp 5 1/2" sciessors	Fisher	13–808–4	
25. One pair of blunt/sharp 5 1/2" sciessors	Fisher	13–808–2	
26. One pair of blunt-point curved 5" forceps	Fisher	08–887	
27. One pair of curved eye dressing 4 1/2" forceps	Henry Schein	101–7739	
27. Four 15-mL centrifuge tubes	Sardstedt	52 554 205	
29. Four divided Petri dishes	Fisher	08–757–110	

Item 21 can be autoclaved in a small box and items 24-27 can be autoclaved together in an autoclaveable box.

Table 7.2. Materials for Elutriation Procedure

Material	*Vendor*	*Catalog number*	*Concentration or amount needed*
1. Beckman Avanti J25 Centrifuge	Beckman		
2. JE6B Elutriation Rotor and accessories	Beckman		
3. Masterflex L/S Variable Speed Pump	Cole Parmer	EW-77521-40	
4. Masterflex L/S Easey Load II Pump Head	Cole Parmer	EW-77200-60	
5. Autoclaved water			
6. Hydrogen Peroxide			6%
7. Catalase	Sigma	C9322	10–15mg/ 100mL
7. Gey's Balanced salt solution			1L

If working with endotoxin, autoclave and bake all glassware before use.

METHODS

The following method is designed for the isolation of Kupffer cells from male Wistar rats that were allowed free access to the Lieber-DeCarli liquid diet (7) containing 36% of calories from ethanol or pair-fed an isocaloric diet that substituted maltose dextrins for the ethanol. The following procedure has been adapted from methods originally developed by the Friedman and Roll, Knock *et. al.*, and Munthe-Kaas *et. al.*

Kupffer cells are isolated as previously described *(2)*. In brief livers are perfused with 0.05% collagenase, and the resulting suspension of liver cells is treated with 0.02% pronase for 15min at 12°C. The resulting cell suspension is centrifuged 3 times at 50 g for 2min, and the supernatant collected after each centrifugation. The pooled supernatant is then centrifuged at 500 g for 7min to collect nonparenchymal cells. Kupffer cells are then purified by centrifugal elutriation *(2)*. The following protocol is for isolation of Kupffer cells from 2 pair-fed and 2 ethanol-fed animals; however, the protocol can be scaled up or down according to the quantity of Kupffer cells required.

Setting up the Peristaltic Pump and Perfusion

All eight channels of the peristaltic pump must be used for isolating Kupffer cells from four rats. Set up the Rainin peristaltic pump per the manufacturer's instructions. It is necessary to check the flow rates of the peristaltic pump before beginning the isolation. For perfusion of two rats, set the flow rate to 25mL/min and to 50mL/min for four rats. To sterilize the tubing, first run sterile water through the pump, and then run approx 200 mL of 6% hydrogen peroxide (H_2O_2) through the pump. Finally, run more sterile water through the pump until all peroxide bubbles are gone from the tubing. While sterilizing the pump, thaw 1 × 10-mL aliquots of 10 mM HEPES and 1 mM ethylene glycol-bis (2-amnioethylether)- N, N, N′, N′-tetraacetic acid (EGTA), and bring 1L of Hanks (pH 7.4) and Williams E ~(pH 7.4) to 37°C.

When the media are at 37°C, oxygenate each medium with a mixture of 95% O_2 and 5% CO_2 for approx 10min, and then add 10mM HEPES and 1 mM EGTA to the Hanks-, mix thoroughly and run through the peristaltic pump. Once the Hanks- is through, stop the pump until you are ready to perfuse the livers. Once the Williams E has been oxygenated, add 200 mg of collagenase, mix by swirling and place in the water bath.

Table 7.3. Materials for Isolation and Culture of Kupffer Cells

Material	*Vendor*	*Catalog number*	*Concentration or amount needed*
1. Waterbath at 37°C			
2. 12°C Shaking Waterbath			
3. Beckman Tabletop Centrifuge			
4. Hanks+(Hank's Balanced Salt Solution (with Ca^{2+}, Mg^{2+} w/o $NaHCO_3$)	Sigma	H6136	10mL/rat
5. CMRL Medium 1066	Invitrogen	11530-037	
6. Antibiotic-Antimycotic	Invitrogen	15240-062	
7. L-glutamine	Invitrogen	25030-081	2mM
7. Fetal bovine serum certified	Invitrogen	16000-044	10%
9. Pronase	Roche	165 921	30mg/2 rats
10. DNase I	Roche	104 159	405mg/rat
11. Gey's Balanced Salt Solution			
12. Nycodenz	Accurate Chemical	ACN1002424	11% and 24.6%
13. 3% Acetic acid			
14. Trypan blue			
15. Thirty-50 mL centrifuge tubes	Sardstedt	62 547 205	
16. Six-15 mL centrifuge tubes	Sardstedt	52 554 205	
17. Two-500 mL polypropylene flasks	Fisher	10–041–10D	
17. Two-250 mL polypropylene flasks	Fisher	10–041–10C	
19. Two plastic funnels	Fisher		
20. One pair straight dissecting dressing 4- 1/2" forceps	Fisher	13–812–39	
21. One pair curved dissecting VWR 4- 1/2" forceps	VWR	25717-005	
22. Two 3 × 3" pieces of Coarse mesh (132 µm)	Sefar Americo (Tetko)	03–132–40	
23. Two 3 × 3" pieces of Fine mesh (224 µm)	Sefar Americo (Tetko)	03–225–42	
24. Two-60-mL syringes	Fisher	13–689–8	
25. Two 18-gauge needles	Fisher	14–826–5G	
26. 5-3/4" Pasteur pipets			
27. 9" Pasteur pipets			
27. Rubber bulb	VWR	56311–062	
29. 1.5-mL Eppendorf tubes			
30. Pipet tips (0.5 – 10 µL, 10 – µL, 10 – 100 µL, 100 – 1000 µL)			
31. Sterile reagent reservoirs	Fisher	07 200 127	
32. Sterile tissue culture-treated plates (96-well, 24-well, 6-well, chamber slides)			

Items l7 – 18 should be autoclaved and items 20 – 23 can be autoclaved together in an autoclaveable box. If working with endotoxin, autoclave and bake all glassware before use.

Setting up the Elutriator

Before using the elutriator to isolate cells, it is necessary to check the flow rates of the peristaltic pump at the desired rate of centrifugation. Before pumping cells through the elutriator, the tubing should be autoclaved. Set up the elutriation chamber and tubing and cool to 4°C. To prepare the elutriator, turn the pump to the 45 mL/min flow rate and run sterile water through the pump and chamber (~100 – 200 mL). Next, recirculate 100 mL of 6% H_2O_2 through the tubing and chamber for approx 5min, with the centrifuge running at 50*g*.

Make sure the pressure gauge stays under 20 psi at all times. Stop the centrifuge, recirculate 100mL of catalase (10 – 15 mg/100mL) through the tubing, and restart the centrifuge at 24*g* for approx 5 more minutes or until all of the peroxide bubbles are gone. Stop the centrifuge again, and then rinse out the tubing and chamber with sterile water. Once the tubing and chamber have been sterilized, run Gey's Balanced Salt Solution (pH 7.4) (GBSS) through the system (let it run all the way through to remove any air bubbles) and then stop the pump until needed.

Setting Up the Tissue Culture Room

It is helpful to set up needed cell culture supplies before beginning the perfusion or while the perfusion is in progress. Place the following in the cell culture hood:

2 – 30-mg aliquots of pronase in a 50-mL centrifuge tube

2 – 4.5-mg aliquots of DNase I in a 15m-L centrifuge tube

1 – 500-mL bottle of CMRL with antibiotic/antimycotic and 2 mM L-glutamine (CMRL).

Fill each 50-mL centrifuge with 50 mL of CMRL and each 15-mL centrifuge with 3 mL of CMRL.

Anesthesia

Rats are weighed and the dose of anesthetic is calculated based on diet and body-weight. In our experiments, we use a mixture of xylazine (20 mg/mL), ketamine (100mg/mL), and acepromazine (10mg/mL), which expires after 1 mo.

Ketamine:	150mg (100mg/mL)	1.5mL
Xylazine:	30 mg (20mg/mL)	1.5mL
Acepromazine:	5 mg (10 mg/mL)	0.5 mL
		3.5 mL

In chow-fed animals, we use a dose of 0.12mL/100g bodyweight. Ethanol-fed animals on the Lieber-DeCarli ethanol diet receive a dose of 0.075 mL/100 g bodyweight, whereas their pair-fed controls receive 0.12mL/100g bodyweight. Animals receive an intraperitoneal injection in the lower middle abdomen (right or left of midline is fine). Allow 5 – 10 min for the animals to reach complete sedation. If there is no movement when the foot or base of the tail is pinched, the animal is completely anesthetized.

Surgery

1. Once the animals are anesthetized, wipe the fur with a Kimwipe wet with water (do not use ethanol) to get rid of any dander, loose fur, or dirt.
2. Using the sharp/blunt scissors and blunt-point curved forceps, peel the abdominal fur by pulling up on the fur at the lower part of the abdomen and cutting up towards the thoracic area, removing enough fur to have a clear surgical field.

3. Remove loose fur from the scissors and animals by wiping with another wet Kimwipe.
4. Using the same sharp/blunt scissors and blunt-point curved forceps, make a horizontal cut into the lower part of the abdomen at midline. Carefully insert the scissors into the opening (do not to puncture the bladder) and make a smooth, V-shaped cut up both sides of the abdomen. Stop the cut at the diaphragm of the animal. Do not puncture the diaphragm.
5. Using a Kimwipe, push the organs to the side and out of the abdominal cavity and place a 15-mL centrifuge tube under the thoracic area to better expose the portal vein.
6. Place two loose ligatures (two different colors of autoclaved buttonhole twist thread) around the hepatic portal vein using the curved eye dressing forceps.
   (*a*) Insert the forceps in the connective tissue near the hepatic portal vein, go underneath the vein and around to the other side. Then, grasp onto one end of the thread and gently pull it through. Pulling too hard on the thread can twist the hepatic portal vein and also cut off circulation to the liver.
   (*b*) After pulling both ligatures through, tie a loose knot in preparation for inserting the perfusion "butterfly" needle and lay the ends of the string out so they are easy to pick up.
7. Inject 0.1 mL of heparin using a 26-gauge, 5/8-inch needle into the posterior vena cava. After the injection, press down on the injection site to stop the bleeding with a Kimwipe, and wait 1 min before proceeding.
   (*a*) If you would like to harvest other tissues (skeletal muscle, adipose), this would be the most appropriate/ideal time to do so. The animal will still be alive, and the area for liver perfusion is ready.
   (*b*) To take a blood sample from the animal, use an 18-gauge needle (smaller needles will shear red blood cells resulting in a hemolyzed plasma/serum sample). Insert the needle into the posterior vena cava and pull back slowly. Try to minimize bleeding by applying pressure to the vena cava. The perfusion should be started as soon as possible after collecting the blood.
7. For the perfusion, start the peristaltic pump at the 25mL/min flow rate, insert the 19-gauge, 3/4-inch perfusion "butterfly" needle behind the rear ligature into the hepatic portal vein and past the front ligature. Tighten the knots of the two ligatures to secure the needle.
9. As each perfusion needle is secured, cut through the diaphragm and cut the portion of the posterior vena cava in the pericardial cavity. This releases the perfusate and allows it to drain into the pan below.
10. First, perfuse with the Hanks- medium. Stop the pump just before all the media is in the tubing, place the tubing into the William's E + collagenase and start the pump again. Do not allow bubbles to get into the liver.

Isolation of Kupffer Cells

1. After the perfusion, using the second pair of sharp/sharp scissors and the blunt-point curved forceps, remove the liver. Take the forceps and grasp onto a portion of the diaphragm near the liver, at midline. Cut a small "tag" of diaphragm to hold onto while cutting the connective tissue that attaches the liver to other organs. When removing the liver, be careful not to puncture the intestines or stomach, as they are a source of potential

bacterial contamination. Once each liver has been removed, place it into a 100mm divided Petri dish. The remainder of the procedure is performed in a laminar flow hood under sterile conditions.

2. Rinse each liver with approx 10 mL of Hanks+.
3. Add 15 mL of CMRL to each liver. Using the straight and curved dissecting forceps, tease each liver slightly.
   (*a*) If the perfusion was successful, the liver will come apart with little effort.
   (*b*) If the liver is hard to tease apart, the perfusion did not work well and Kupffer cell yields will be low.
4. Once the livers have been teased apart, carefully pour the cell suspension from each Petri dish into its appropriately labeled (P or E) 500-mL flask through the plastic funnel. Rinse each Petri dish with 20 mL of CMRL adding the rinse to the flask. Then, add the 50 mL of CMRL + pronase to the flask.
   (*a*) If hepatocytes are required, for experiments, remove them prior to adding the pronase.
5. Pour the 3mL of CMRL + DNase I into each flask and cover with aluminum foil. At this point, there should be approx 100mL of media in each flask.
6. Shake the 500-mL flasks for 15 min at 12°C in a shaking water bath (100rpm).
7. After shaking, filter each suspension through the coarse mesh (224 μm) into each 250-mL flask.
7. Next, filter the suspension from each 250 mL flask through the fine mesh (132μm) into 4 × 50-mL centrifuge tubes labeled P or E.
9. Centrifuge at 500 g for 7 min (4°C).
   (*a*) This step pellets all cell types: hepatocytes, red blood cells, Kupffer cells, etc.
10. Carefully aspirate off most of the supernatant (leave 1–2mL over the cells as the pellet is very loose).
11. Resuspend each of the pellets from two tubes into 10-15mL of CMRL with antibiotic/antimycotic, 2mM L-glutamine, and 10% fetal bovime serum (FBS; CMRL 10% FBS) and bring each one to 50 mL. Repeat for the other two tubes, and then the other treatment group.
12. Centrifuge at 50 g for 2min.
13. Carefully pour approx 25–30mL of the supernatant from each tube into two sterile centrifuge tubes labeled PA (for pair-fed tube A) or EA (for ethanol-fed tube A) and place them in ice.
    (*a*) The volume of media poured off will depend on the size of the pellet. At this step the pellet is very soft and easily disrupted. Pour off as much supernatant as possible without displacing the pellet. The nonparenchymal cells are in the supernatant.
14. Resuspend the pellet in the remaining media in each tube and take the volume up to 50mL using CMRL 10% FBS.
15. Repeat steps 12 – 14 two more times.
    (*a*) This step is to separate the Kupffer cells from hepatocytes. After repeating there will be 6 × 50-mL centrifuge tubes for each treatment group.

16. Centrifuge all 12 centrifuge tubes at 500 g for 7 min to pellet nonparenchymal cells.
17. Aspirate off most of the supernatant (the pellet is soft and easily disrupted).
17. Combine the pellets of the six centrifuge tubes from each treatment group into one centrifuge tube with GBSS and then bring the volume up to 50mL.
19. Once the cells are resuspended in 50 mL of GBSS, load the cells into a 60-mL syringe with an 18-gauge needle attached.
    (*a*) First carefully remove the plunger from the syringe, turn it over, and set aside to avoid contamination. Next place the 60-mL syringe (without plunger) into an empty 50-mL centrifuge tube.
    (*b*) Carefully, but quickly, pour the cell suspension into the syringe, insert the plunger (some of the suspension will be lost into the centrifuge tube), and invert the syringe so the needle is facing upward.
    (*c*) Get rid of as many bubbles as possible and push out any air left in the syringe. Pour the lost cell suspension into a reagent reservoir. Invert the syringe and collect the remaining cells by pulling back on the plunger.
    (*d*) Again, remove all the air from the syringe. Replace the cap of the needle onto the syringe and place the syringe in ice.

It is now time to separate the Kupffer cells from other cell types within the cell suspension in the elutriation chamber. Depending on the centrifuge, make sure that the correct rotor is selected on the instrument panel (JE6B) and check that the centrifuge is at 4°C.

20. Turn on the peristaltic pump allowing the GBSS to flow through the tubing into the elutriator and out into the collection bucket. Run the elutriator for approx 1 min and then turn the pump to the 10mL/min flow rate.
    (*a*) Some bubbles may have collected in the tubing near the pressure gauge. Allow these bubbles to pass out of the elutriator into the collection bucket.
21. Once the bubbles are gone, start the centrifuge at 24 g. If there are any bubbles left in the chamber allow them to pass out of the elutriator. If no bubbles are present and the pressure gauge is reading close to 0 psi, increase the speed to 96 *g*, check again for bubbles and high pressure, and increase the speed finally to 602 g.
22. Load the cells into the elutriation chamber at 10mL/min. Take one cell-loaded syringe from the ice and invert it to make sure the cells are in suspension. Remove the 18-gauge needle from the end and discard. Invert the syringe and twist it into place on the stopcock making sure it is secure. Turn the valve on the stopcock so the pump starts to pull the cell suspension into the tubing. Once the cells have started to move into elutriator tubing, place the outlet tubing into a 50-mL centrifuge tube.
23. Allow the majority of the cell suspension to be taken into the tubing, return the stopcock to its original position allowing the GBSS to rinse the cells through the elutriation chamber. Remove the syringe, discard it, and allow the first 50-mL centrifuge tube to fill.
24. Change the pump setting to 21 mL/min, and collect another 50 mL of cells; this will elute off any remaining red blood cells, hepatic stellate cells, and endothelial cells.
25. Change the pump setting to 45 mL/min and collect 2 × 50mL of cells; this will elute off the Kupffer cells.

26. Once you have collected the Kupffer cells, return the outlet tubing to the collection bucket and stop the centrifuge. The cells that were not eluted out of the chamber will wash out into the collection bucket at this time.

27. Allow 100-200 mL of GBSS to wash the elutriation chamber before repeating steps 21 – 26 for the other cell suspension.

    If there is debris left in the chamber, disassemble the chamber and clean out the debris. If left inside the chamber, it may increase the pressure during the next round of elutriation.

27. Centrifuge the Kupffer cells (2P and 2E 50-mL centrifuge tubes) at 500 g for 7 min.

29. Under the tissue culture hood, prepare the nycodenz (ND) gradient for the final purification step.

    (*a*) Add 4mL of 24.6% ND into 2 × 15-mL centrifuge tubes for each treatment group,

    (*b*) Slowly layer (drop by drop initially) 4mL of 11% ND onto the –24.6% ND. An interface will be visible between the layers if done correctly.

30. Take the pelleted cells from step 28, aspirate off the GBSS, and resuspend in 10 mL of fresh GBSS.

31. Take 5 mL of cells and slowly layer them onto each ND gradient.

32. Centrifuge the gradients at 1644 g for 15 min WITH NO BRAKE. Using the brake will disrupt the separation of cells within the gradient.

33. After spinning, pair-fed Kupffer cells will be suspended in the 24.6% ND. Anything that pellets at the bottom should be discarded. With cells from rats that were ethanol-fed, two layers of cells can be seen within the gradient. The top layer, within the 11% ND, will be fatty hepatocytes. The bottom layer within the 24.6% ND, is Kupffer cells. Any cells that pelleted through the gradient should be discarded.

34. Under the tissue culture hood, take a 9-inch Pasteur pipet and attach a rubber bulb. Squeeze the rubber bulb and carefully insert the Pasteur pipet through the ND to the Kupffer cells and carefully-remove the cells within the 24.6% ND layer.

35. Remove the Pasteur pipet and place the collected cells into a new 50-mL centrifuge tube. Repeat step 34 to remove all of the Kupffer cells from the gradient.

36. Take the 50-mL centrifuge tube with the Kupffer cells (step 35) and fill to 50mL with GBSS.

37. Centrifuge the Kupffer cell suspensions at 500 g for 7 min.

37. Aspirate off the GBSS and resuspend the pellet with 10 mL of CMRL 10% FBS.

39. Into a 1.5-mL sterile Eppendorf tube, pipet 70 μL of 3% acetic acid, 10 μL of trypan blue, and 20 μL of Kupffer cells for counting.

40. Resuspend the cells to your desired density for plating. In our lab, we resuspend the cells to 2×10^6 cells/mL.

    (*c*) We use a plating density of 6.25×10^5 cells/cm^2 or 0.625 mL/cm^2, maintaining a consistent plating density across different types of cell culture plates.

41. Wash within 1 h of plating to remove any nonadherant cells.

Isolation of RNA From Primary Kupffer Cells

Analysis of gene expression in isolated Kupffer cells from pair- and ethanol-fed rats is integral to our understanding of the mechanisms that contribute to the pathology associated with chronic ethanol exposure. Using real-time reverse transcriptase polymerase chain reaction (rtPCR), the effects of ethanol on the expression of several genes can be examined rapidly, even with a small number of Kupffer cells. The brief protocol below describes a method to isolate total RNA and synthesize cDNA from rat Kupffer cells for gene expression analysis by real-time rtPCR.

1. Seed six wells of a 96-well plate at a density of 6.25 × 10^5 cells/cm^2 for each experimental condition.
2. The RNeasy Micro kit (Qiagen, Valencia, CA) is used to isolate total RNA from Kupffer cells. The total number of cells per experimental condition is not to exceed 5 × 10^5. Fifty-eight microliters of the lysis buffer, provided with the kit and supplemented with 2-mercaptoethanol by the user, is added directly to each well using a multi-channel pipet. This results in a total of 355 μL of total lysis buffer for a maximum of 5 × 10^5 cells. Wells of the same experimental condition are combined into a single DNase-RNase-free 1.5-mL microcentrifuge tube before nucleic acid precipitation.
3. The remainder of the RNA isolation, including on-column DNA digestion, proceeds according to the manufacturer's instructions.
4. RNA yield is small (250-500 ng/sample) when compared with isolating RNA from, for example, whole liver. Therefore, quantify the RNA and test purity using the least amount of sample possible. The NanoDrop instrument (NanoDrop Technologies, Wilmington, DE) is ideal as it only requires 1 μL of undiluted RNA per sample.
4. We routinely reverse transcribe the RNA immediately to maximize the amount of sample for subsequent analyses. Total RNA is reversed transcribed using Ambion's RETROscript kit (Austin, TX) according to the manufacturer's instructions, using random decamer as primers according to manufacturer's instructions.
5. Finally, cDNA can be diluted 1:2 with sterile, RNase- DNase-free water. This will only increase threshold cycle time by one, but will double the starting material and therefore maximize the number of genes one is able to examine per sample.

NOTES

1. We make use of two surgical pans (similar to a 9 × 13 baking pan), and two wire mesh racks on which to place the four animals. One wire mesh rack is placed on each baking pan, and one pair-fed and one ethanol-fed animal are placed on top of each wire mesh rack. The purpose of this set up is to allow the 2L of perfusate (1L of Hanks- and 1L of Williams E) to drain from the animal into the surgical pan. This allows for a more sterile field to be maintained around the liver because it is not covered with perfusate and also allows for easy clean up of the area after the perfusion is complete.
2. It may be helpful when first learning the perfusion procedure to insert the butterfly needle into the hepatic portal vein before turning on the peristaltic pump. However, starting the pump before inserting the needle generally allows for a better perfusion of the liver.

   If the needle is inserted before starting the pump, many red blood cells may be left at the edges of the liver, which are hard to separate from the Kupffer cells.
3. It is also important to test different batches of collagenase before starting experiments. We use Collagenase type IV from Sigma (Catalog C5138). If you plan on doing multiple experiments, it is a good idea to call the Reserves Department at Sigma and request a 100-mg tester batch of a lot of collagenase with sufficient reserves to completer all of our

experiments. Chow-fed control animals can be used to test the new batch of collagenase. It is important to look at the success of the perfusions, as well as, the yield of Kupffer cells. If there is a problem with either of these, it is likely that the lot of collagenase used will not work for other animals. Call Sigma for another tester batch, if necessary.

4. Centrifuging the cells at 4°C and keeping them in ice, starting at Subheading 3.6., step 13, is important for good quality and yield of Kupffer cells.
5. When taking the ethanol cells off the ND gradient, be careful not to remove the fatty hepatocytes. One trick is to insert the Pasteur pipet just past the fatty hepatocyte layer and then gently squeeze out any remaining air from the rubber bulb. This will push any cells that were in the tip of the pipet out before inserting the pipet down to the Kupffer cell layer.
6. The 3% acetic acid will cause any remaining red blood cells to lyse, and the trypan blue will allow for enumeration of the viable Kupffer cells.
7. The total number of cells from which RNA can be isolated using the RNeasy micro columns is 5×10^5. If you exceed this number of cells, it is likely that RNA yields will suffer in quantity as well as quality. Therefore, determine the number of cells per well and ensure the total number of cells per experimental condition does not exceed 5×10^5. Cell counts can be done immediately before experiment initiation. If the cell culture medium is to be changed before initiation of the experiment, count cells after the media change. If needed, adjust experimental protocol accordingly (*i.e.*, use fewer wells/ experimental condition).
7. DNase I used for the on-column DNA digestion step of the RNeasy micro procedure can be dispensed in 10 µL aliquots in 600 µLof RNase- DNase-free tubes and stored at –20°C until use. One of these aliquots is used per sample during RNA isolation.
9. It is also possible to perform the reverse transcription reaction without prior RNA quantification. This is due to the fact that the reagent concentrations for the reverse transcription reaction, using Ambion's RETROscript kit, are suitable for up to 2 µg of RNA, far above what is typically isolated using the small numbers of Kupffer cells described here. Input sample amounts can be normalized within and between samples by comparing the expression of a house keeping gene to the gene of interest.

CHAPTER 8 Biochemical Defense Mechanisms

Our bodies are under constant attack by viruses, bacteria, protozoa, and metazoan parasites. Persons born without an immune system adequate to fight off these invaders die very quickly unless heroic measures are taken. We have learned to cooperate with our immune systems by immunizing ourselves against some bacteria and viruses. At other times we may fight a stubborn battle with our own defense systems against allergic reactions and a variety of autoimmune responses.

A. LOCATIONS AND ORGANIZATION OF THE IMMUNE SYSTEM

The immune system has many components, many of which are dispersed throughout the human body (Fig. 8.1). Lymphocytes, which are a foundation of the immune system, constitute little more than 1% of the body's mass. However, this represents ~10^{12} cells of several types, about 10 times more than there are neurons in the brain. These 10^{12} cells make antibodies and T-cell receptors, both of which are thought to have ~10^{15} different peptide sequences. And this is only the beginning of the complexities.

Immune responses have often been described in terms of *humoral* and *cellular* components. The humoral response involves the small circulating *B lymphocytes* (B cells), the *antibodies* (immunoglobulins), and proteins of the *complement* system. The cellular response is mediated by another group of small lymphocytes, the *T lymphocytes* (T cells). They resemble B cells in appearance but have quite different functions. However, newer knowledge has provided a somewhat different description of the body's defense systems, which can be classified into three levels. (1) The skin and internal mucous membranes, which are resistant to infection and have antibacterial properties, provide the first level of defense. (2) A fast-acting *innate immune system* can respond within a few minutes to breaches in the barriers provided by the tough outer skin and the glycoproteins of mucous surfaces and provides a second level. (3) A slower *adaptive* (acquired) part of the immune system leads to synthesis of antibodies and to long-term immunity, providing the third level (Table 8.1). Both B and T lymphocytes together with *antigen-presenting cells (APCs)* are necessary for the selection and development of immunoglobulin structures appropriate for attack on an invading organism.

The innate immune system utilizes *phagocytic cells* including neutrophils, monocytes, and macrophages to ingest and kill invading organisms. Basophils, mast cells, eosinophils, and other cells release inflammatory mediators, which attract additional lymphocytes and affect their development. Specialized T lymphocytes called *natural killer (NK) cells* may also attack foreign cells (Table 8.2). The innate system is ancient and has apparently evolved to recognize molecular structures that are foreign to the host but are characteristic of pathogens. These structures, which are described as *pathogen-associated molecular patterns (PAMPs)*, include those of lipopolysaccharides of bacterial cell walls (Fig. 8.30), mannans, other carbohydrates of surface layers, oxidized phosphatidylcholines, bacterial flagellins, various posttranscriptionally modified proteins, teichoic acids, etc. However, they do not include patterns characteristic of host cells. Those have been avoided during evolution of the system. Although T lymphocytes are major mediators of the innate immune response they are under control of the *dendritic cells* (*DCs*), which are found in "immature" forms in tissues throughout the body.

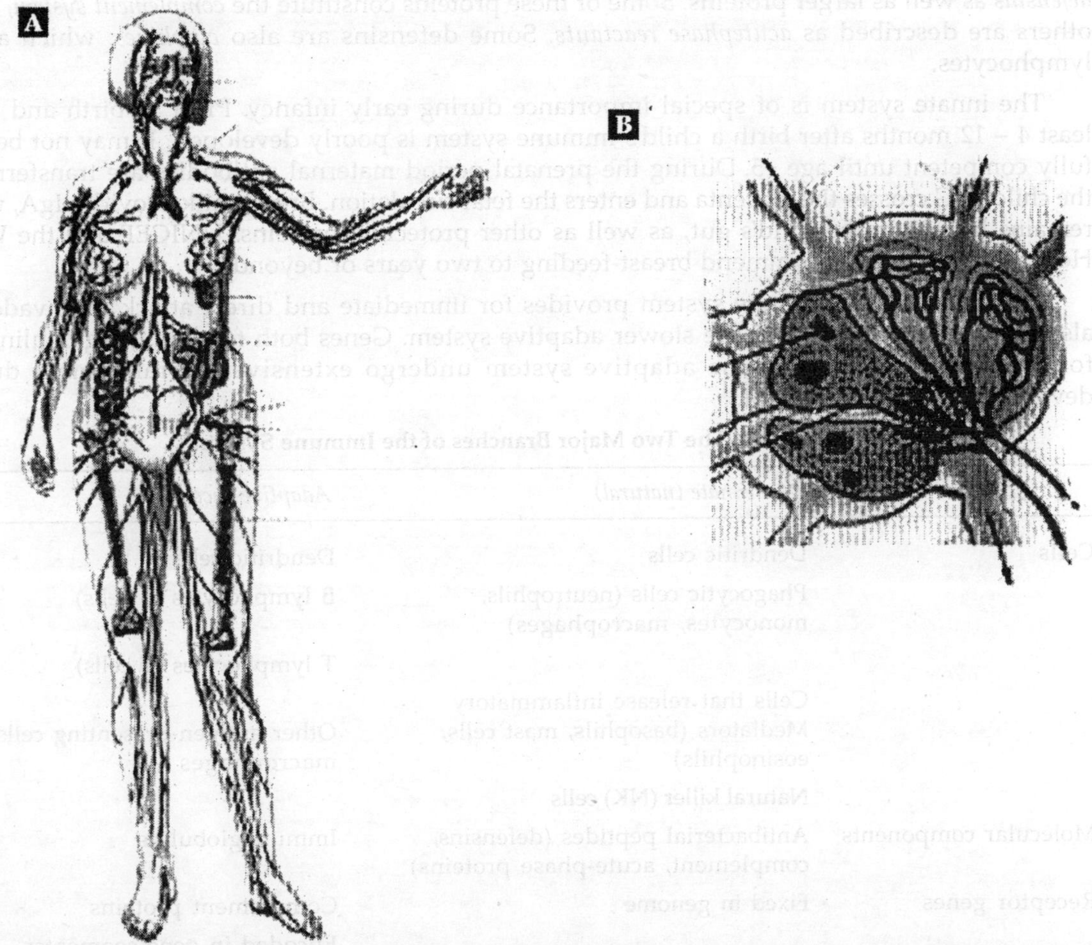

Fig. 8.1. (A) Locations of the primary and secondary tissues of the immune system. The primary lymphoid organs are the thymus, which makes T cells, and the bone marrow, which forms B cells. After moving from these organs into the blood circulation the cells reach one of the secondary lymphoid organs, which include lymph nodes, spleen, tonsils, and Peyer's patches on the small intestine. Immature dendritic cells are found in body tissues including skin and mucous membranes and respond to foreign proteins by inducing attack by T lyphocytes and antibody formation by B cells. (B) Schematic drawing of a lymph node.

The immature DCs are phagocytic cells that act as "immunological sensors." They recognize various PAMPs, which act as *danger signals*, using what are known as *toll-like receptors* (*TLRs*). They are also the most active APCs. Their proteasomes cleave proteins, both of the host and of invading organisms, into short peptides. These peptide fragments are displayed on the APC surfaces for recognition by T lymphocytes and for activation of adaptive immune responses. Some "autoreactive" B cells are also part of the innate immune system as are the IgA antibodies present in mucous membranes. The innate immune system of insects resembles that of vertebrates. The toll-like receptors of the latter are named for their resemblance to the Toll receptors of *Drosophila,* which are utilized in resistance to fungi. In both mammals and insects the innate immune system activates responses via the transcription factor NF-κB. However, many details of the signaling pathways differ. The innate system also provides for synthesis of small antibiotic peptides called *defensins* as well as larger proteins. Some of these proteins constitute the *complement system,* while others are described as *acutephase reactants.* Some defensins are also *cytokines,* which attract lymphocytes.

The innate system is of special importance during early infancy. Prior to birth and for at least 4 – 12 months after birth a child's immune system is poorly developed. It may not become fully competent until age ~5. During the prenatal period maternal antibodies are transferred to the child. IgG crosses the placenta and enters the fetal circulation. Breast milk provides IgA, which remains largely in the child's gut, as well as other protective proteins. UNICEF and the World Health Organization recommend breast-feeding to two years or beyond.

While the innate immune system provides for immediate and direct attack on invaders, it also provides information to the slower adaptive system. Genes both for immunoglobulins and for the T-cell receptors of the adaptive system undergo extensive rearrangement during development of an individual.

Table 8.1. The Two Major Branches of the Immune System

	Innate (natural)	*Adaptive (acquired)*
Cells	Dendritic cells	Dendritic cells
	Phagocytic cells (neutrophils, monocytes, macrophages)	B lymphocytes (B cells)
		T lymphocytes (T cells)
	Cells that release inflammatory Mediators (basophils, mast cells, eosinophils)	Other antigen-presenting cells, *e.g.*, macrophages
	Natural killer (NK) cells	
Molecular components	Antibacterial peptides (defensins, complement, acute-phase proteins)	Immunoglobulins
Receptor genes	Fixed in genome	Complement proteins
		Encoded in gene segments; rearrangement necessary
Recognition	Conserved molecular patterns	Small molecular groups (epitopes)
Immunogenic memory	Absent	Present
Self–nonself discrimination	Perfect	Imperfect
Action time	Immediate	Delayed

This provides potential defensive proteins directed at almost every imaginable invader. It also ensures that every individual has a set of proteins that labels its own cells as "self," and that virtually every individual on earth has cell surface proteins different from those of every other person. In both the innate and adaptive responses the immune system must carefully distinguish "self" from "nonself." In the innate system this discrimination developed during evolution of the host and its pathogens. In the adaptive system it depends upon interaction of the T cells with surface molecules, primarily those of the *major histocompatibility complex* (MHC).

Another basic characteristic of immune responses is the development of *immunologic memory.* This is exemplified by the fact that vaccination can sometimes impart immunity for a person's lifetime. If a foreign protein is injected into an animal, after a lag period of 2 – 5 days the animal will synthesize antibodies against this foreign antigen. This is called a *primary adaptive immune response.* If after a few days or weeks a second injection of the same protein is made, a much more rapid synthesis of additional antibodies occurs. This *secondary immune response* may take place within hours and will last longer than the primary response. It is a manifestation of immunologic memory.

1. Development of Lymphocytes and other Specialized Cells

Both the B cells and T cells arise in the fetal liver or bone marrow (Fig. 8.1) from pluripotent stem cells. In birds the B cells develop in a special organ, the bursa of Fabricius. Mammalian B cells complete their differentiation into mature *lymphocytes* within the bone marrow. However, the T cells must travel to the *thymus,* where they complete their maturation. The T lymphocytes include the previously mentioned NK cells as well as the somewhat similar *cytolytic T cells* and *immunoregulatory T cells.* The latter are further characterized as *helper T cells* or *suppressor T cells.* The adaptive response requires cooperation of helper T cells in many instances. The mature B and T cells leave the bone marrow and thymus, which are known as the *primary lymphoid tissues,* and enter the blood circulation. Following "homing" signals they take up residence in a variety of locations in the lymph nodes, spleen, adenoids, tonsils, and Payer's patches. The last are small clusters of lymphoid cells in the wall of the intestine. All of these tissues, which are referred to as *secondary lymphoid tissues,* are the sites in which the adaptive immune system is developed.

An important component of the immune system that was neglected until recently is located in the mucous membranes and the skin. The mucosal surfaces of airways and the gastrointestinal tract provide the point of entry for many diseases. Both internal and external body surfaces are protected by dendritic cells, whose immature forms in skin are called Langerhans cells.

Eight different types of cells of the immune system (Table 8.2) develop by differentiation of pluripotent stem cells as indicated in Fig. 8.2. Dendritic cells (which are not shown in this figure) may be formed from monocytes but may also arise by other routes. The development of the various cells takes place under the influence of a number of *hemopoietic regulators.* Among these are the protein hormone *erythropoietin* and various interleukins and colony-stimulating factors.

2. Triggering an Immune Response

When a foreign antigen enters the body the B cells, with receptors of appropriate specificity and present in the lymph nodes, are stimulated to divide repeatedly and to produce a large clone of *plasma cells.* These contain a highly developed ER and actively synthesize and secrete immunoglobulins. One activated B cell may produce 10 million antibody molecules per hour. This B-cell response occurs within a network of *follicular dendritic cells* in the *germinal centers* of the lymph nodes (Fig. 8.1B). The antibodies are "adaptor" molecules, which bind to antigenic proteins often on surfaces of invading microorganisms. Another part of the antibody binds to one of several *effectors* systems. These immobilize microorganisms, induce phagocytosis, activate the complement system, carry antibodies across placental membranes, etc.

Table 8.2. Cells of the Immune System[a]

Type	*Functions*
B lymphocytes	
Plasma cells	Antibody synthesis
Memory B cells	Immunologic memory
T lymphocytes	
Cytolytic	Destroy infected and malignant cells
Helper cells	
Type 1 (T_H1)	Participate in activation of B cells
Type 2 (T_H2)	
Memory T cells	Immunologic memory
Natural killer (NK) cells	Destroy infected and malignant cells with pore-forming protein perforin and cytotoxic granules
Dendritic cells	
Interdigitating	Antigen recognition and processing
Follicular	Antigen presentation
Microglia	Defensive network in brain[b]
Cells that release Inflammatory mediators	
Mast cells and basophils	Posses high-affinity receptors for IgE May secrete histamine, prostaglandins, leukotrienes Important in allergies
Eosinophils	Weakly phagocytic, secrete cationic proteins, reactive reduced oxygen species, leukotrienes, prostaglandins, cytokinins
Phagocytic cells	
Neutrophils	Acute inflammatory response
Monocytes	
Macrophages	Carry receptors for carbohydrates not normally exposed on surfaces of cells in vertebrates *e.g.*, mannose; kill engulfed organisms with $^{\bullet}O_2^-$, HOCI, NO, cationic proteins and peptides, lysozyme Antigen processing and presentation

Induction of a T-cell response is more complex and is very demanding. Antigenic peptide fragments from infected or malignant cells anywhere in the body must be recognized by an appropriate receptor on a T cell that is circulating in the blood. Only a few T cells with receptors for any given antigen specificity exist. The foreign antigen fragment must bind to a protein of the *major histocompatibility complex (MHC)* while present within a dendritic cell or other APC. The resulting MHOantigen complexes pass through the ER and the Golgi to the outer cell surface in a rather complex process. Recognition of the antigen that is "presented" in this manner on the APC surface is accomplished with the aid of ~10^{15} different receptor proteins *(T-cell receptors)* on the T-cell surfaces. When a T-cell receptor is occupied by an HMC •antigen complex of appropriate specificity the T cell is activated to participate in adaptive immunity. However, some T cells, notably σδ8 T cells, like antibodies bind to antigen directly. The recognition process occurs in an *immunological synapse,* which has elements of similarity to neurological synapses.

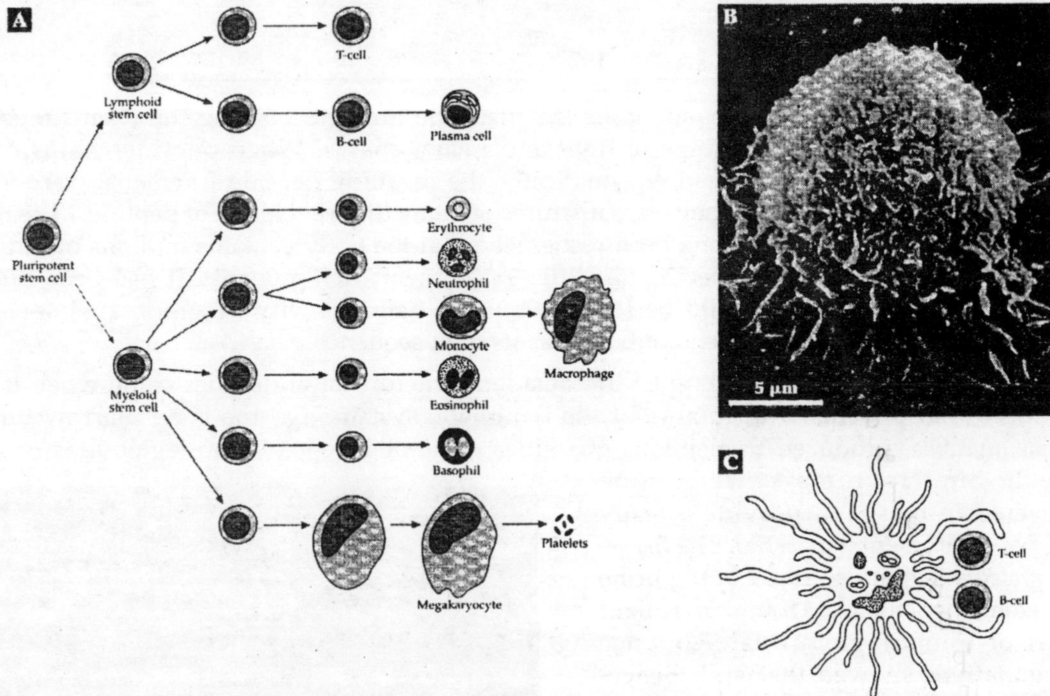

Fig. 8.2. (A) Development of eight types of blood cells including those of the immune system from pluripotent (multipotential) stem cells. The cells develop under the influence of a variety of protein growth factors. Some steps, *e.g.*, maturation of B cells, involve complex rearrangements in the DNA of the cell. (B) Scanning electron micrograph of a macrophage, a large motile cell that plays a key role in the immune system. It moves by means of its surface "ruffles." It actively phagocytoses both pathogens and waste materials and is also one of the cells that releases the hormones known as colony-stimulating factors. (C) Schematic drawing of a dendritic cell. Both macrophages and dendritic cells present antigens for recognition by T cells and synthesize cytokines, which affect lymphocyte development.

B. THE IMMUNOGLOBULINS (ANTIBODIES)

1. Molecular Structures

There are five classes of antibodies or immunoglobulins. The first three, IgG, IgM, and IgA, are quantitatively the most significant, but IgD and IgE are also important. For example, the content of IgE is elevated in allergic persons. The basic structure of all of the immunoglobulins is that of a quasi-symmetric dimer composed of a pair of light chains and a pair of heavy chains whose lengths vary among the different classes of immunoglobulins. Two classes of *light chains,* κ and λ are found in human antibodies. The *heavy chains* are designated γ, μ, α, δ and ε (see accompanying tabulation). Both IgM and IgA contain an additional J chain.

Symbol	Mass (kDa)	Formula
IgG	150	$\kappa_2\gamma_2$ or $\lambda_2\sigma_2$
IgM	950	$(\kappa_2\mu_2)_{2,5}$ or $(\lambda_2\mu_2)_{2,5}$
IgA	320	$(\kappa_2\alpha_2)_n$ or $(\lambda_2\alpha_2)_n$
IgD	180	$\kappa_2\delta_2$ or $\lambda_2\delta_2$
IgE	190	$\kappa_2\varepsilon_2$ or $\lambda_2\varepsilon_2$

Treatment with mercaptoethanol splits the disulfide linkages holding the chains together, permitting preparation of monomeric light and heavy chains. When peptide chains of the immunoglobulins were hydrolyzed enzymatically, the resulting peptide fragments were found to be extremely heterogeneous. They were mixtures of many different kinds of peptides. The result was not unexpected, for it had long been recognized that the body contains millions of different antibodies, with binding sites specific for different antigenic determinants. It had been unclear how different binding sites could be formed, but the heterogeneity in amino acid sequence suggested the correct answer: Each antibody has its own sequence.

Progress toward understanding of the detailed structure of antibodies came when it was recognized that patients with tumors of the lymphatic system, *e.g.*, the bone marrow tumors *multiple myeloma,* produced tremendous quantities of homogeneous immunoglobulins or parts thereof. Similar tumors were soon discovered in mice and provide a ready source of experimental material. The *Bence-Jones proteins* that are secreted in the urine of myeloma patients were found to be light chains of immunoglobulins. Sequence determinations showed that each Bence-Jones protein was homogeneous, even though no two patients secreted the same protein. Later, intact myeloma globulins and macroglobulins (IgM) of a homogeneous kind were also obtained.

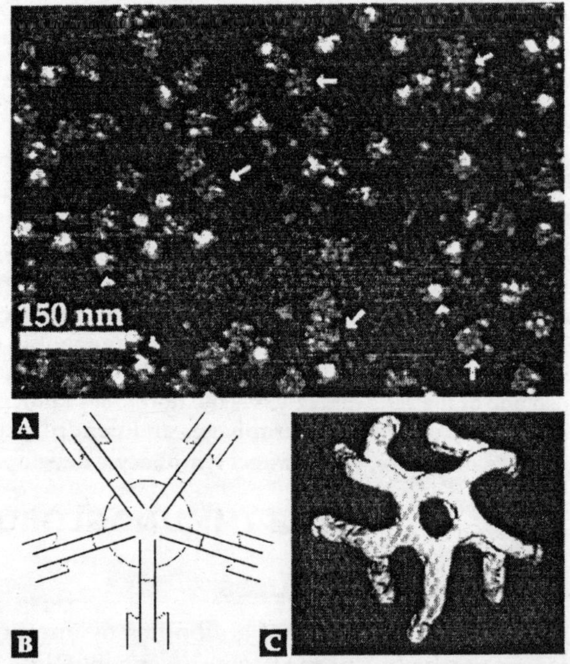

Fig. 8.3. (A) Cryo atomic force (AFM) micrograph of molecules of the human immunoglobulin IgM. Courtesy of Zhifeng Shao, University of Virginia. (B) Schematic diagram. One-fifth of this structure is shown in greater detail in Fig. 8.4A. (C) Model based on earlier electron microscopic images.

The first complete amino acid sequence of an IgG molecule was announced in 1968. The protein contained 446 amino acids in each heavy chain and 214 in each light chain. The longer heavy chains of IgM molecules contain 576 amino acids. In all of the immunoglobulins the heavy and light chains are held together by disulfide linkages, and the chains are folded into loops to form compact domains. The IgM molecule is polymerized through additional disulfide linkages to form a pentamer readily visible with the electron microscope (Fig. 8.3). The heavy chains also carry oligosaccharide units. In IgM there are five of these, as indicated in Fig. 8.4 A. They contain man-nose and *N*-acetylglucosamine units linked to asparagine. Other immunoglobulins (IgA, IgE, and IgG) contain fucose, galactose, and N-acetylneuraminic acid as well. In fact, almost all of the most important macromolecules that participate in innate and adaptive immune responses are glycoproteins.

Digestion of an intact molecule of IgG with papain cleaves both heavy chains in the hinge region near the interchain disulfide bridge. This splits the molecule into three parts; two *Fab* (antibody-binding) *fragments*, each containing the N-terminal end of a heavy chain together with a linked light chain, and an *Fc fragment*. Even before it was known that IgG could be split into two Fab fragments, the antibody was known to be divalent, *i.e.*, capable of binding with two different antigens (Fig. 8.4). The shape and overall structure of IgG molecules have been verified by electron microscopy and numerous X-ray diffraction studies.

Sequence determinations showed that in some regions of immunoglobulin molecules there is extreme variation in the amino acid sequence between one homogeneous antibody and the next; other regions have a constant sequence. The molecule can also be divided into domains. The *variable regions*, which occupy the N-terminal ends of the chains, are designated V_L and V_H for the light and heavy chains, respectively. The *constant regions* are C_L and C_H. Examination of the C_H region showed that much of the sequence is repeated after ~110 residues. In the IgG molecule the constant region of the heavy chains is made up of three such homologous domains (C_H1, C_H2, and C_H3). A fourth C_H domain is present in IgM. These facts suggest that duplication of a smaller gene coding for about 110 amino acids took place in the evolution of the immunoglobulins. Within the variable regions of immunoglobulin chains are *hypervariable regions* that form the antigen binding sites. These regions are located at the ends of the Fab fragments and involve both the light and heavy chains.

Within all of the domains each of the two peptide chains is folded in a similar way. Seven extended lengths of chain form two mostly antiparallel sheets between which hydrophobic side chains are packed. The overall size of the unit is ~4.0 × 2.5 × 2.5 nm. An S–S bridge links the two sheets in the center of each domain. The folding patterns in the variable domains are somewhat more complex. Different domains are linked by segments of extended peptide chain known as hinge or switch regions. These impart a segmental mobility, which seems to be important for functioning of the molecules.

The exact mode of binding to Fab fragments has been established for several specific *haptens*. Haptens are small molecules having the binding properties of antigenic determinants but unable by themselves to induce formation of antibodies when injected into animals. Binding of the hapten phosphocholine to one Fab fragment and of vitamin K to another involves the hypervariable regions of both the heavy and light chains. The same is true for the binding of lysozyme and a bacterial oligosaccharide. The binding sites for the haptens or for the antigenic determinants on larger antigens are largely within the ninestranded elliptical barrel or β sandwich formed from the two p sheets (Fig. 8.5). Four strands come from the V_L domain and five from the V_H domain. The barrel forms the bottom and sides of the antibody-binding site, which can also be viewed as consisting of six separate loops of peptide chain. IgA molecules are similar to those of IgG but have structurally different hinge regions as well as an extra 18-residue tail-piece at the C terminus of each heavy chain.

As shown in Fig. 8.5, which provides a three-dimensional view of the variable domain of an Fab fragment of an immunoglobulin, there is a conserved hydrogen-bonded network even in this region. There are also "framework residues," which are highly conserved. The antibody-binding site is provided largely by the three hypervariable regions present in each of the V_L and V_H

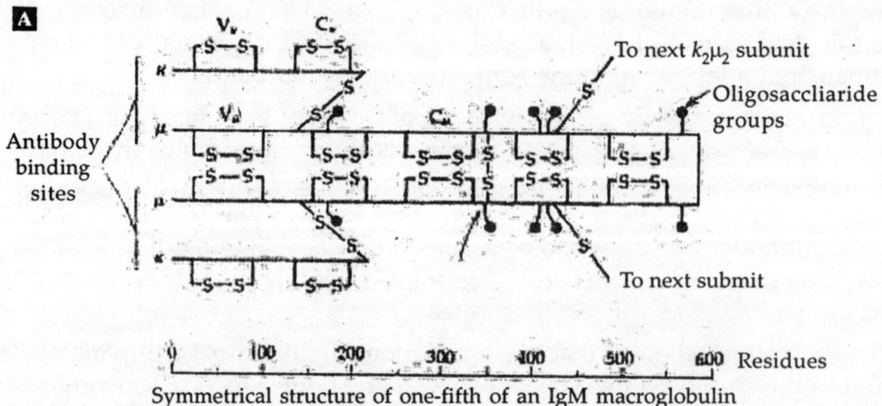

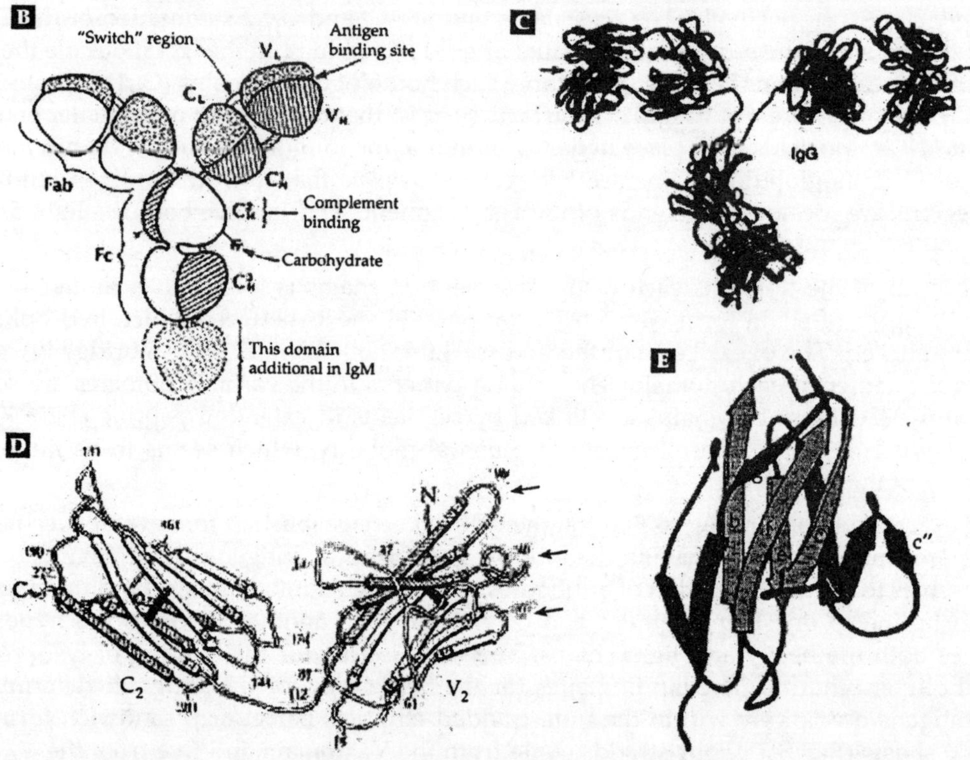

Fig. 8.4. Schematic structure of one-fifth of an IgM molecule. From Putnam *et. al.*, (A) Covalent structure. (B) Schematic three-dimensional representation. (C) Ribbon diagram of an IgG molecule. From Cochran *et. al.*, (D) Folding patterns of one chain in a constant and a variable domain of a Bence-Jones protein. From Schiffer *et. al.* Green arrows indicate hypervariable regions. (E) MolScript drawing of the common core structure of Ig-like domains. The lighter shaded strands (*b, c, e, f*) form the core common to all Ig-like domains, which is surrounded by structurally more varied additional strands (darker). The front sheet has up to five strands (*a, f, c, e, c"*) and the back sheet up to four (*a, b, e, d*). Strand *c"* is very flexible and is not always a part of the β sheet.

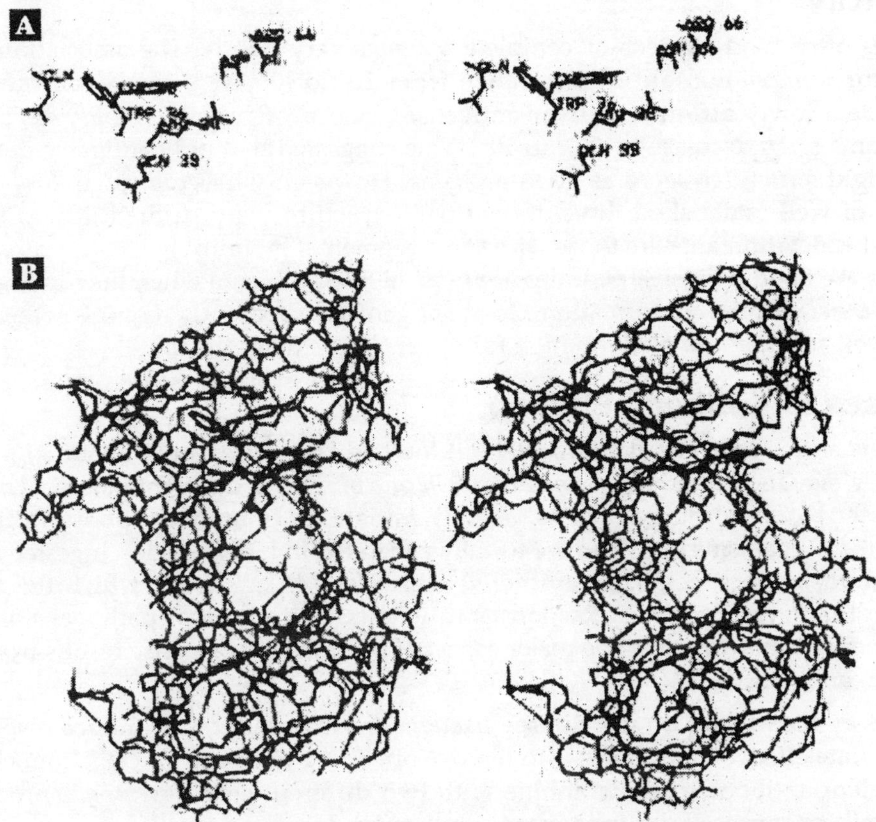

Fig. 8.5. The extensive conserved hydrogen-bonding pattern in an immunoglobulin variable domain provided by polar residues buried inside the V_L and V_H domains. (A) To facilitate orientation, prominent side chains are displayed here and identified by names and numbers in the same orientation as in (B). (B) Polypeptide chain backbones of both domains are denoted by heavy lines and hydrogen bonds by light lines. In addition to the regular interbackbone hydrogen-bonding network characteristic of antiparallel (β-sheets, there are hydrogen bonds provided by side-chain atoms. Note the two hydrogen bonds of Gln-38 (V_L) and Gln-39 (V_H) that span the domain-domain interface.

domains. These are usually referred to in current literature as *complementarity-determining regions (CDRs).* Each pair of heavy and light chains is held together by a conserved disulfide bridge. Three-dimensional structures of a substantial number of different Fab fragments have provided precise knowledge about the antibody-binding cavities and about the forces involved in binding. Among the established structures are those of Fab fragments specific for the following antigens: the haptens *p*-azophenylarsonate and phencyclidine, a sweet-tasting hapten, triple-stranded DNA, a DNA photoproduct, creatine kinase, staphylococcal nuclease, an HIV capsid protein, and an EGF receptor. Structures are also known for single-domain antibodies from camels and llamas. These antibodies are naturally lacking in heavy chains but have single chains that fold back to mimic two-chain Fab fragments. Similar single-chain antibody domains have also been created artifically.

Not all proteins bind to antibodies in the usual binding cavity. *Protein G,* a cell surface protein from *Streptococcus* bonds to IgG molecules from many different species. Its binding site is on the outer surface of the heavy chain C_H1 domain.

2. Antigenicity

Antibodies often bind haptens or complete antigens very tightly. The association constants K_f observed for monoclonal antibodies range from 10^6 to 10^{12} M^{-1}. However, most natural antibodies have a lower affinity for their antigens. When protein antigens are denatured, the binding constants often decrease by 10^{-4} to 10^{-5}. This suggested that only antigenic determinants of relatively rigid structures serve as good antigens. However, when the reaction of antibodies with proteins of well-established three-dimensional structure were studied, it was found that the best antigenic determinants are those with some segmental mobility. Furthermore, while some small peptides are good antigens, peptides are most highly antigenic when they can readily fold into a bend or other definite conformation. Good antigenicity apparently requires some segmental flexibility as well as a definite conformation for the antigenic determinant.

3. Responses to Antibody Binding

Both B cells and T cells circulate throughout the body, spending only about 30 min during each cycle. They may meet and bind to an antigen in one of several different places. Lymphocytes, which encounter blood-borne pathogens, usually initiate an immune response in the spleen. Responses to microorganisms in tissues are usually generated in lymph nodes. Ingested pathogens activate lymphocytes in specialized epithelial *microfold (M) cells* from which the antigen is transported to the Peyer's patches. Responses to inhaled or intranasal pathogens arise in the tonsils and adenoids. In every case one major aspect of the immune response results from binding of antibodies to antigens.

Antibodies by themselves do not destroy bacteria or viruses, but they induce responses that do. One immediate effect of antibodies is to remove offending materials or cells from circulation. When multivalent antibodies each combine with two different cells *agglutination* occurs. The agglutinated cells or multicellular organisms can then be destroyed by phagocytes. The coating of a cell surface by IgG is one form of a process called *opsonization,* a process that marks the cell as foreign and a target for phagocytosis. Antibody-antigen interactions trigger several other responses as well. One of these results from the binding of protein C1q, a component of complement. Complement consists of a series of blood proteins that is poised to respond and to *complement* the action of antibodies in a variety of ways. It has been established that it is the C_H2 domain of the Fc region of IgG that binds to C1q. The binding occurs only after antigen (but not a small hapten) binds to the immunoglobulin.

Complement C1q is only one of several types of Fc receptor. Others are involved in antigenic stimulation of B and T lymphocytes, macrophages, polymorphonuclear lymphocytes, and mast cells. Binding of the antibody-antigen complex to the recep-tors on phagocytic cells induces phagocytosis and release of oxygen metabolites, leukotrienes, prostaglandins, and other mediators of inflammation. The Fc domain mediates the uptake of antibodies from the mother's milk by young rats. It also is the binding site of antibodies to *protein* A, a constituent of the cell wall of *Staphylococcus aureus,* which is also widely used as a tool in immunological studies. The neonatal Fc receptor, which is related structurally to Class I MHC antigens is one of three major types of Fc receptor. The other two are the receptors for Fcγ (of IgG) and Fcε (of IgE). They (like their ligands) are members of the immunoglobulin superfamily. An exception is FcεRII (also called CD23), which resembles a C-type lectin. Some Fc receptors, *e.g.,* FcγRI (CD64) and FcεRI, have a high affinity for their ligands with K_d ~10^{-8} to 10^{-10} M. Others, such as FcγRII (CD32) and FcγRIII (CD16), have lower affinities with K_d ~10^{-5} to 10^{-7} M. Three-dimensional structures of several Fc receptor fragments, some in complexes with Fc fragments (Fig. 8.6), are known. These include both IgG and IgE receptors.

It may be worthwhile to recall that many quite different proteins are members of the immunoglobulin structural family (Fig. 8.16). These include proteins encoded by 64 genes of *E. elegans,* an organism that doesn't form antibodies.

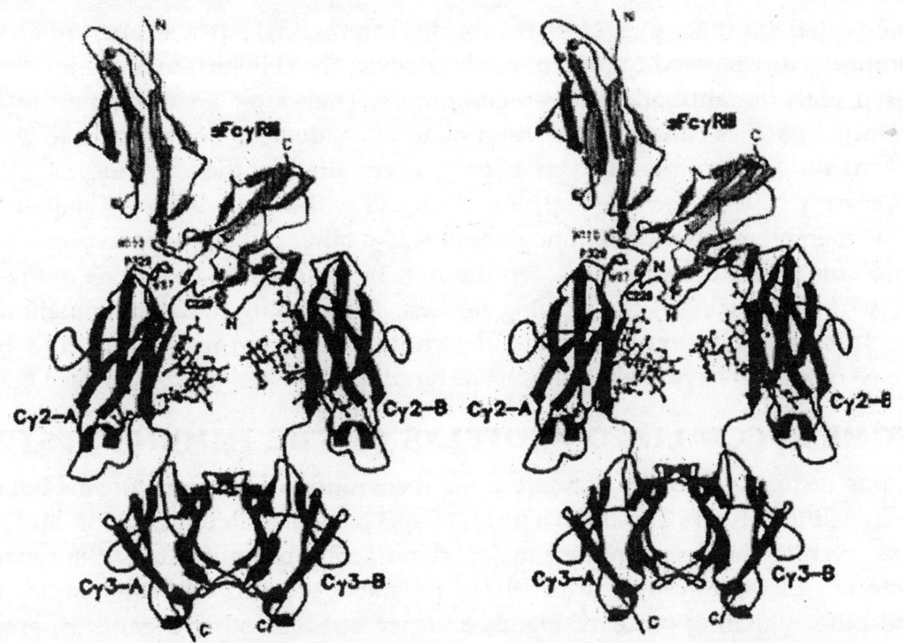

Fig. 8.6. Three-dimensional ribbon representation of the structure of a complex of a soluble Fc fragment of a human IgGl molecule. Pro 329 of the IgG and Trp 87 and Trp 110 of the Fc-receptor fragment form a "proline sandwich," which is shown in ball-and-stick form. The oligosaccharide attached to the Fc fragment of the antibody and the disulfide bridge between the two Cys 229 residues (at the N termini of the C2 domains of the heavy γ chains) are also shown. The small spheres on the Fc receptor fragment are potential sites for *N*-glycosylation.

4. Clonal Expansion of B Cells; Plasma Cells

The immunoglobulins are synthesized both by plasma cells and by their precursors, the B cells. Each B lymphocyte makes antibodies of specific sequence in two forms, secreted and membrane-bound or *antigen receptor.* Mature "virgin" B cells, which are responsible for the primary immune response, make largely monomeric IgM and some IgD. It isn't clear why IgD should be the predominant surface immunoglobulin on most B cells. The secreted and membrane-bound (receptor) antibodies differ in their C-terminal sequences but are otherwise the same. A B cell responds to the binding of an antigen with a shape complementary to that of its IgM antigen receptor by multiplying and differentiating. Some of the progeny B cells start to divide and begin to differentiate into clones of plasma cells that secrete IgG or into cells of the gut that secrete IgA. Some of the B cells give rise to *memory cells,* long-lived lymphocytes that can be triggered into rapid proliferation many years later if the same antigen is encountered. The B cells also undergo a shift to secretion of pentameric IgM rather than to synthesis of membrane-bound antibody.

5. Help from T Cells

The maturation of B cells is a complex process that requires the cooperation of helper T cells. The B cells must process some antigen and present the peptide fragments for recognition by the MHC complex and T-cell receptors. If the antigen is recognized as foreign, the differentiation goes forward with the activated T cells secreting lymphokines that promote B-cell growth.

Before activated lymphocytes start to divide, interesting surface phenomena occur. If fluorescent antigens are allowed to bind to a lymphocyte, the cell surface is seen to be relatively evenly covered with the antibody-antigen complexes. Then after a short time the antibodies aggregate to form "patches" and begin to migrate to one side of the cell, where they eventually form a "cap." At still longer times the cap material is engulfed by the lymphocytes. Perhaps this phenomenon simply reflects the oriented flow of liquid within membranes (Chapter 8). On the other hand, the membrane-bound immunoglobulins, like other cell surface receptors, are integral membrane proteins whose cytoplasmic C termini may be attached on the inside of the cell to the cytoskeleton, which may control the capping process. The binding of lectins sometimes triggers lymphocytes into antibody synthesis, but it is not clear how the binding of a lectin to a carbohydrate receptor can have the same effect as binding of an antigen to a surface IgM.

C. SOME SPECIALIZED PROTEINS OF THE IMMUNE SYSTEM

The immune response depends not only upon recognition of foreign antigens but also upon an extensive signaling network and upon a series of specialized antibiotics, toxins, and hormones. Some of these, such as the defensins, are major, rapid-acting components of the innate system. Both the innate and adaptive systems utilize the complement proteins, and both employ numerous cytokines and other signaling proteins. Plants, and perhaps also other organisms, employ gene silencing by small RNA molecules as part of their defense against viruses.

MONOCLONAL ANTIBODIES

A mouse may make over 10 million different antibodies. Because of this heterogeneity it was impossible to learn antibody structures until the discovery of the myeloma proteins (Fig. 8.4D). These were produced in the bone marrow by clones of specific immunoglobulin-forming malignant cells. However, it was still not possible to obtain homogeneous antibodies to any desired antigenic determinant. The discovery of a method of forming such *monoclonal antibodies* by Milstein and Kohler in 1975 provided a new tool with many biochemical and medical applications. What Milstein and Kohler did was to immunize mice against an antigen of interest. They then fused B cells from the spleen of the immunized mouse with cultured myeloma cells. The resulting *hybridomas* grow vigorously and produce antibodies of the type dictated by the B cells. Since each hybridoma cell is derived from a single B cell, it makes a single kind of antibody. By plating out and selecting clones of hybridoma cells it is often possible to find a monoclonal antibody that binds well to a specific antigenic determinant. The hybridoma can be cultured indefinitely, producing its monoclonal antibody in any desired quantity.

A major application of monoclonal antibodies is in clinical assays for drugs, bacterial and viral products, tumor antigens, hormones, and other circulating proteins. Their use in conjunction with immunoassays has provided increased specificity and sensitivity. Another major application is to observe binding of antibodies to specific proteins by electron microscopy. The location of specific receptor proteins can be established as can the locations of ribosomal proteins and many other cellular components (Fig. 8.1). Monoclonal antibodies to acetylcholine receptors have been shown to induce symptoms of myasthenia gravis, supporting the autoimmune origin of that disease. Monoclonal antibodies specific for such a small hapten as mercuric ion have been isolated.

Several problems have limited the wider use of monoclonal antibodies created by the hydridoma method. The antibodies are those of a mouse and are antigenic to humans. This long prevented many medicinal uses. Years of effort have gone into attempts to "humanize" the antibodies. One approach is to introduce human immunoglobin genes into mice. Another is to use recombinant DNA techniques to clone genes for immunoglobulin fragments and to introduce these into cells of *E. coli* in which additional genetic diversity in the antibodies arises. Selection of antibody fragments is often accomplished using bacteriophage display systems (Fig. 8.16). After selection gene fragments can be reassembled into a final form. Recently, using clon-ing of large pieces of the several Mbp of human immunoglobin genes into yeast artificial chromosomes, it has been possible to prepare purely human monoclonal antibodies.

Many attempts have been made to link monoclonal antibodies specific for antigenic determinants on cancer cells to protein toxins such as ricin (Box 29-A). It is hoped that this may provide an effective way of carrying toxins into cancer cells. Therapeutic human monoclonal antibodies are already in use as antirejection drugs for kidney transplantation, for treatment of rheumatoid arthritis, Crohn disease, and for some types of cancer.

CATALYTIC ANTIBODIES

Both enzymes and antibodies are proteins. Antibodies consist of subunits with multiple domains, just as do some enzymes. Both enzymes and antibodies have binding sites for small molecules between domains or subunits. In view of such similarities it isn't surprising that some antibodies have catalytic properties. The possibility was suggested in 1969 by Jencks.[a] He also proposed that injection of a mouse with a hapten, that resembled a transition state for an enzyme, might induce formation of antibodies complementary to the transitionstate structure. These might be catalytic. By the early 1980s such antibodies were discovered.[b - d] Some of the first catalytic antibodies (also dubbed *abzymes*) had esterase activity. The haptens used to induce antibody formation were phosphorates such as the following.

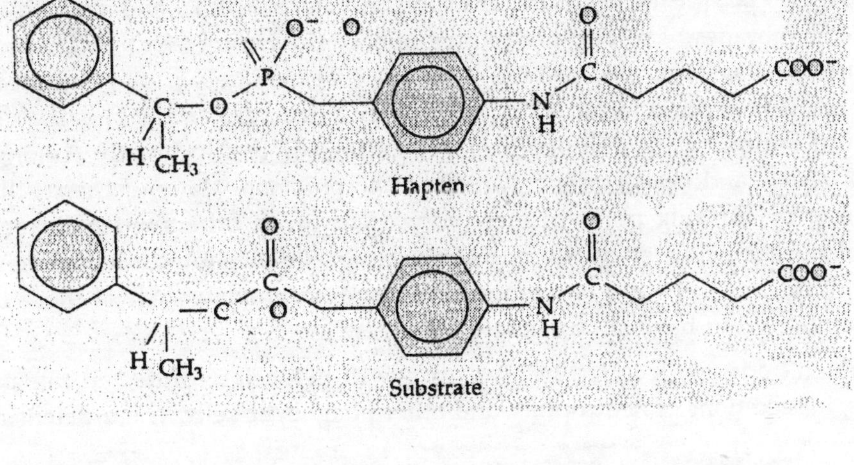

Hapten

Substrate

Using the transitionstate analog shown on a catalytic antibody with chorismate mutase activity was isolated. Many antibodies catalyzing additional reactions have also been found. Although they are usually less active than natural enzymes, in some cases they approach enzymatic rates. Furthermore, they may catalyze reactions for which no known enzymes exist.

Catalytic antibodies, like enzymes, must be isolated and purified to homogeneity before they can be studied. Initially this was done by using the hybridoma technique for isolation of monoclonal antibodies. After induction of antibody formation by injecting a selected hapten into a mouse, large numbers of monoclonal antibodies had to be tested for catalytic activity. Even if several thousand different monoclonal antibodies were tested, only a few with catalytic properties could be found. Newer methods have incorporated recombinant DNA techniques and use of combinatorial libraries and phage display. Incorporation of acidic or basic groups into the haptens used to induce antibody formation may yield antibodies capable of mimicking the acid-base catalysis employed by natural enzymes.

A sample of the types of reaction for which catalytic antibodies have been discovered or designed include following: ester hydrolysis, transesterification, amide hydrolysis serine protease-like hydrolysis elimination, aldol cleavage, decarboxylation, deiodination by a selenium-containing antibody, pericyclic rearrangements, and the Diels–Alder reaction. Like natural enzymes catalytic enzymes can be mutated and engineered and can be used to study fundamental aspects of catalysis. Fluorescent probes incorporate near active sites may provide information about mechanisms or may signal information of diagnostic significance.

In science we must always expect the unexpected. Do antibodies all catalyze the reaction of singlet molecular oxygen $^1O_2^*$ with H_2O to form H_2O_3 and H_2O_2? How?

1. Defensins and other Antibacterial Polypeptides

Only higher vertebrates have an adaptive immune system with circulating antibodies. However, from bacteria to higher plants and human beings all of us utilize defensive polypeptides for protection. More than 500 have been identified. Many have a broad specificity, attacking both bacteria and other pathogens. Among these peptides are more than 200 bacterially produced antibiotics such as gramicidin, tyrocidines, and colicins. More recently discovered are the 37- to 59-residue *bacteriocins,* formed by lactic acid bacteria. Like colicin E1 and alamethicin they disrupt cytoplasmic membranes of some other groups of bacteria. *Helicobacter pylori,* which is associated with stomach Ulcers, forms a 38-residue antibiotic that may help protect infected persons from other bacteria. This peptide forms a simple two-helix structure and is one of a large number of simple helical antimicrobial polypeptides 40 residues or less in length. Among them are the *cecropins* of insects and *magainins and buforins* of amphibians. Many of these kill by disrupting membranes or by forming pores in membranes. However, others enter bacteria and disrupt functions of nucleic acids, enzymes, etc. Many antibacterial peptides have been isolated from insects, scorpions, spiders and horseshoe crabs, and amphibians. All of these organisms lack adaptive immunity but have strong innate immunity.

The human body is protected by two groups of defensins formed in the skin, in mucous membranes, in secretions of neutrophils, and other phagocytic cells. The α-defensins (Fig. 8.7) are

29 – 35 residues in length and are active against both gram-positive and gram-negative bacteria as well as fungi and enveloped viruses including HIV. The β-defensins are mainly active against gram-negative bacteria and yeast. They also possess immunostimulatory activity that is important in activating the adaptive immune response. Various tissue-specific defensins have been discovered. Peptides of the *trefoil family* protect the gastrointestinal tract. Eosinophils, leukocytes, and neutrophils make additional protective proteins. One leukocyte defensin is a macrocyclic peptide, whose gene may have arisen by fusion of two segments encoding nonapeptide segments of α-defensins, Neutrophils form, in addition to defensins, α-helical peptides called *cathelicidins,* which protect skin from invasive bacterial infection. Their synthesis is greatly increased after wounding. They may be among the proteins whose absence after severe burning is likely to be fatal.

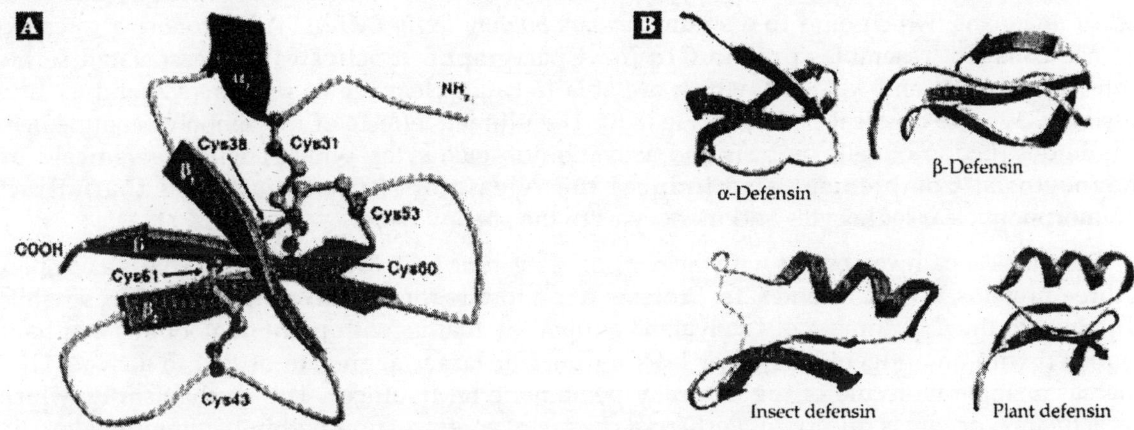

Fig. 8.7. Ribbon structures of some defensins. (A) Structure of a human β-defensin showing the three disulfide bonds. (B) Comparison of the folding patterns of four types of defensins. Mammalian α- and β-defensins are all β sheets with somewhat different arrangements of disulfide bridges. Insect and plant defensins have an a helix joined to the β sheet. Mammalian and insect defensins have three disulfide bridges, while plant defensins have four.

Both α and β-defensins consist largely of β strands (Fig. 8.7) and are linked by three disulfide bridges. Some scorpion and insect defensins resemble scorpion toxins (Fig. 8.16) and have four S–S bridges. Fungi and green plants also form antimicrobial peptides. A 30-residue fungal protein is highly knotted and contains four S–S bridges. Some polypeptides from the oleander and related plants are 29- to 31-residue macrocyclic structures with two S–S bridges in a *cysteine knot* structure (Fig. 8.7). They are exceptionally stable and protease-resistant and may have defensive activity against insects. Defensins are small polypeptides, but larger proteins are also part of the innate defense system. For example, a 93-residue protein from onion seeds resembles plant in lipid-transfer proteins. Some frog skins contain a 60-residue trypsin inhibitor. Ribosome-inactivating proteins are well known.

2. Complement

Complement is a group of more than 30 proteins found in blood serum, which are activated in a cascade mechanism when antibody and antigen combine (Eq. 8.1). This *classical pathway* for activation of complement is outlined in Fig. 8.8. The proteins involved in the cascade are designated C1 to C8. Many of them undergo proteolytic cleavage, the

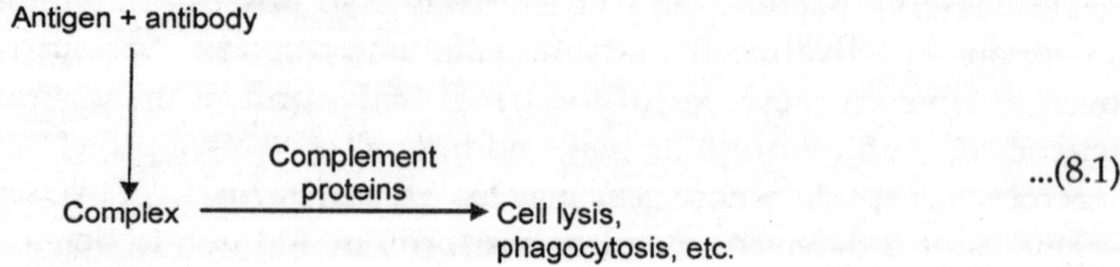

cleavage products being designated by a or b, *e.g.*, C3a and C3b. The *b* fragment is usually the larger of the two. There is also an *alternative pathway* that is part of the innate system. It is activated by such foreign surfaces as lipopolysaccharides of bacterial cell walls. Its special proteins are called *factors, e.g.*, factor B, factor D. A third pathway, the *lectin pathway,* is activated by microbial surface mannans, which bind to a serum *mannan-binding lectin (MBL).* This protein, a so-called defense collagen, resembles protein C1q (next paragraph). It activates two associated serine proteases (MASP1 and MASP2), which are able to cause cleavage of proteins C4 and C2 and possibly C3 in the classical pathway (Fig. 8.8). The ultimate effects of the action of complement include destruction of cells by lysis and activation of leukocytes, which engulf foreign cells by phagocytosis. Complement also induces the release of *chemotactic factors* that attract polymorphonuclear leukocytes and monocytes to the site involved.

The classical pathway begins with the *recognition component* C1 of complement. This is a complex of three proteins, C1q, C1r, and C1s. Proteins C1r and C1s form a mixed tetramer $C1r_2 s_2$ while C1q binds to the C_H^2 domain of "activated" antibodies, that is, with immunoglobulins that have combined with an antigen. It takes at least a dimer or larger aggregate of IgG to activate C1q, whereas a single molecule of the naturally pentameric IgM suffices. The mechanism by which this activation occurs is uncertain. Perhaps a change of conformation within the immunoglobulin acompanies antibody binding and is responsible for generation of a binding site for C1q. It may seem strange that haptens cannot cause complement binding, and that they do not cause detectable conformational alterations in Fab. Only multivalent antigens able to bind to more than one antibody induce complement binding. However, as we have learned in recent years, many biological responses involve transient assembly of large aggregates of different protein components. In this context, the requirement for two or more antibody molecules doesn't seem so strange.

The 400-kDa C1q consists of a central portion of diameter 3 – 6 nm and length 10 – 12 nm to which are attached six very thin connecting strands. These are ~14.5 nm long and ~1.5 nm in diameter and terminate ~135-residue globular ends of ~6 nm diameter, which are thought to be the sites of combination with the immunoglobulins. The thin connecting strands have, for most of their length, a collagenlike structure with a high content of hydroxyproline and hydroxylysine. The latter is glycosylated by glucosylgalactosyl disaccharides as in collagen itself. The reason for this unusual structure is not obvious. We do know that the binding of antigens activates the complement-binding regions of antibodies, and that the activated antibodies then bind C1q. This binding in some manner activates C1q, which in turn activates C1r subunits of the $C1r_2 s_2$ tetramer. The latter is thought to bind at the center of C1q, while the antibodies bind at the outer ends. We don't know how the activation message is carried from the outer arms to the center. Clq is a member of *a* group of collagen-like proteins that includes protein MBL (also designated MBP) and surfactant protein A.

Activated C1r is one of five different serine proteases involved in activation of complement. The substrate for the trypsinlike C1*r* is C1s, a proenzyme which is converted by the action of C1*r* into another trypsinlike serine protease C1*s*. Through a rather elaborate cascade mechanism, depicted in Fig. 8.8, the important proenzyme C2 is activated. Its active form C2*a* is a serine protease, which cleaves proteins C3 and C5 to the active forms CSb and C5b. Protein C4 is also cleaved to C4b by activated Cl. C4 and C3 are also activated, and protein C5 is cleaved to C5b. The 200-kDa C4 consists of three chains, all derived by proteolytic cleavage of a precursor. It is also glycosylated and sulfated. Both C4 and C3 contain internal thiol ester linkages and act as "molecular mousetraps". They react to fix the proteins covalently to the assembling complement complex. Protein C5b interacts with C6, C7, C8; and six molecules of C9 to generate an "attack complex," which inserts donutlike rings into the cell membrane being attacked (Fig. 8.9). Although there has been some uncertainty about the mechanism of lysis, it seems likely that it is at least partly a result of loss of ions through the holes in the donut.

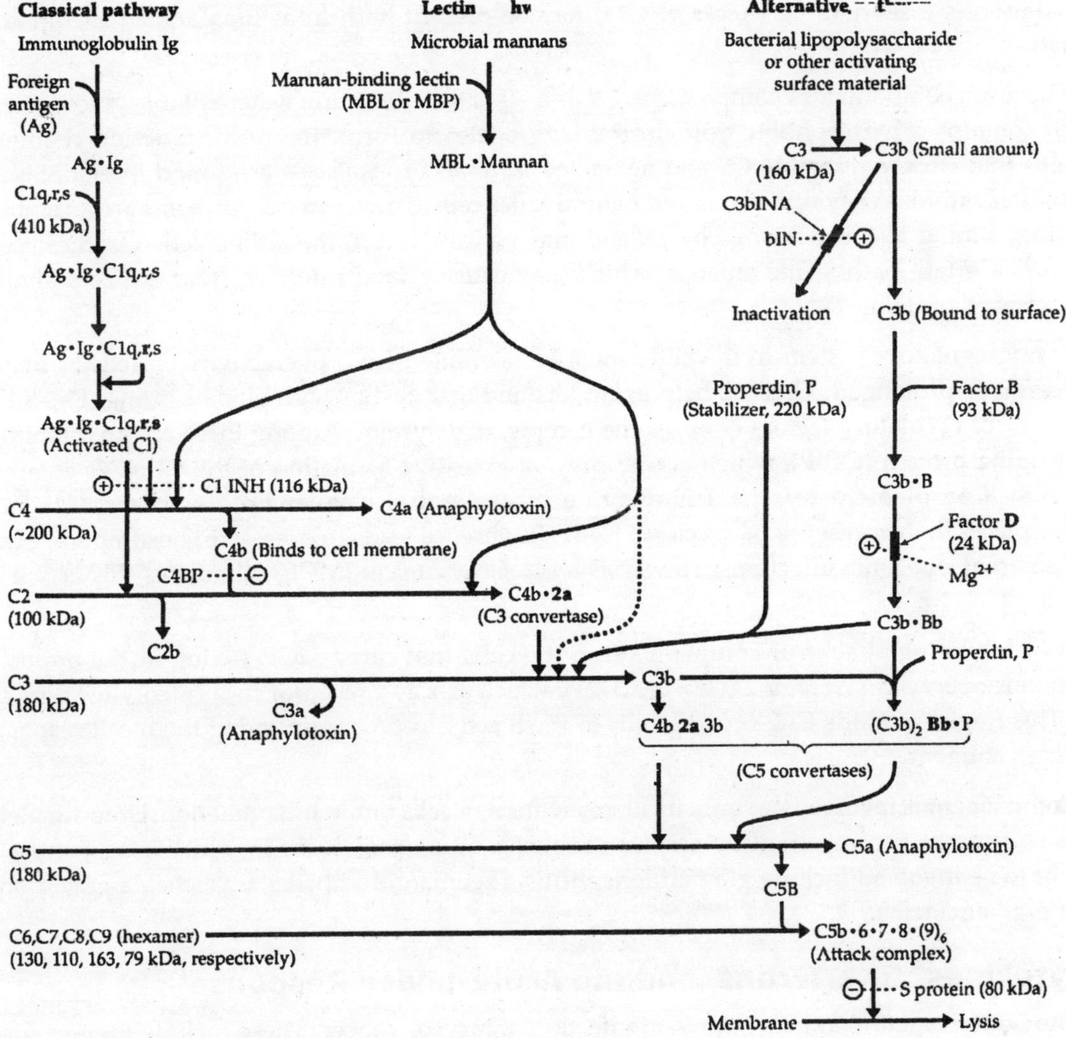

Fig. 8.8. Pathways for activation of the complement system. Active proteases are by abbreviations in boldface.

In the alternative pathway of activation a small amount of C3b is formed and becomes bound to the cell surface. This binds another proenzyme factor B, which is converted by protease factor D to active protease **Bb**. The latter in its complex with C3b is the enzyme that cleaves C3 in large amounts and permits a rapid formation of more **Bb** and also of the complex $(C3b)_2$ **Bb•**P, which attacks C5 (Fig. 8.8). These complexes are stabilized by the abundant serum protein *properdin* (P).

Other components of complement are the plasma C1 protease inhibitor, which prevents accidental activation of the system, and protease C3bINA, which inactivates-C3b. The latter depends upon accessory protein bIN. Another component, serum carboxypeptidase B (SCPB), inactivates anaphylotoxins C3a, C4a, and C5a. These small ~80-residue pieces have a variety of powerful biological activities. They are chemotactic factors for leukocytes and induce release of histamine from mast cells. In excess, they can cause anaphylaxis; hence their rapid degradation is essential. An excess of C5a may be present both in asthma and in rheumatoid arthritis.

The ninth component of complement, C9, is a 70-kDa 537-residue water-soluble glycoprotein, which contains a hydrophobic domain that aggregates to form the ion conducting channels. Proteins that closely resemble C9 and are called *performs* or *cytolysins* are found in cytoplasmic granules of cytotoxic T-lymphocytes and natural killer cells. These ~66-kDa proteins are assembled into rings similar to those formed by C9 and may be involved in the killing action of these cells (Fig. 8.9). Certain pathogenic amebas, which may cause a fatal infection, also utilize a similar pore-forming protein.

Every regulatory system in the body must be prevented from overactivity or activity that is unnecessarily prolonged. This can help us understand that, just as with blood clotting (Fig. 8.17), a network of regulatory factors controls the complement system. Among these are an inhibitory C4b-binding protein (C4BP), which acts to prevent excessive formation of the C4b•C2*a* complex (Fig. 8.8). Complement *cofactor* I is a sefine protease that cleaves both C3b and C4b into smaller pieces in the presence of *cofactor H* C4BP. Its absence leads to excessive consumption of C3 and recurrent pyogenic infections. The *membrane cofactor protein* (MCP) stimulates this action of cofactor

I in inhibiting attacks of complement on the cells that carry MCP. Acting in the opposite direction is *complement receptor* 2 (CR2 or CD21), which acts as a receptor for proteolytic fragment C3d. This fragment binds CR2-bearing cells to the B cell receptor, amplifying the B cell response to foreign antigens.

Complement is involved not only in attacking foreign cells but in inflammation. Unfortunately, this is sometimes accompanied by serious problems. Human diseases in which complement is thought to be involved include glomerulonephritis, rheumatoid arthritis, myasthenia gravis, and lupus erythematosus.

3. Cytokines, Interferons, and the Acute-phase Response

The body responds in many ways to infection, injury, or cancer. These include the secretion of cytokines, interferons, and proteins of the acute phase response. These proteins, many of which are quite small, are involved in communication with other cells, often with specialized cells of the immune system. *Cytokines,* are small hormonelike molecules. They may stimulate, inhibit, or exhibit other effects on cells of the immune system. They often have pleiotropic effects, not acting in the same way on all types of cells. The cytokines known as *interleukins* (IL-1, etc.) are produced by leukocytes. *Lymphokines* are formed by lymphocytes and *monokines* by monocytes. Based on their functions there are four categories of cytokines.

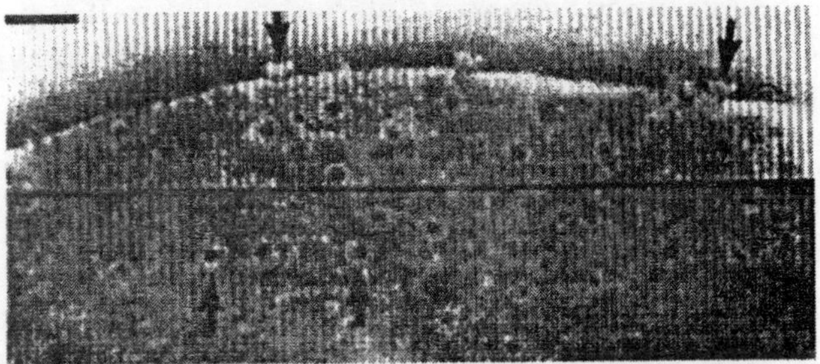

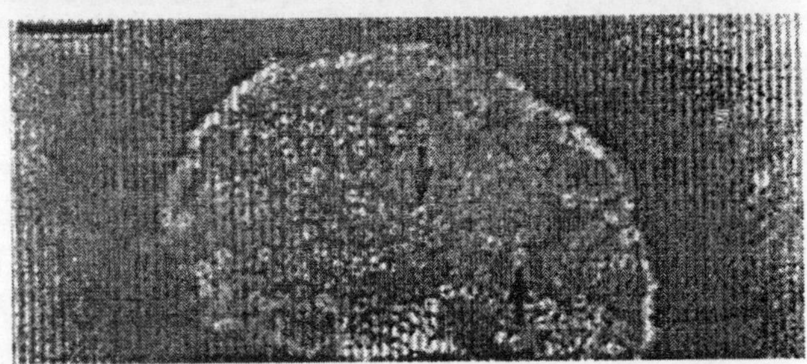

Fig. 8.9. Electron micrograph of a negatively stained sheep erythrocyte lysed with human complement. The cylindrical "attack complex" embedded in the membrane is seen in the upper left frame in side projection and in the lower frames in axial projection. The top views are of a proteolytically "stripped" ghost, the lower view from a freshly lysed ghost. The inner diameter of the cylinders is 10 nm; scale bars 50 nm. From Bhakdi and Tranum-Jensen.

1 Mediation of natural immunity: type I interferons, IL-1, IL-6, and more than 40 *chemokines* (small highly basic chemotactic proteins).

2. Regulation of lymphocytes, activation, growth, and differentiation of B and T cells: IL-2, IL-4, IL-21, TGF-β.

3. Regulation of immune-mediated inflammation: Interferon-γ, tumor necrosis factor (TNF), IL-5, IL-10, IL-12, and migration inhibition factor (MIF).

4. Stimulation of hematopoeiesis (IL-3, IL-7), colony-stimulating factors.

Cytokines all function using a group of transmembrane receptors embedded in the plasma membranes of target cells. The receptors have no tyrosine kinase activity but associate with and activate kinases known as *Janus kinases (JAKs).* These kinases phosphorylate tyrosine side chains in their receptors, and the phosphorylated receptors activate transcription factors of the *STAT* (signal transducer-activators transcription) group. The specificity of cytokine action results from a combination of receptor recognition and recognition of the various STAT molecules by different JAKs. Cytokines have a variety of structures. Many are helix bundles or have β sheet structures.

Interferons

The interferons (IFNs), which were discovered in 1957, are proteins secreted by leukocytes, fibroblasts, and activated lymphocytes. They inhibit replication of viruses as well as the growth of host cells and also have antitumor activity. Interferons are classified as α (from leukocytes), α (from fibroblasts), β and γ (from lymphocytes). According to their affinities for the two types of known interferon receptors, interferons IFN-α, IFN-β, and the less well known IFN-ω and IFN-τ are designated type I, while interferon γ (IFN-γ) is type II. At least 15 homologous 166-residue human α interferons are known.

The binding of interferons to their receptors induces a rapid increase in the transcription of particular genes and synthesis of corresponding proteins. One of the proteins induced is a *double-stranded RNA-activated* 2′ – 5′ *A synthase,* which polymerizes ATP to a series of 2′ – 5′ linked oligonucleotides containing triphosphates at the 5′ termini. Double-stranded RNA is uncommon except in replicating viruses, and it is thought that the activation by dsRNA is related to establishment of an antiviral state. Another interferon-induced enzyme is the small subunit of eukaryotic protein synthesis initiation factor eIF-2. This is converted to an inactive phosphorylated form by a dsRNA-dependent protein kinase (Fig. 8.10). The protein kinase also appears to be an interferon-induced protein as is the oligo(2′ – 5′ A)-activated RNAse indicated in Fig. 8.10. Interferons have effects other than inducing the antiviral state. Thus, human IFN-β_2 is identical to a B-cell differentiation factor. Both IFN-α and IFN-β have antigrowth activity and are currently in use for treatment of some forms of cancer as well as for viral infections.

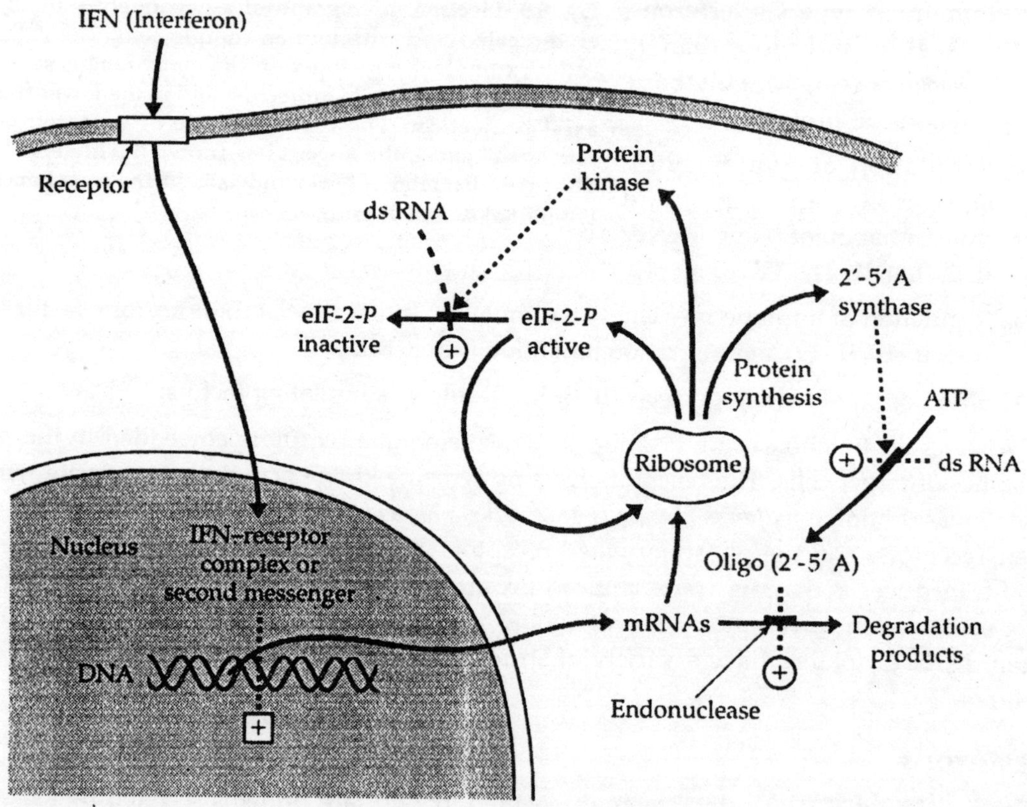

Fig. 8.10. Some responses to the binding of an interferon to a cell surface receptor.

Interleukin-1 (Il-1) plays a key role in the body's response to microbes and to tissue injury. It actually consists of three similar proteins, *Il-1 α, IL-1 β, and Il-1 receptor antagonist.* The first two are the active cytokines with a wide range of effects among which are induction of inflammation and pain. Il-1β is thought to be most active in promoting inflammation but only after it is cleaved by *interleukin-1 β-converting enzyme*. Blocking of Il-1 receptors provides a potential new method for control of pain. *Interleukin*-6 (IL-6) is also needed for an optimal immune system. Its effects overlap those of Il-1, and it has a potent activity in inducing the acute-phase response. Like Il-1, it is a four-helix cytokine.

IMMUNOASSAYS

Among the important techniques that have permitted rapid progress in studies of hormone action is the use of specific antibodies formed against hormones, hormone-protein conjugates, or other molecules.[a–c] The first of these techniques to come into general use was the *radioimmunoassay* (RIA),[d–f] which was devised by Yalow and Berson.[d] In one form of RIA various amounts of a sample containing an unknown quantity of hormone, *e.g.*, insulin, are placed in a series of tubes. Additional tubes containing known amounts of the hormone are also prepared. Then a standard quantity of radiolabeled hormone (often iodinated with a γ emitter such as I) is added to each tube together with a standard quantity of the specific antibody to the hormone. The solution is incubated for minutes or hours to obtain equilibrium between hormone (the antigen) and antibody-hormone complex.

The antibody-hormone complex is then separated, *e.g.*, by gel filtration or ammonium sulfate precipitation, and the radioactivity of the complex is measured. In the tubes containing higher concentrations of hormone, the labeled hormone has been diluted more, and the amount bound to antibody is less than in tubes with lower concentrations of hormone. The tubes of known concentrations are used to construct a standard curve from which the unknown concentrations can be read. As little as a femptomole of hormone (*i.e.*, the amount present in 1 ml of a 1CT^{12} M solution can be detected)[f] Methods are available for virtually every pure hormpne.[c] The RIA methods were made more convenient by adsorbing either the antibody of antigen to the plastic surface of a tube or depression plate. This facilitates separation of the antibody-ligand complex and washing. A variety of other immunoassays techniques have been devised. For example, in *enzyme-linked immunoabsorbent assays* (ELISA),[C] the amount of adsorbed antibody-ligand complex is measured by treating the washed surface with a second antibody, which is directed against the first. The second antibody is linked covalently to an enzyme, whose activity can then be measured by a suitable colorimetric procedure. The reactions involved are as follows.[g, h]

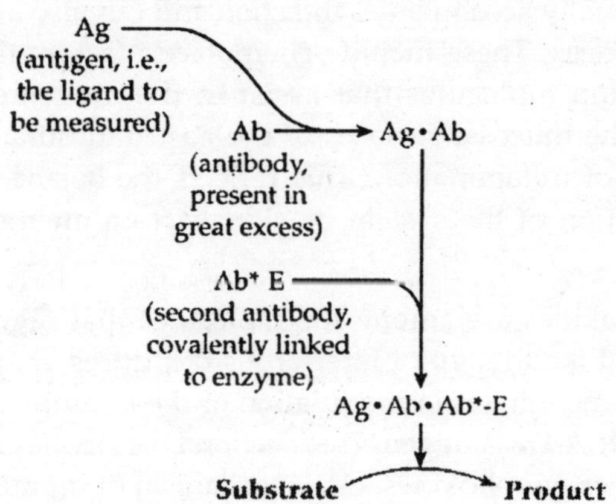

Variations, which avoid the use of radioisotopes, are replacing RIA. Some utilize stable isotopes. However, ^{14}C at such low levels that there is no radioactive waste can be coupled with accelerator mass spectrometry to provide very sensitive immunoassays.[i] A great variety of other procedures are available. Some involve coupling to antibodies that carry fluorescent labels. Many are now automated. Often protein A from *Staphylocaccus nureus* is utilized in various ways that take advantage of its ability to bind to the Fc portion of IgG from virtually all mammals. For example, it may fix antibodies to a surface or to a label.[j]

Also required by the immune response are the numerous *chemokines.* Chemoattractant molecules provide concentration gradients that direct the movement of B and T cells and other leukocytes. Chemokines bind to seven-helix receptors, often without rigid specificies. However, some such as *eotaxin* (Fig. 8.6E) are more specific. Eotaxin attracts primarily eosinophils and basophils during allergic reactions. Interleukin-8 (IL-8) is a proinflammatory cytokine and a powerful attractant for neutrophils. Neutrophils are attracted into affected tissues, where they undergo a respiratory burst and generate toxic compounds from O_2.

The second group of cytokines regulate B and T lymphocytes. Among them interleukin-2 (IL-2) stands out as the major promoter of growth and differentiation of T cells. It was the first hormone of the immune system to be recognized. Both Il-2 and Il-4 have short four-helix structures (Fig. 8.6A). Il-2 is synthesized by activated T cells and binds to a multisubunit receptor complex. The latter associates with tyrosine kinases of both the Src family and with Janus kinases and also activates phosphatidylinositol 3-kinase. Il-2 promotes growth and differentiation, and clonal expansion of T cells, a key aspect of the cellular immune system. It also acts as an immunomodulator of B cells, macrophages, and NK cells. Considerable excitement has accompanied the possibility of activating lymphocytes with IL-2 produced from cloned genes in bacteria to increase their ability to kill cancer cells. However, IL-2 is toxic, and this is limiting its use.

Interleukin-4 (IL-4), cooperating with IL-21, stimulates growth of activated B cells, T lymphocytes, and mast cells, induces formation of cytotoxic $CD8^+$ T cells, and enhances formation of IgG. The *transforming growth factor*-β (TGF-β) is another cytokine that modulates the development of the immune system.

Inflammatory Influences

Inflammation, which usually accompanies infection and can also arise from allergic responses, is affected by many substances. These include chemotactic factors that attract neutrophils and monocytes and the adhesion molecules that assist in the movement of lymphocytes. Some epithelial tissues, such as the mucosal surfaces of the gastrointestinal tract, are maintained in a continuous very low level of inflammation. This reflects the balance between activation of the immune system and inhibition of the system by signals from microorganisms both pathogenic and commensal.

The third group of cytokines are among the molecules that regulate inflammation. One of these is *interferon*-γ (IFN-γ). Like the type I interferons it induces an antiviral state. However, its most important biological function lies in modulation of the immune system. It induces synthesis of both class I and class II (HLA-DR) antigens (see Section D), activates macrophages, and regulates synthesis and activity of other lymphokines. One mechanism of immunomodulation may involve induction of an aminopeptidase that participates in "trimming" of antigenic peptides that participate in antigen presentation. This may directly alter the T-cell response. IFN-γ has a major effect on skin cell-mediated immune responses. INF-γ exists in solution as a symmetric dimer, which binds to two molecules of its major surface receptor. The antiviral activity of IFN-γ is largely a result of induction of large amounts of *guanylate-binding proteins,* large 60- to 100-kDa GTPases.

Occasionally a well-established cancer regresses and disappears spontaneously. In the late 1800s it was observed that this sometimes happened, when a person had a concurrent severe bacterial infection. W.B. Coley pursued this lead for many years, treating cancer patients with extracts of killed bacteria, which, although highly toxic were safer than live bacteria. In the laboratory filtrates from cultures of gram-negative bacteria were shown to kill some experimental mouse tumors. The active ingredient was identified as a highly toxic and pyrogenic lipopolysaccharide. This lipopolysaccharide has a powerful activating effect on macrophages. More recently it was found that the activated macrophages produce a protein known as *tumor necrosis factor* (TNF-α) that can destroy tumor cells and also acts together with interferons in inducing resistance to viruses. A similar *lymphotoxin* (TNF-β) is secreted by lymphocytes. Although highly toxic there has been hope of obtaining engineered forms of these proteins more specifically toxic to tumors.

TNF-α is identical to *cachetin,* a protein that suppresses completely the lipoprotein lipase of adipose tissue and is believed to be responsible for *cachexia,* a condition of general ill health, malnutrition, weight loss, and wasting of muscle that accompanies cancer and other chronic diseases. Nevertheless, TNF-α may be overproduced in obesity as well. It has been suggested that abnormal production of TNF-α may induce cachexia while abnormal action of the cytokine may cause obesity. Some TNF receptors have "death domains" and trigger apoptosis, while other receptors promote proliferation and differentiation via transcription factor NF-κB.

Other cytokines with lymphocyte-regulatory functions are IL-5, IL-10, IL-12, and the *macrophage migration-inhibition factor* (MIF). IL-10 is secreted by B cells, T cells, keratinocytes, monocytes, and macrophages. It suppresses synthesis of many cytokines but stimulates growth and activity of activated B cells. IL-12 is formed by monocytes, macrophages, neutrophils, and dendritic cells. It activates T cells and NK cells, is a very potent stimulator of INF-γ forma-tion, and also inhibits angiogenesis in tumor cells. It stimulates defenses against a wide range of infectious diseases caused by bacteria, fungi, protozoa, and worms. The 115-residue MIF is formed not only by the immune system but also by many other tissues. The first lymphokine to be discovered, MIF, inhibits migration of macrophages and is also a mediator of toxic shock. MIF is also an enzyme, a *phenylpyruvate tautomerase.*

The fourth group of cytokines are involved in hematopoiesis and control the developmental steps portrayed in Fig. 8.2.

The *acute-phase response* consists of increased production of a group of plasma proteins in response to tissue injury or inflammation. Important acute-phase reactants are the *C-reactive protein, serum amyloid A, haptoglobin, hemopexin,* α1-*acid glycoprotein,* and α2-*macroglobulin.* The C-reactive protein precipitates pneumoccocal polysaccharides in the presence of Ca^{2}+. It is present in primitive invertebrates and may serve as a rudimentary immunoglobulin. Serum amyloid A is one of the apolipoproteins associated with high-density lipoproteins. Its concentration may increase as much as 1000-fold during the acutephase response, and during prolonged stress it may precipitate as extracellular amyloid fibers. This secondary amyloidosis is sometimes a severe pathological problem. The level of the general protease inhibitor α_2-macroglobulin can increase several hundredfold.

D. ORGANIZING THE IMMUNE RESPONSE

A person's immune system must be able to respond to a large variety of foreign antigens without reacting against the individual's own tissues. The huge variety of antibodies that can be formed arise from the existence of B cells with millions of different sequences in their antibody genes. When an immune response occurs only a few B cells are stimulated to proliferate, and it is these selected clones that provide the needed specific antibodies and memory cells. However, it is not immediately obvious how we avoid a disastrous attack of the immune system triggered by the many antigenic determinants *(epitopes)* present in our own cell surfaces and macromolecules. Part of the answer is that the immune system "learns" early in life what is self and what is nonself. Thus, while foreign tissues cannot usually be grafted without rejection, cells of two immunologically incompatible embryos can be mixed at a very early stage of development, and an animal tolerant to both types of cell will develop.

A full understanding of self-discrimination is not yet available. The adaptive system, as generally understood, is outlined in the following pages. A current view of the innate system is presented by Medzhitov and Janeway. However, an alternative description, the *Danger model,* is being developed by Matzinger. Her view is that the immune system is designed not so much to recognize *nonself* as to send *alarm signals* from injured tissues. Most of the basic mechanisms of the adaptive immune system are not in dispute, but many hard-to-explain phenomena remain uncertain.

1. Coreceptors and the B-Cell Response

Early in life most B cells that would produce anti-bodies directed against a person's own tissues (auto-reactive B cells) are eliminated or altered to reduce their reactivity. When functional B-cell receptors do bind an antigen, the B cell will not be activated unless *coreceptors* also bind to the antigen-bearing particle. The transmembrane glycoproteins known as CD22, CD21, CD72, and FcγRIIb are among the many coreceptor molecules. Coreceptors often induce tyrosine phosphorylation of internal receptor domains and attract other molecules to form a signaling complex that may release cytokines. The coreceptors ensure that an immune response doesn't take place without at least two signals. They also help to localize the immune response.

Among the most important factors in B-cell activation are the effects of T cells. B cells can independently mount an attack using IgMs against surface antigens. However, B-cell responses to many antigens, *e.g.,* those present on flagella or inserted into membranes, are also dependent upon assistance from helper T cells (T_H cells). These cells also have a major role in determining the longer term fate of B cells. Upon activation B cells may survive or die via apoptosis. They may proliferate (clonal expansion) and differentiate into plasma cells or may become unreactive *(anergy).* They may become long-lived memory cells.

2. The Leukocyte Differentiation Antigens

Before discussing T-cell responses it seems appropriate to mention the nomenclature of molecules (largely glycoproteins) that have been recognized as antigens present on leukocyte surfaces. These same molecules are found on other cells, but the designation of the antigens by a *cluster of differentiation* number, such as CD1, CD4, or CD8, has provided a convenient way of distinguishing different types of leukocytes. For example, helper T_H cells are usually $CD4^+$, carrying predominantly CD4. Cytotoxic T cells are predominantly $CD8^+$. Both CD4 and CD8 consist largely of Ig-like domains. CD4 is a 55-kDa transmembrane protein with tyrosine kinase activity. It is a monomer containing four Ig-like domains, but CD8 is a disulfide-linked αβ dimer. Not all CD molecules are related to IgG. Proteins are often designated by a specific name followed by a CD number, *e.g.*, FcγRII/CD32, ICAM-1 (CD54).

3. Functions of T Cells

T cells carry the responsibility of identifying antigens as foreign or as belonging to self. They do this in immunological synapses (Fig. 8.11) in conjunction with the major histocompatibility complex MHC. T cells circulate through the body searching for antigens that indicate danger to the body. To avoid being swept through the bloodstream too rapidly and to be able to enter the lymphoid organs lymphocytes form tethers with adhesion molecules such as the *selectins*. They then roll more slowly to their destination. Within the lymphoid tissues the T cells may form synapses with activated B cells, dendritic cells, and macrophages. Within these cells proteosomes generate a stream of peptide fragments, some of which arise from phagocytosed pathogens. These foreign peptide fragments are displayed on the cell surfaces as complexes with type II proteins of the major histocompatibility complex (MHC). The complexes are checked by $CD4^+$ T cells, some of whose *T-cell receptors* (TCRs) will probably be complementary to the surfaces of the complex of the class II MHC protein and the foreign peptide. The T cell will recognize two things about this complex: the MHC protein is of *self* origin but the antigen is *foreign*. The CD4 on the T cell must also bind to the MHC on the surface of the antigen-presenting cell. Other co-stimulatory interactions may be needed as well. Both $CD4^+$ and $CD8^+$ T cells tend to bind to oligomeric *activation clusters* of receptors within the immunological synapses. Other proteins also participate in assembly of these activation complexes. Of particular interest is the recognition of an antigenic peptide produced by a B cell. The B cell has probably already recognized and phagocytized a foreign protein and is displaying peptides from that protein on its MHC I molecules. Recognition of this peptide complex by a $CD4^+$ T cell will stimulate the cell to become a T_H helper cell, which will in turn stimulate the B cell to proliferate and differentiate into a clone of as many antibody-forming plasma cells. The essential nature of the costimulation by CD4 is emphasized by the fact that infection by HIV-1, which is mediated by CD4, leads to loss of CD4 from plasma membranes and to the weakening of the immune response toward various pathogens that is characteristic of AIDS.

In a similar manner $CD8^+$ T cells recognize peptide fragments displayed on MHC class I molecules. These fragments arise via a somewhat different pathway that forms fragments of viral proteins or proteins of other intracellular pathogens. Recognition by a $CD8^+$ T cell usually converts it into a *cytotoxic (killer) T cell,* which will kill the infected cell. This type of immune reaction was first recognized by the phenomena of *delayed hypersensitivity* and of *transplantation immunity, i.e.,* the rejection of transplanted tissues. Both phenomena are caused by cytotoxic T cells. In delayed hypersensitivity they appear to be confused and to attack host cells.

Fig. 8.11. The T lymphocytes seen here are forming synapses with the large flat macrophage in the center. The macrophage is displaying antigenic peptide fragments bound to molecules of the major histocompatibility complex (MHC). Most T cells carry their own specific type of receptor. If it is complementary to a displayed antigen fragment it will bind, and the T cell will respond. Depending upon what other coreceptors are activated, it may become a T_H helper cell, or a cytotoxic T cell, or it may become inactive.

Some very hydrophobic antigens are presented by neither a class I nor a class II MHC molecule but by members of the CD1 family, leukocyte surface proteins that are not encoded in the MHC gene region.

4. Natural Killer Cells

An additional line of defense is provided by *natural killer cells* (NK cells), a type of circulating lymphoid cell able to kill cancer cells, to participate in antiviral defenses, and to help control immune responses. NK cells, which utilize their own signaling pathways, are also able to use MHC class I molecules to recognize and to spare the lives of normal, healthy cells. Partial deprivation of a night's sleep can reduce NK cell activity, damaging the cellular immune response.

5. Identifying Self: The Major Histocompatibility Complex

Proteins encoded by a single cluster of genes are known as the *major histocompatibility complex* (MHC). These proteins, which are essential to T-cell function, were first recognized as the primary determinants of the compatibility of grafted tissues with the host's immune system. A lack of histocompatibility can be disastrous. Not only are grafted tissues rejected but T lympocytes from the grafted tissues sometimes proliferate, attack, and kill the host. The MHC of mice is usually referred to as the *H-2 complex* and that of humans as *leukocyte locus A (HLA)*. Although the MHC is the most important determinant of histocompatibility, differences in other genes may also lead to a slow rejection of transplanted tissues. Since there are many different MHC genes, transplantation is successful only within inbred lines.

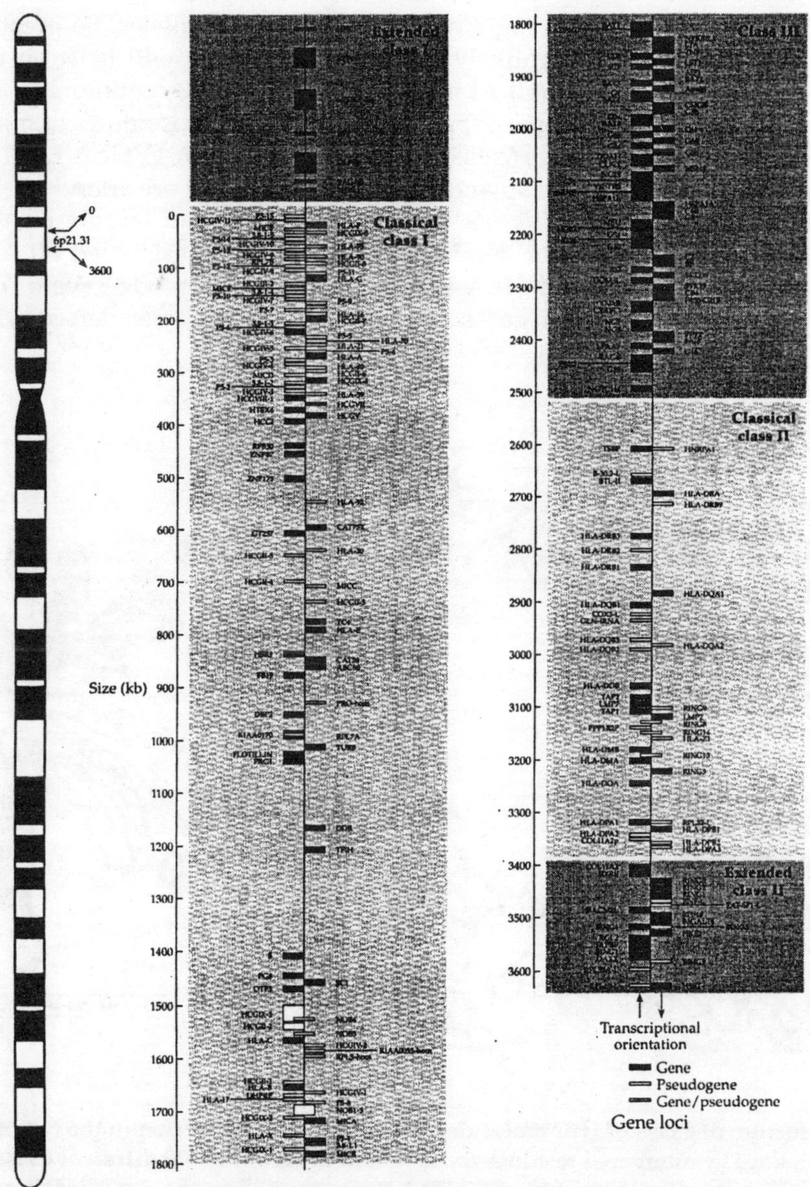

Fig. 8.12. (Opposite page) Arrangement of genes of the human major histocompatibility complex (MHC). Left: Banding pattern of a stained chromosome 6 with the MHC region marked. Center and right: locations of all genes and psedogenes in this region. The MHC molecules can be divided into three classes on the basis of their structure and function. The class I antigens constitute a single class structurally but fall into two functional groups. The first of these contains the "classical" class I antigens, first discovered as the transplantation antigens and now known to function as target antigens in the recognition and destruction of virus-infected cells by cytotoxic T lymphocytes. They are expressed on virtually all somatic cells. The class II antigens are expressed largely on B lymphocytes and macrophages of the immune system and are essential for presenting antigen to the helper and suppressor T cells that regulate the immune response. Many class III products are components of the complement system. These maps are based on serological and biochemical data, together with complete sequences.

Some of the MHC genes have a large number (50 – 100) of alleles. So great is this genetic polymophism that it is extremely unlikely that two individuals will have an identical set of histocompatibility genes. The MHC (HLA) genes are located in a 2-centimorgan (~3.6 kb) region of the short arm of human chromosome 6 (Fig. 8.12) and on chromosome 17 of mice. These genes are of at least three classes. *Class I genes* (called HLA-A, HLA-B, HLA-C, HLA-E, HLA-F, and HLA-G in humans; encode the major transplantation antigens, which are found on the surfaces of nearly all cells of the body. *Class II genes* encode proteins found largely on the membranes of B lymphocytes, macrophages, and dendritic cells. They are designated HLA-DP, HLA-DN, HLA-DM, HLA-DQ, HLA-DR, and HLA-DO. *Class III genes* encode several components of the complement system. Many other genes and pseudogenes are interspersed with those of the MHC.

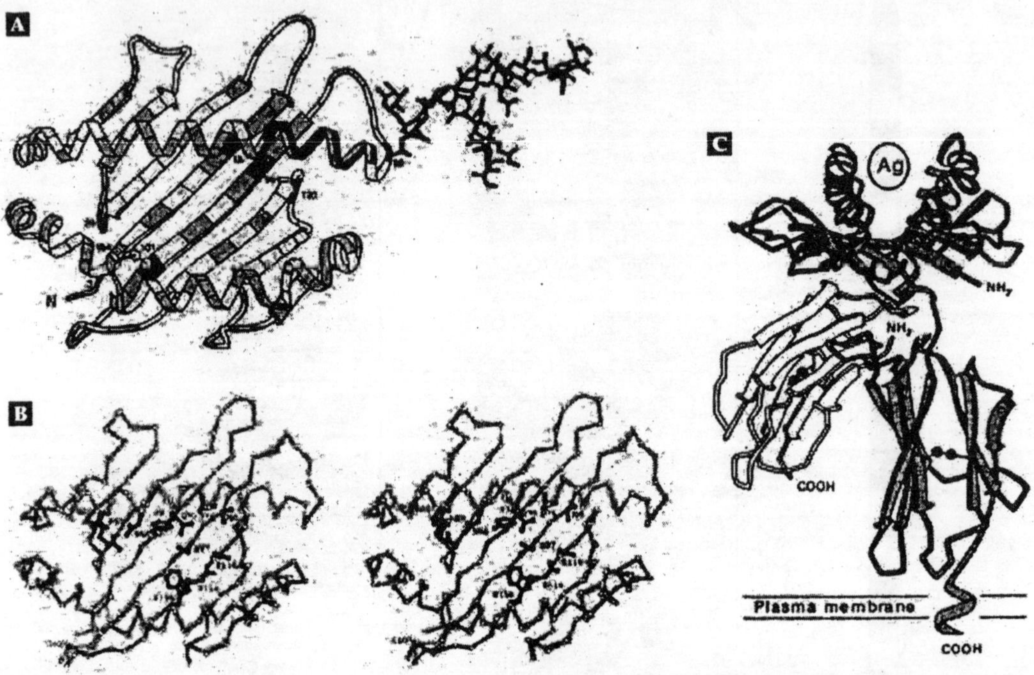

Fig. 8.13. The structure of Class I MHC molecules. (A) The specificity pocket in the N-terminal part of the ~360-residue a chain. The numbered residues are invariant in all of the ~20 different Class I molecules. An oligosaccharide is shown on the invariant Asn 86. (B) A stereoscopic view of a similar MHC molecule showing some of the polar residues that protrude into the peptide-binding groove and may form hydrogen bonds with the peptide. (C) Side view of a complete MHC molecule with an antigenic peptide (Ag) bound into the peptide-binding groove. The C terminus of the long ~360-residue α chain is in the cytoplasm of the displaying cell. The small 99-residue β chain (unshaded) is a molecule of β microglobulin, which is also a constituent of blood plasma.

All type I MHC molecules are integral membrane glycoproteins each of which is composed of a 45-kDa heavy chain of about 350 residues together with a noncovalently linked 17-kDa light chain. The genetic variation occurs in the heavy chain between residues 43 and 195 in the human proteins. This chain appears to consist of three ~90-residue domains protruding from the outside of the cell, about 25 residues embedded in the membrane, and a short C-terminal tail in the cytoplasm (Fig. 8.13). The light chain has an invariant composition and is identical to the plasma protein β_2-microglobulin, whose gene is located on a different chromosome. Its structure closely resembles that of a single immunoglobulin domain. The MHC Class II antigens (Fig. 8.14) are also $\alpha\beta$ dimers, the α chains being 34-kDa glycoproteins and the 28-kDa β chains being larger than in the type I antigen. While the MHC of humans and mice have been studied the most, all vertebrates possess similar self-identification systems. Although both the sequences and the folding patterns of the MHC antigens (Figs. 8.13, 8.14) are somewhat similar to those of immunoglobulins, there are many differences. Furthermore, the cause and significance of the polymorphism is quite different in the two cases. Each individual has millions of antibodies with different variable regions but only one set of HLA antigens, which are largely the same on germ, embryonic cells, and adult cells.

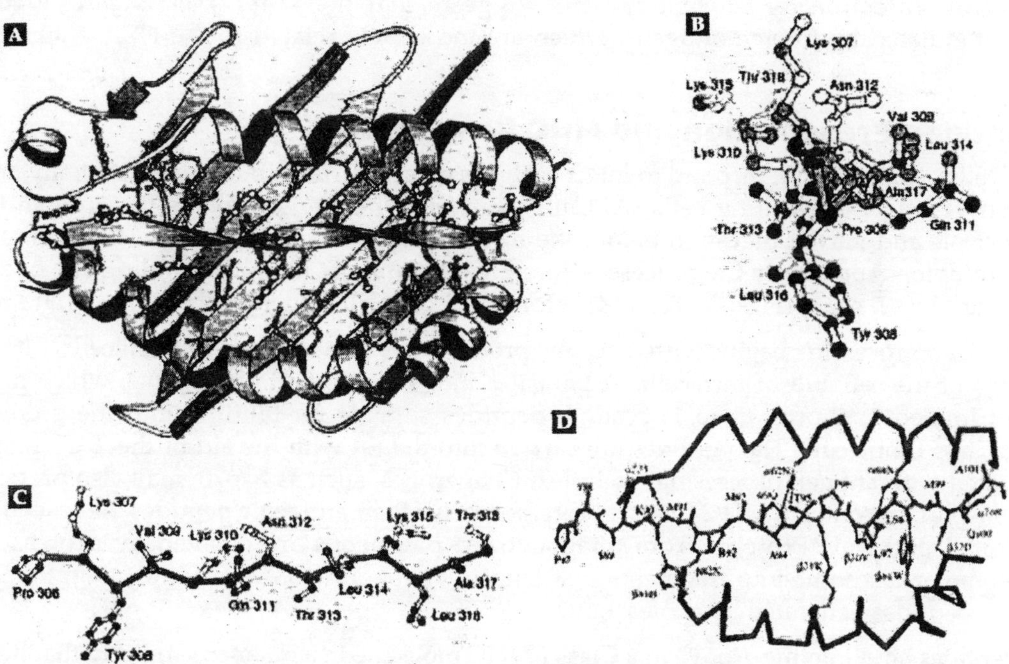

Fig. 8.14. Illustration of the binding of a short peptide fragment (central ribbon) into the antigen-binding groove of an MHC type II molecule. (A) The binding groove of molecule HLA-DR1 with a bound peptide (HA) derived from an influenza virus. (B) End view of the bound peptide. (C) Side view of the same peptide. (D) A similar HLA molecule (HLA-DR3) with the peptide CLIP (class II associated invariant chain peptide) bound into the antigen-binding groove. The binding is almost identical to that in (A). Notice the specific hydrogen bonding to side chains of the HLA molecule.

Serological tests allow tissue types to be defined by the HLA genes. Thus the commonest HLA type in Caucasian populations is HLA-A1/B8/Dw3, whereas A1/B17 is common among Asian Indians. In every case subtypes can be defined, and this fact together with the polymorphism in other genes leads to a unique HLA type for nearly every individual. As is indicated in Fig. 8.12, complete nucleotide sequences are known for typical alleles of all of these genes. It is of medical interest that the susceptibility of an individual to many degenerative diseases is determined in part by the HLA type. Thus among patients with a kind of arthritis, *ankylosing spondylitis* that affects 1 or 2 per 1000 men of Caucasian origin, 96% have the HLA-B27 antigen. Of patients with *celiac disease,* a type of intolerance to gluten, 60% have the HLA-B8 antigen. Persons with HLA-Bwl7 and B13 have an increased susceptibility to *psoriasis* and those with HLA-DRw4 an increased tendency to develop *rheumatoid arthritis.* Susceptibility to *autoimmune insulin-dependent (Type I) diabetes* is strongly correlated with the presence of the neutral residues Ala, Val, or Ser at position 57 of the HLA-DQ β chain. However, aspartate in position 57 protects against the disease. It may prevent development of a dangerous autoantibody to this cell surface protein. The presence in populations of both humans and apes of a balanced polymorphism among the residues Ala, Val, Ser, and Asp at this position suggests an essential evolutionary origin to this disease susceptibility. HLA-B53 protects against severe malaria in Africa. Other diseases with a strong association with HLA type include multiple sclerosis, *Crohn disease* (inflammatory bowel disease), and several diseases induced by infections with viruses, bacteria, trypanosomes, etc. For example, arthritis can follow infection by *Salmonella.* This suggests that the killer T cells can be confused when stimulated by foreign antigens, which are too closely related to the HLA antigens of the host.

6. Antigen Presentation and MHC Restriction

T cells usually do not respond to intact antigens on cell surfaces but only to partially degraded antigens. Antigen-presenting cells (APCs) of various types must process the antigen through endocytosis and partial digestion before the foreign antigen can bind as an MHC complex to a T-cell receptor. Apparently the processed foreign antigen must lie in the binding site of an MHC Class I or Class II chain (Figs. 8.13C, 8.14) before the T cell will recognize the complex and respond.

Antigen processing begins with cytosolic proteasomes that are present in all cells. They cleave proteins of the cell and of intracellular parasites into short peptide fragments, which may need to be trimmed to shorter 8- to 11-residue peptides suitable for binding into the groove in an MHC Class I molecule. The peptides are carried into the ER with the aid of the *TAP* (transporter associated with antigen processing) complex. Chaperones, such as hsp70, may also participate in the transport. In contrast, MHC Class II proteins receive their antigenic peptides via an endosomal-lysosomal pathway. Proteins from phagocytized pathogens are cleaved by proteases in an endosome or lysosome into fragments that tend to be longer (13–25 residues) than the 9- to 11-residue peptides generated by proteasomes.

Peptides bind into the groove in a Class I MHC molecule in a manner similar to that illustrated in Fig. 8.14 for a Class II MHC molecule. However, in the Class II complex the longer peptides extend from the two ends of the binding groove. The peptide, which assumes a polyproline II helical conformation, is held by hydrogen bonds from the Class II MHC molecule to the peptide backbone. A single peptide may shift and bind in a different register with the possibility of being recognized by a different T cell receptor, when it is displayed. In contrast, MHC Class I molecules bind best to 8- or 9-residue peptides, which are held by an array of hydrogen bonds to the $-NH_3^+$ and $-COO^-$ termini (not shown in Fig. 8.13, which displays the empty binding groove). The central part of this groove contains a deep pocket, which together with smaller pockets near the ends provides specificity. However, there is a puzzle. Because of the great genetic variability in the MHC genes there will be great differences (polymorphism) in the shapes of the binding pockets among different people. However, an individual has at most six different kinds of MHC molecules. Yet, a single MHC molecule has been estimated to be able to bind more than 10,000 different peptides. Essential to this process is a final trimming of the peptides at their N termini to provide better fits.

Peptides are loaded onto Class I MHC molecules, while they are together in the ER. They move as tightly bound complexes through the Golgi and into the external plasma membrane, where they remain tethered via the MHC molecule (Fig. 8.13C). In contrast, binding to Class II MHC molecules occurs in the endosomes or lysosomes. The process is somewhat complex. Class II MHC molecules are chaperoned from the cytosol into late endosomal/lysosomal organelles with their antigen-binding grooves occupied by a peptide fragment known as the Class II-associated invariant chain peptide (CLIP; Fig. 8.14D). This is cut from the end of an invariant chain known as Ii.

The MHC proteins HLA-DO and HLA-DM (Fig. 8.12) are resident in the lysosome-like organelles and chaperone the class II molecules, until they are ready for loading with peptides. HLA-DM assists in removal of the CLIP peptide when loading occurs. An asparaginyl endopeptidase may also be required as well as a disulfide reductase. After loading the Class II MHC•peptide complexes, like the Class I MHC•peptide complexes, are exported to the plasma membrane. There they may be recognized by a T cell, which utilizes its T-cell receptor to recognize an antigen and its CD4 or CD8 proteins to distinguish Class I from Class II MHC complexes. The binding of CD4 and CD8 to their cognate MHC molecules has also been described at the molecular level. As mentioned previously, some hydrophobic antigens are presented by CD1 molecules. They also have an MHC-like fold with a large hydrophobic binding groove.

An interesting approach to the treatment of autoimmune diseases is design of peptide mimics that bind into the antigen-binding groove of specific MHC proteins. For example, a protease-resistant pyrrolinone-peptide hybrid has been designed to bind to the rheumatoid-arthritis-associated HLA-DR1.

An important distinction between B- and T-cell responses is that T cells recognize a foreign antigen only when associated with an MHC antigen of the same type as is carried by the T cell. This "MHC restriction" limits the actions of T lymphocytes. The function of cytotoxic T cells appears to be primarily one of killing virus-infected cells and perhaps cancer cells. MHC restriction ensures that the T cell is attached by its rereceptor to a Class I MHC molecule belonging to self. The dual recognition ensures that the cell probably contains a foreign antigen and should be killed. The Class I MHC antigens are found on almost all body cells. Therefore, cytotoxic $CD8^+$ T cells can attack infected cells of all types. They may kill their target cells by injecting their cell membranes with the complement C9-like cytolysins.

Regulatory $CD4^+$ T^- cells recognize the Class II MHC molecules, which are found primarily on lymphocytes. Thus, the attention of regulatory T cells is directed towards other lymphocytes. In this case, too, MHC restriction enables helper T cells to recognize B lymphocytes as self. If foreign antigen is present so that B lymphocytes have been activated by the binding of a foreign antigen, they will be stimulated by the T_H cells to proliferate and make and release immunoglobulins. This is accomplished in part by secretion of the lymphokine interleukin-2 (Fig. 8.6) and B-cell growth factors. Some T cells become suppressor T cells.

7. T-Cell Receptors

T cells both mediate the recognition of self and also participate in the immunologic response to foreign antigens. Their surface receptors function much like the immunoglobulins that are attached to the surfaces of B cells. The T-cell receptors are αβ disulfide-linked heterodimeric glycoproteins (Fig. 8.15) consisting of 40- to 45-kDa α subunits and 42- to 44-kDa β subunits. They are associated in the T-cell membrane with a larger complex called CD3, which contains additional 26- to 28-kDa γ, δ, ε, and ζ chains. The polypeptides of the CD3 complex have C-terminal cytoplasmic tails that contain tyrosine residues within several immune system tyrosine-based activation motifs (ITAMs; Fig. 8.15). As is illustrated in this figure, the activating antigen is cradled in the binding groove of an MHC molecule attached to the APC (top) with some side-chain and backbone atoms of the peptide available for bonding to the T-cell receptor. Notice that in the synapse the T-cell receptor also makes direct contact with the MHC molecule (Fig. 8.15).

Signaling by an activated T-cell receptor is quite complex. The ITAMs are sites of tyrosine phosphorylation by kinases of the Src family. Another tyrosine kinase, Zap-70 (Fig. 8.15), associates with the C-terminal tails of the disulfide-linked dimer of subunits ζ. It recognizes the phosphotyrosines groups via its SH2 domains (Fig. 8.30). Zap-70 appears to act synergistically with the Src. kinases. The nature of the APC is also of importance. For example, some dendritic cells secrete 11-12, which favors formation of T_H1 helper cells. A second type of dendritic cell induces formation of T_H2 helper cells. Dendritic cells may also control growth and proliferation of T cells by regulating the availability of cysteine, which is a nutritional essential for lymphocytes. Other effects may result from endocytosis by T cells of occupied T-cell receptor•MHC complexes.

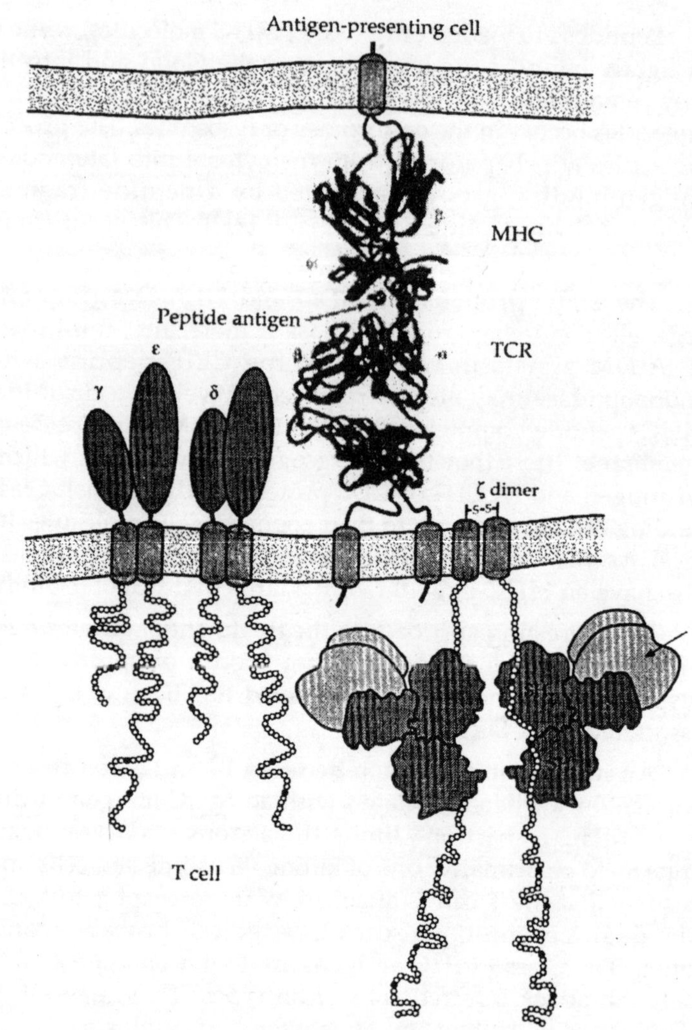

Fig. 8.15. Interaction between an MHO • peptide complex on an antigen-presenting cell (APC) with a T-cell antigen receptor (TCR) that is attached to the plasma membrane of a T cell. Structures of the α- and β-subunits of the MHC molecule and of the T-cell receptor are based on crystallographic data. The detailed structures of the disulfide-linked σε and γε modules of the T-cell receptor are not shown. The dimeric ζ_2 subunit has large cytoplasmic domains that are thought to be involved in signaling the protein ZAP-70 (zeta-associated protein of M_r 70,000). The σ, ε, and y subunits may also undergo phosphorylation of their tyrosyl groups (gray spheres), which are found in immune system tyrosine-based activation motifs (ITAMS).

A second type of T-cell receptor, the γδ receptor, is carried by a small subgroup of T cells. It may have a distinct role in generating an immune response to certain microorganisms including *Mycobacterium tuberculosis.* Like immunoglobulins, T-cell receptors have a great variety of amino acid sequences. They have C-terminal constant domains and N-terminal variable and hypervariable regions. Thus, T cells can bind to a variety of foreign antigens. However, except as a result of autoimmune diseases, they do not attack cells recognized as self unless these cells are infected with a virus or for some other reason carry foreign surface antigens.

Certain bacterial immunostimulatory molecules known as *superantigens* are able to stimulate MHC Class II molecules to activate large numbers of T cells without any assistance of an antigenic peptide, sometimes with disastrous results. Superantigens are a group of related proteins that includes enterotoxins from species of *Staphylococcus* and *Streptococcus,* a staphylococcal exfoliative toxin, and *toxic shock syndrome* toxins. Superantigen molecules don't occupy the peptide-binding groove in the MHC molecule but bind as intact proteins at an external site. They also bind to the variable region of the TCR in the MHC-TCR complex. Attempts are being made to design decoy molecules that prevent binding of a particular superantigen.

8. Self-Tolerance

The immune system is flexible enough and powerful enough to protect us from a great variety of dangers, even from viruses and organisms that may be entirely new. However, it is hard to understand how the immune system completely avoids fatal damage to our own bodies. The answer is complex. It has baffled generations of investigators and is still not fully understood.

Since the discovery of vaccination in 1796 immunology has claimed the attention of many scientists. However, it was not until 1891 that the German bacteriologist Emil von Behring proposed the term antibody for the protective materials in blood. By about 1900 Paul Ehrlich and Svante Tiselius, who wrote the first immunochemistry book, initiated serious investigations. Ehrlich proposed that binding of an antigen to a surface receptor would induce the cell to make additional identical surface receptors, which would be released to become antibodies. The concept was correct, but it would be many years before knowledge of the structures and biosynthesis of antibodies became available.

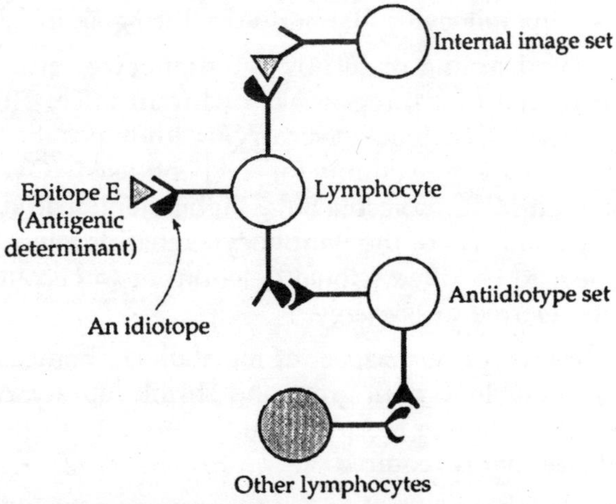

Fig. 8.16. Schematic depiction of lymphocyte receptors forming anti-idiotype and internal image sets as proposed by Jerne.

Nevertheless, immunological tolerance interested Ehrlich and other immunologists. One proposal, offered by Niels Jerne in 1974, was that self-tolerance depends upon *immunological networks.* Consider a lymphocyte bearing a bound immunoglobulin receptor or a bound T cell receptor. It will be specific for some epitope E. The receptors are shown in Fig. 8.16 as having V-shaped antigen-binding sites. Jerne pointed out that the variable region of this receptor will itself carry epitopes that can be recognized by other appropriate antibodies. These epitopes on the receptor are called *idiotopes* and as a group define the *idiotype* of that receptor or immunoglobulin. There will usually be other lymphocytes with receptors that recognize the idiotype of the first lymphocyte (Fig. 8.16). These constitute an *antiidiotype set.* In addition there will be lymphocytes, whose idiotopes resemble those of epitope E and which will therefore be recognized by the first lymphocyte as foreign. These lymphocytes constitute an *internal image set.* There will be other sets of lymphocytes that recognize the lymphocytes of the antiidiotype set or of the internal image set. Thus, there will be an elaborate network of clones of interacting lymphocytes. When an immune response occurs many members of this network will respond. A B lymphocyte will recognize a particular antigen and gives rise to a clone of plasma cells making antibodies against that antigen. The body will then make new antibodies against the first antibodies formed, etc. A whole segment of the network will respond in this fashion. Jerne suggested that the overall effect would be to limit and suppress the immune response. Anti-diotype antibodies as well as anti-anti-idiotypic antibodies have been prepared and have been used in studies of receptors. However, Jerne's theory is generally regarded as incorrect or at least a great oversimplification.

A process of *clonal selection* is now thought to be basic to self-tolerance. Credit for the theory, developed in the 1950s, is usually given to F. M. Burnet. However, Ehrlich, Jerne, and David Talmage were also prominent contributors. An essential postulate of the clonal selection theory is that each B lymphocyte is predetermined to make antibodies of only a single specificity. The mechanism of *allelic exclusion,* which makes this possible, is described in Section E. Clonal selection can occur because the B cells carry their antibodies as surface receptors. Binding of an antigen provides a signal for clonal expansion. However, during development in the thymus the progenitor B cells carrying self-reactive antibodies are killed by apoptosis. Later, peripheral B lymphocytes also undergo selection by a complex network of signaling and apoptosis. In a similar manner excess lymphocytes that build up during an immune response must be removed.

The learning of self by T lymphocytes also happens in the thymus early in development, during the first three weeks of life in mice. This again involves selection against potentially autoreactive lymphocytes carrying idiotypes that are also present on the body's own tissues and which have a high affinity for self peptide•MHC complexes. However, T cells with a weaker affinity for a self MHC molecule but a high potential affinity for a nonself peptide are allowed to develop. Only about 1% of the lymphocytes that develop in the thymus emerge as mature T cells. Others appear to be killed (clonal deletion) or to become unreactive toward antigen. The latter enter a state referred to as *anergy.*

As with every aspect of metabolism, homeostasis is essential to the immune system, which must be able to both grow and shrink rapidly. Antigens, cytokines, apoptosis-inducing signals, immune inhibitor receptors, and receptor tyrosine kinases all participate in preserving the delicate balance that is required.

8. Immunologic Memory and Vaccination

In 1796, Edward Jenner carried out the first human vaccination. Attempting to protect a teenaged boy from disfigurement and possible death from smallpox, he vaccinated him with material from a cowpox lesion on the hand of a milkmaid. (She had contracted the disease from a cow named Blossom, whose hide hangs in St. George's Hospital in London). Six weeks later he inoculated the boy with virulent smallpox. Fortunately, the boy didn't contract the disease. Today vaccination is in use for more than 70 bacteria, viruses, parasites, and fungi, and the results have been impressive. Poliomyelitis has been almost eliminated. Smallpox has not been seen for many years, and the decreases in diphtheria, measles, mumps, whooping cough, and rubella have been impressive.

A nagging question is "How long will vaccination last?" One unplanned experiment resulted from two epidemics of measles in the remote Faroe Islands. The first outbreak was in 1781 after which the islands remained free of measles for 65 years. The second outbreak in 1846 affected 75-

95% of the population, but according to a physician who investigated the epidemic not a single one of the many aged people living who had had measles in 1781 contracted the disease a second time. This bit of history confirms that immunological memory is sometimes very long-lived (although it doesn't prove that the smallpox vaccination older people received is still good!).

Vaccines have been prepared traditionally by use of viruses or organisms killed by compounds such as formaldehyde or by attenuated viruses or live organisms. These are selected for a low degree of virulence after repeated passages through live animals or cell cultures. Newer methods utilize purified viral proteins, bacterial capsular polysaccharides, or DNA. In the future edible vaccines may be produced in plants. Nevertheless, it is often very difficult to devise effective vaccines. In spite of 80 years of effort better vaccines against tuberculosis are needed. All efforts to produce an AIDS vaccine have failed. A satisfactory vaccine must activate both B cells and T cells. Activation of the latter may be especially difficult. Continuous development of new strains of bacteria is a problem for vaccination against tuberculosis and has been an insurmountable barrier to vaccination against AIDS. However, in the latter case it may be possible to use vaccination to prevent an HIV infection from progressing to AIDS with complete destruction of $CD4^+$ T cells. As the immune system and also diseases become better understood, it is possible that new strategies for induction of specific cytolytic T cells can be devised, *e.g.*, for AIDS and even for cancer. A current obstacle to development of new vaccines is that pharmaceutical companies view vaccines as unprofitable.

We need a better understanding of how memory B and T cells are formed and selected for long-term survival. After differentiation and selection in the bone marrow, B cells move to the spleen. These *transitional B cells* within germinal centers undergo further selection to become mature B cells and B memory cells. Interactions with cytokines and with coreceptors play important roles. A subset of B lymphocytes may remain in the germinal centers, serving as a kind of stem cell providing new memory B cells continuously. *Naive T cells,* which have not yet encountered antigen, travel throughout the body but apparently don't enter nonlymphoid tissues. However, after being presented with antigen by dendritic cells within lymph nodes, some T cells move to the skin and other peripheral locations. After a pathogen is destroyed most of these T cells die, but a few remain as long-lived memory cells. These are able to respond to a second encounter with a pathogen. Apparently continuous new exposure to antigen is not needed for long-term immunity, but the slowly dividing $CD8^+$ memory T cells may require continuous stimulation by 1l-15 to counteract inhibition by Il-2.

E. THE REARRANGING GENES FOR IMMUNOGLOBULINS AND T-CELL RECEPTORS

An impressive example of the kind of permanent changes that can occur in the genome of specialized cells is provided by the genes of immunoglobulins and T-cell receptors. B lymphocytes make tens of millions of antibodies of differing sequence, and T lymphocytes make a similar number of different T-cell receptors. This diversity is established in major part in the DNA but also by alternative splicing and editing of RNA.

1. Rearrangements of Germline DNA

Each ~110-residue domain of an immunoglobulin is encoded by a single exon, but the exon for the amino-terminal or variable domain is assembled from two or three small genes (or segments) selected from a large family of such genes present in the germ cells and in the lymphoid progenitor cells. Within the V region the three short hypervariable segments alternate with four framework segments that have a more nearly constant structure. The V region of the light-chain genes (of either the κ or λ type) is put together in part from a V gene that encodes an approximately 95-residue sequence making up the first three framework regions plus two hypervariable regions and part of the third hypervariable segment. There are ~100 different V_κ genes and ~30 V_λ genes in the light-chain, family. The similar arrangement of genes in the mouse is indicated in Fig. 8.17. These V genes are spaced at intervals of 14-30 kb within the DNA.

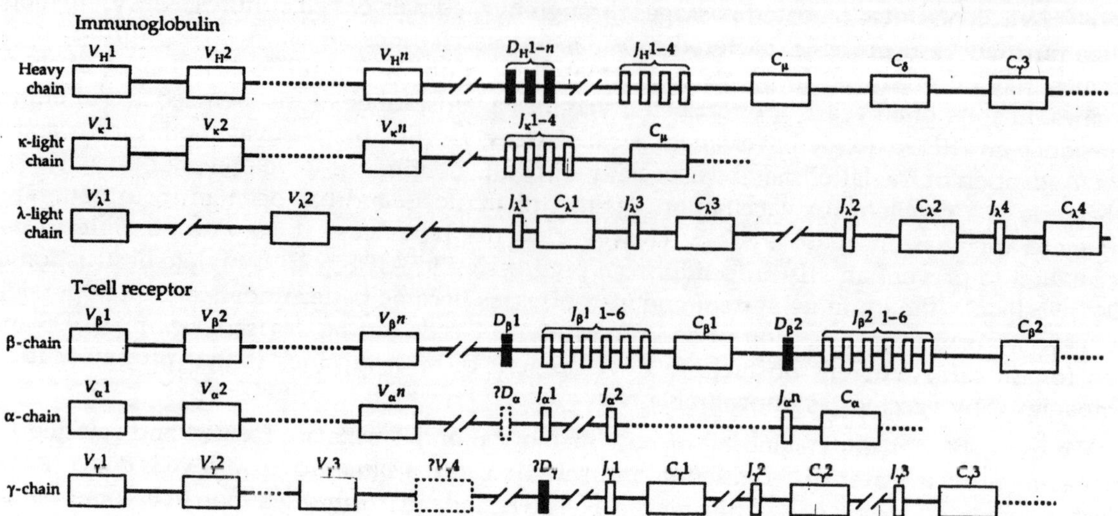

Fig. 8.17. Organization of the immunoglobulin and T-cell receptor gene families of the mouse. The human γ-gene pool is larger as is the λ light-chain immunoglobin gene pool. All six gene pools contain separate gene segments encoding the variable and constant regions of the antigen receptors of lymphocytes. In the course of lymphocyte development, one of the V segments is juxtaposed by chromosomal rearrangement with one of the J segments and, where applicable, a D segment to form a complete variable-region gene. Each V segment has two regions of hypervariability, which are known in the case of the immunoglobulins to contribute to the antigen-binding site in the folded molecule. A third hypervariable region, which also contributes to the antigen-binding site, is generated by the junction of the V segment with the J or the D and J segment(s) at recombination. There is more germline diversity in the T-cell receptor than in the immunoglobulin gene pools in the J segments. The β-gene pool is organized in a way that also allows more combinatorial diversity. The β-gene pool contains fewer J regions than the gene pool, but it has two distinctive features that allow for exceptional diversification during somatic rearrangement. First, the 'rules' for recombination allow in principle the joining of both D to D segments and V segments directly to J segments, neither of which is possible in the immunoglobulin heavy-chain pool. The D segments of the β genes can be read in any of the three possible reading frames, so that varying the site of the V – D junction alone can make a substantial contribution to the diversity of the third hypervariable region.

The rest of the third hypervariable region and the fourth framework region of the κ and λ chains are encoded by short *J (joining) genes* that specify ~15 residues. There are about five J_κ and three J_λ genes in the mouse. The first mechanism for creating antibody diversity lies in the large number of V genes, which are especially diverse in their hypervariable regions. The second mechanism is the joining of any one of these V genes with any one of the J genes, the joining taking place within the third hypervariable region. There are also at least six different constant (C_λ) genes in the human genome.

The heavy-chain genes are more complex. There are over 200 V_H genes located at 14- to 16-kbp intervals, which are followed by ten -15-17 bp *D (diversity) genes* and 4-6 J_H genes. During the differentiation of the lymphoid stem cells pre-pro-B cells can be identified in which the V, D, and J genes are still separate. Later there are pro-B cells containing joined DJ_H segments, then pre-B cells with joined VDJ_H segments, then B cells with a VJ_L (either κ or λ) segment also joined. The overall process of gene rearrangement, mRNA splicing, and immunoglobulin synthesis is outlined in Fig. 8.18. The joining of the gene segments occurs by recombinational mechanisms that involve 7-bp and 9-bp recognition sequences known as *recombination signal sequences* (RSSs).

5′-CACTGTG • 3′-GTGACAC
heptamer

5′-GGTTTTTGT • 3′-CCAAAAACA
nonamer

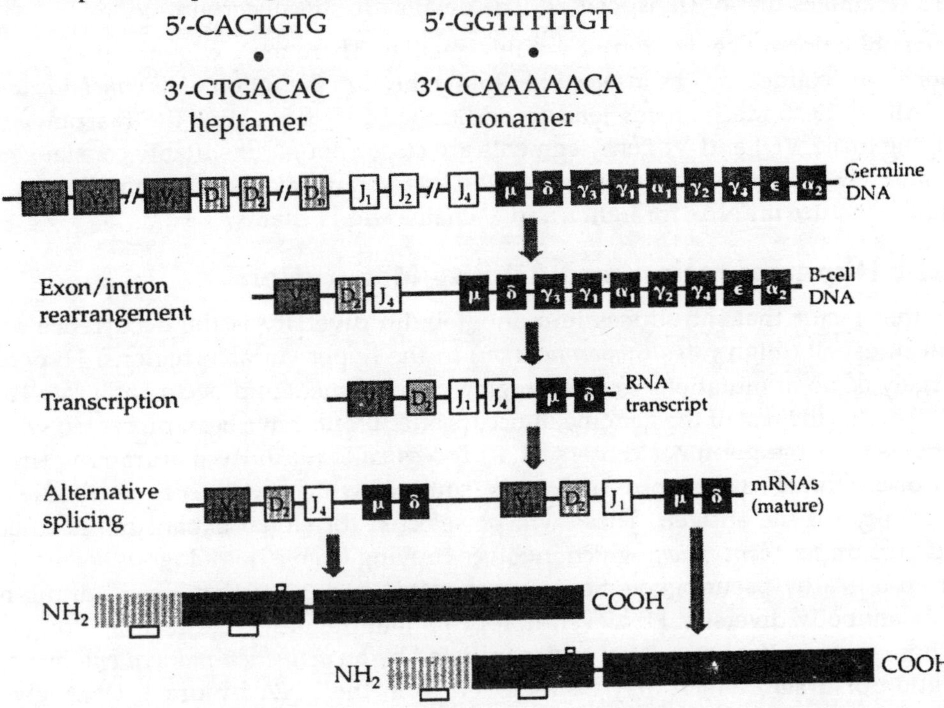

Fig. 8.18. Human immunoglobulin heavy-chain gene structure and gene processing. Exons of the heavy-chain genes that encode the variable regions of the immunoglobulin molecule are labeled V_1, V_2, ... V_n, Selection from these V exons during embryonic development produces the unique sequences of each B-cell clone. The germline genes for the immunoglobulin heavy chains also contain diversity exons, labeled D_1. D_2 .... D_n. Recombination between the V and D regions occurs more frequently than that between the V and J exons in the light-chain exons. Introns between the V and D and between the D and J exons contain signal sequences that regulate synthesis of the Rag1/Rag2 recombinase. This enzyme is responsible for the efficient recombination that gives rise to the epitope-specific B-cell clones with their individual Ig genes. The heavy-chain genes contain exons that encode all of the isotype heavy chains. Class switching, *i.e.*, the change in chain expression that occurs during antibody synthesis after B-cell activation, results from alternative splicing between the J exons and the exons for the various heavy-chain isotypes.

The heptamers have twofold rotational symmetry. The recognition sequences adjacent to one or both sides of the coding segments are separated by 12- or 23-bp spacers, and an empirical rule states that joining can occur only when one pair of recognition sequences are separated by 12 bp and the other by 23. Recombination involves cleavage and rejoining of the DNA, a process also used for reorganization of the T-cell receptor genes. The cleavage is catalyzed by a complex of two proteins RAG1 and RAG2, which are encoded by the adjacent *recombination activating genes RAG*1 and RA1G2. Cleavage is assisted by HMG chromatin proteins and by the level of histone acetylation in an associated enhancer. In the presence of a divalent metal such as Mg^{2+} or Mn^{2+} a single-stranded nick is made between the 3′-end of the DNA coding region and the heptamer sequence. The released 3′-OH group then attacks the phosphodiester bond on the opposite strand, cleaving it in a transesterification reaction. This leaves the coding sequences capped by a hairpin end and a blunt cut end on the RSS. The cut ends are apparently held until synapsis with a nonamer RSS further in the 3′ direction and rejoining can occur. Rejoining requires DNA ligase IV. The process resembles that of transposition. The details are still uncertain.

The recombinational events are not entirely regular. There are excisions of pieces of DNA, and additional nucleotides may be inserted randomly through the action of *terminal deoxynucleotidyl transferase.* All of these mechanisms lead to additional diversity. After the rearrangements are completed, the fused VDJ and VJ gene segments are close enough to suitable constant-region (C) sequences that when the genes are transcribed the intervening sequences in the RNA are spliced out to yield the mature mRNAs for light κ and λ chains and H chains.

2. Somatic Hypermutation and Affinity Maturation

Yet another factor that introduces immunoglobulin diversity is the occurrence of somatic mutation at unusually high rates (*hypermutation*) in the hyper-variable regions. Hypermutation consists largely of point mutations in the V-region gene segment and occurs at a rate 10^5- to 10^6-fold higher than for the rest of the genome. It occurs after B cells have been presented with antigen by dendritic cells in the germinal centers in a process called affinity maturation. The result is generation of an enlarged repetoire of B cells, some of which synthesize antibodies with an increased affinity for the antigen. These will be selected for clonal expansion. Further genetic alteration occurs by *gene conversion,* which involves copying from a homologous segment of DNA, perhaps from a nearby pseudogene. Some experiments suggested that RNA editing may also contribute to antibody diversity. However, it appears that the observed deamination of cytosine to uracil rings occurs at the DNA level and is initiated by an *activation-induced cytosine deaminase.* The generation of uracil, which may be removed from the DNA by uracil-DNA glycosylase, apparently triggers both somatic hypermutation and gene conversion. Both processes also depend upon DNA repair via homologous recombination or nonhomologous end joining. The error-prone DNA polymerase τ is also needed. Class-switching recombination, discussed in the next section, is apparently also initiated by the activation-induced deaminase. These are affected by *RNA editing. Receptor editing,* gene rearrangements that occur in the peripheral immune system, also contributes to affinity maturation.

It is important that the genes as finally assembled maintain correct reading frames so that a potentially useful antibody can be made. Although lymphocytes are diploid and therefore contain two sets of immunoglobulin genes, a single cell produces only one kind of immunoglobin (*allelic exclusion*). When the genetic rearrangements produce a light chain able to combine with a heavy chain to form a functional immunoglobulin, a signal may be sent that stops the rearrangements in the other chromosome. However, recent findings suggest that one allele may be marked for inactivation early in development, just as one X chromosome becomes inactive in females.

3. Immunoglobulin Class (Isotype) Switching

A newly matured B cell produces initially IgM bound to its surfaces. The difference between the C_H domains of bound and secreted antibodies of a given type seems to lie in alternative splicing of the mRNA. More mysterious are the consecutive switches from IgM to other types in the following sequence:

$$IgM \rightarrow IgG \rightarrow IgE \rightarrow IgA$$

This is the same sequence in which the C_H genes lie (Fig. 8.18). However, the C_δ gene, which codes for IgD, is not utilized in this sequence. A newly matured B cell transcribes the Cμ gene to give IgM. Later, the class switch occurs, apparently by a looping-out recombinational mechanism, allowing a C_γ gene (for IgG) to be expressed. The switches are mediated by tandemly repeated DNA sequences in 1- to 10-kbp switch regions and are controlled in part by cytokines. Other complexities are involved in synthesis of the J chain of IgM and in synthesis of IgD. Although the latter is a major surface immunoglobulin on B lymphocytes, its exact functions have been hard to understand. The 8 exon of its heavy chain gene is joined to a J exon by alternative splicing of the mRNA (Fig. 8.18). In a similar manner, the difference in the surface-bound and soluble forms of IgM arises by alternative mRNA splicing.

The T-cell receptor gene families are also indicated in Fig. 8.17. Their development is remarkably similar to that of the immunoglobulin genes and involves most of the same mechanism of diversification with the exception of somatic hypermutation. The same recombinase may cut the DNA to initiate rearrangements of all of these gene families. A single T-cell precursor may give rise to 1000 or more clones with unique P-chain sequences. With a total of ~10^6 different β chains there are potentially ~10^{15} unique T-cell receptor structures that could arise from the 42V and 61J segments of the α-chain gene and the 47V, 2D, and 13J segments of the human β-chain gene. Allelic exclusion is observed, as with the immunoglobulins.

F. DISORDERS OF THE IMMUNE SYSTEM

Many things can go wrong with a system as complex as the human immune system. In immunodeficiency disease some component is missing or has been inactivated. In autoimmune diseases the immune system attacks some component of the body. Of the known problems none is more common than *allergy,* which may be described as the inappropriate activation of the immune system by environmental antigens *(allergens).*

1. Allergy

One in 10 persons, ~22 million people, in the United States have allergies. Ten million of these suffer from the nasal discomfort of "hay fever" and six million from the more serious *asthma.* Substantial numbers of people in the United States die of allergic reactions to insect stings (more than 30 per year) or to injections of penicillin (300 per year in 1970). Foods, drugs, pollens, mold spores, mites in house dust, and even heat or cold can evoke serious allergic reactions. Among these *eczema* (atopic dermatitis) is very common. A major cause of allergic reactions has been traced to molecules of immunoglobulins IgE, which bind to the *basophils* in the blood and to the related *mast cells* of tissues. Binding of an antigen to these IgE molecules activates them. These activated antibodies bind to the α subunits of the $F_{C\varepsilon}RI$, a transmembrane receptor on basophil or mast cell surfaces. If two or more IgE molecules bind to a mast cell, they may aggregate and activate the mast cell to release its histamine-containing granules. The granules also release cytokines and arachidonate, which is converted primarily into prostaglandin D_2 and into products of the 5-lipoxygenase pathway. The products include the criemo-tactic leukotriene B_4 and leukotrienes C_4 and D_4. The latter two constitute the slow-reacting substance of anaphylaxis. The result is a rapid inflammatory response with dilation of blood vessels, increased vascular permeability, infiltration of leukocytes, and destruction of tissues.

What is the normal function of IgE and the mast cells? These cells are located at places where parasites might enter tissues. IgE is involved in killing of schistosomes, and elevated IgE levels are seen in patients infected with various parasites. The killing of schistosomes seems to be mediated by blood platelets as well as by neutrophils and eosinophils with the help of mast cells. Allergic persons often have an IgE level over ten times normal. This high level makes the individual especially sensitive to IgE-mediated reactions, a condition called *atopy* (meaning "strange disease"). Allergies may also be accompanied by increased B cell levels. This can sometimes'be responsible for the sudden and sometimes fatal systemic reaction of *anaphylaxis*. T-cell responses may also cause anaphylaxis.

Most allergy-inducing antigens are proteins, but proteins vary widely in their antigenicity. Only a few natural proteins are major allergens. Many of these are relatively small, with molecular masses of 5-50 kDa. Most are soluble, and some are glycoproteins. Mites are the closest animals to human life and carry allergens that are among the most important causes of asthma and allergic dermatitis. The allergens are 125- to 129-kDa proteins crosslinked by three disulfide bridges. Cockroache and other insects also form many allergens. Among them are the hemoglobins of small flies of the chironomid family. Studies of the latter suggest that antigenicity may arise both from flexible regions and also from the presence of a preponderance of amino acids with polar side chains.

Proteins of cat saliva dry and flake off as dander, which contains major indoor allergens linked to asthma. Dogs, horses, cattle, and other animals also provide several allergens, some of which are lipocalins. Other close-to home allergens are provided by fungi that live on skin or nails.

Plants provide a host of allergens. Major allergens are found in pollen of rye-grass, of many other grasses, of ragweed, and of olive trees. Natural rubber latex would appear to be a harmless high polymer, but it contains antigenic proteins, which have been blamed for 1100 anaphylactic attacks with at least 15 deaths, between 1988 and 1992.

Food is a major source of allergens, which are often overlooked. Food allergies may be hard to diagnose and symptoms such as headache, diarrhea, itching, and asthma may be attributed to other causes. However, the occasional rapid death from anaphylactic shock, *e.g.*, from exposure to peanuts, is a reminder that unrecognized food allergies exist. About 100 – 200 persons die annually of food allergies. About 90% of recognized food allergies involve milk, eggs, fish, crustacea, peanuts, tree nuts, soybeans, and wheat. There are usually only a few allergenic proteins in any one food. Many of these are resistant to digestion in the stomach. Some but not all are compact proteins with multiple disulfide crosslinkages. However, no structural generalization can be applied to all food allergens.

The increasing use of genetic engineering of foods poses both risks and hopes. The accidental incorporation of an allergenic protein into a plant or animal product can make a food previously safe for a person deadly. For this reason, attempts are now made to identify likely allergens and to avoid transferring their genes. However, this process can't be completely reliable, partly because we each have our own personal immune system. This is one reason for requiring accurate labeling of foods. On the positive side recognized major allergens can probably be eliminated by genetic engineering. The muscle protein tropomyosin is a well-known allergen whose allergenicity varies among different sources. Tropomyosin from beef, pork, and chicken is usually not highly allergenic, but that from shrimp often is. Perhaps safer shrimp can be created.

Allergies are treated in various ways, often with antihistamines or corticosteroid ointments. Injection of adrenaline is an emergency treatment of anaphylaxis or asthma. Also important is *specific immunotherapy,* better known as "allergy shots." Small amounts of allergen in increasing amounts are injected subcutaneously at intervals to desensitize the patient. Use of purer antigens, which may be engineered to decrease their antigenicity, may provide advances in this technique.

MYASTHENIA GRAVIS

One of the best-understood autoimmune diseases is myasthenia gravis, a condition associated with a decrease in the number of functional post-synaptic nicotinic acetylcholine receptors in neuromuscular junctions. The resulting extreme muscular weakness can be fatal. Myasthenia gravis is not rare and affects about one in 10,000 people. An interesting treatment consists of the administration of physostigmine, diisopropyl-phosphofluoridate or other acetylcholinesterase inhibitors (Box 12-E). These very toxic compounds, when administered in controlled amounts, permit accumulation of higher acetylcholine concentration with a resultant activation of muscular contraction. The same compounds are widely used in the treatment of glaucoma.

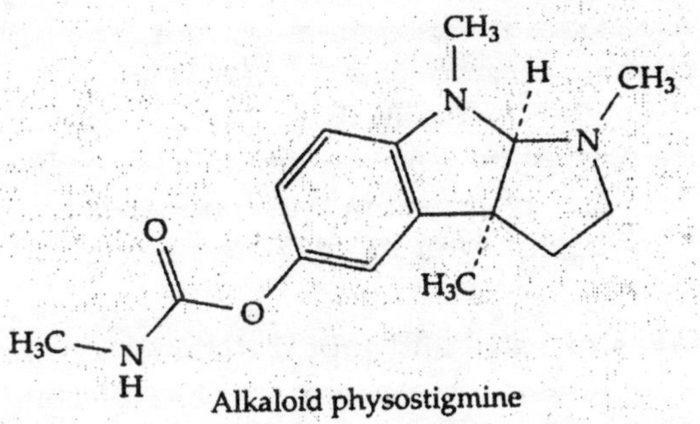

Alkaloid physostigmine

More than 90% of patients with myasthenia gravis have circulating antibodies directed against a subunit of the acetylcholine receptor. Immunosuppressive drugs and steroids help to cut down on these autoantibodies, and many patients are benefitted by removal of the thymus. Newer approaches involve specific immunotherapy aimed at increasing tolerance to either T cells or to B cells.[c, d] For example, oral ingestion of purified acetylcholine receptors to desensitize the body's response or inhibition of production of Il-2.

A possible cause for the production of the damaging antibodies may be the sharing of common antigenic determinants between the receptor protein and surface proteins of bacteria such as *E. coli*.[f]

Asthma is one of the most common chronic diseases in industrialized countries, affecting 10% or more of young children in some countries. An atopic disease with high IgE levels, asthma is induced by small particles of antigen, which are able to penetrate deep into the lungs. About 80% of asthmatic children are allergic to house mites. Animal danders are another major cause. The incidence of asthma appears to be increasing in many modern societies, but the reasons are unclear, and the increase doesn't appear to be linked to air pollution. A chronic atopic condition usually precedes an acute attack of asthma. In addition to IgE and eosinophils there are excessive numbers of neutrophils with high-affinity IgE receptors in the airway tissues. Prostaglandin D_2 released from mast cells may play a role in triggering an attack. Cytokines, nitric oxide, and nerve growth factor may also participate in the response. The presence of a high concentration of glutathione and of glutathione peroxidase, whose concentration increases in asthmatic lungs, may reflect the action of the antioxidant system in combatting inflammation. The surfactant proteins SP-A and SP-D are Ca^{2+}-dependent lectins, which serve as regulators of the innate immune response. Their concentrations also increase in asthma.

Treatment of asthma has depended upon inhaled glucocorticoids, quick-acting bronchodilators usually p-adrenergic antagonists, and long-acting beta agonists such as theophyllins and leukotriene antagonists.

2. Autoimmune Diseases

There are numerous *autoimmune diseases* in which the body makes antibodies against its own cells. In *myasthenia gravis* antibodies attack the acetylcholine receptors in postsynaptic membranes. In *Graves disease* aberrant antibodies are directed against receptors for thyrotropin. They have a stimulatory rather than an inhibitory effect and cause hyperthyroidism. *Childhood onset (Type I) diabetes* results from destruction of insulin-secreting cells by an autoimmune reaction triggered by environmental factors in genetically susceptible persons. The principal autoantigen appears to be the 65-kDa form of glutamate decarboxylase (GAD). While GAD has an essential function in the formation of γ-aminobutyrate in the brain, its role in the pancreatic islets is not clear. What is established is the presence of specific GAD65-stimulated T cells in diabetic individuals. The stimulating autoantibodies, which may appear in the blood years before diabetes is evident, carry HLA-DR4 type surface MHC antigens. After the activated T cells kill enough of the pancreatic β cells, diabetes appears.

Myasthenia gravis, Graves disease, and type I diabetes are organ-specific autoimmune diseases. Another group of autoimmune diseases are systemic, affecting many tissues. For example, in the severe *systemic lupus erythematosus* there are often antibodies against the victim's own DNA. The antibodies may then attack any tissue, *e.g.*, the red blood cells. Antibodies against a variety of other nuclear constitutents such as histones, ribosomal protein L7, ubiquitin, enzymes, cardiolipin, and small nuclear RNAs are also made. The primary defect appears to be an intolerance to chromatin and in particular to nucleosomes. Antibodies to nucleolar and other components of nuclei are also present in progressive systemic sclerosis (*scleroderma*). In *rheumatoid arthritis,* a chronic inflammation of joints, the serum and joint fluids contain abnormal complexes, which appear to consist entirely of immunoglobulins. They may be antibody – antiidiotype antibody complexes. Immunization of animals with type II collagen induces a very similar arthritis, but this collagen probably doesn't supply the offending human antigen. Susceptibility to rheumatoid arthritis is linked to HLA-DR4 class II MHC genes. Molecules of the class II DR4 subtypes may associate with antigenic peptides of uncertain origin to induce a T-cell response. $CD4^+$ T cells are thought to drive inflammation in the disease. Monocytes are attracted from the blood and become inflammatory macrophages.

Another autoimmune disease in which antibodies attack collagen is *Goodpasture disease.* It is mediated by B cells, which form antibodies directed at the N-terminal domain of the α3 chain of collagen IV. The antibodies attack the glomerular basement membranes causing a rapidly progressing glomerulonephritis and also lung hemorrhages. Primary *glomerulonephritis,* a major kidney disease, may be caused by a cross-reaction between the membrane of streptococci and the glomerular basement membranes.

Table 8.3. Some Autoimmune Diseases

Addison disease	Adrenal glands
Ankylosing spondilitis	
Celiac disease	Upper intestines
Crohn disease	Intestines
Diabetes, type I	Pancreatic islets
Glomerulonephritis	Kidney
Goodpasture disease	Kidney
Graves disease	Thyroid gland
Guillain-Barre syndrome	Gangloisides
Multiple slerosis (MS)	Peripheral myelin
Myasthenia gravis	
Paroxysmal cold hemoglobinuria	Red blood cells
Primary biliary cirrhosis	
Psoriasis	Skin
Polymyocystits	
Rasmussen encephalomyelitis	Cerebral cortex
Rheumatoid arthritis	Joints
Scleroderma	Skin
Sjögren syndrome	
Systemic lupus erythematosis	Many tissues
Thyroiditis	Thyroid gland
Ureitis	

In *Sjögren syndrome* autoantibodies are directed against α fodrin. In primary *biliary cirrhosis* they are directed at mitochondria and specificially to a pyruvate dehydrogenase subunit. In the inflammatory muscle disease *polymyocystitis* autoantibodies are often directed against cytoplasmic proteins including aminoacyl-tRNA synthetases. In the rare *paroxysmal cold hemoglobinuria* autoantibodies attack red blood cell membranes only when the temperature of an extremity is lowered. *Paroxysmal nocturnal hemoglobinuria,* a serious complement-mediated condition, results from deficiency in the complement decay accelerating factor. This is a result of a defect in the PGI tail on this factor.

Celiac disease (celiac sprue) is an allergic inflammatory condition caused by poorly digested prolinerich sequences of wheat gluten and related proteins. The disease is usually not recognized, but it may occur in 3% or more of the United States population. A T-cell response that causes destruction of the smaller intestinal mucosa, celiac disease is characterized by malabsorption and diarrhea. It can cause death by starvation. A primary target of the autoantibodies is a transglutaminase.

Most cells of the immune system are ordinarily kept apart from those of the nervous system by means of the blood-brain barrier. However, allergic encephalomyelitis, in which T cells attack the myelin sheath of brain neurons, can easily be induced in mice. A similar autoimmune process is thought to be involved in human *multiple sclerosis*. High levels of circulating IgM are found in some demyelinating diseases of peripheral neurons. In *Rasmussen's encephalitis,* which causes brain inflammation and epilepsy, serum antibodies attack a glutamate receptor subunit *GluR*3.

The causes of autoimmune disease doubtless lie largely in the difficulty of developing a repetoire of immunoglobulin-forming B cells and of T-cell receptors that will always reliably distinguish self from a foreign antigen. The problem can lie either with B-cell recognition or with the T-cell receptors. Extensive medical use is made of *immunosuppressants* in treatment of persistent allergic reactions, autoimmune problems, and rejection of transplanted tissues. Among these compounds are the steroidal anti-inflammatory agents such as prednisone and the folate antagonist methotrexate. A new vision of the possibilities for immunosuppression came, however, with the discovery of *cyclosporin A.* This fungal metabolite inhibits lymphokine formation by helper T cells. It alleviates rejection of grafted tissues and prevents graft-versus-host disease. The use of cyclosporin and FK506 has permitted organ transplantation to the extent that by 1987 surgeons had transplanted in one year 1200 livers, 1500 hearts, and 9000 kidneys with one-year survival rates of 80% for hearts and over 90% for kidneys. Cyclosporins bind very tightly to *cyclophilins,* which have peptidylprolyl *cistrans* isomerase activity and perhaps other independent functions.

EVADING THE IMMUNE SYSTEM

Parasitic species always have a problem with the antibodies and killer T cells of their hosts, and the chemical makeup of the external coats of parasites tends to reflect this fact. An example is provided by *trypanosomes,* which cause sleeping sickness and which make much of Africa unsuitable for cattle grazing. Trypanosomes in the blood-stream evade the immune system by covering the outer surface of their plasma membrane, flagella and all, with a dense 12- to 15-mm thick monolayer of an ~60-kDa *variable surface* glycoprotein. The glycoprotein molecules are anchored in the cell membrane by C-terminal glycosylphosphatidyl-inositol (GPI) anchors (Fig. 8.13). The glycoprotein layer protects the parasite but is soon attacked by the host's immune system. However, the parasite has perhaps 1000 different genes for the variable surface protein, and every ten days or so new clones of trypanosomes appear with new coats that the immune system is not prepared to attack. To accomplish this cells occasionally copy one of the previously unused variable surface glycoprotein genes and place it into a new location in the genome, where it is expressed.

Parasitic *nematodes* shed the outer layers of their external cuticle and like trypanosomes reveal a new layer with different antigenic proteins. *Giardia* protects itself in a similar fashion. *Schistosomes,* tiny parasitic flatworms, evade a host's immune system by shedding complex glycoproteins from specialized double outer membranes. Antigenic determinants including MHC antigens characteristic of the mouse have been identified in the membrane of schistosomes from infected mice. Thus, one aspect of the parasite's defense may be to hide behind surface recognition markers stolen from its host. Schistosomes also secrete the peptide Thr-Lys-Pro, which inhibits macrophages, as well as a small molecule that inhibits T lymphocytes.

The malaria parasite *Plasmodium* has a complex life cycle with several forms and spends much of its life hiding within red blood cells. It may also suppress the immune system. The unicellular sporozoites, which are injected into the bloodstream by mosquitos, are protected by an external coat protein that is unusual in containing many short repeated sequences. For example, that of *P. falciparum,* which causes the most deadly form of malaria, contains the sequence Asn-Ala-Asn-Pro repeated 37 times. These coat proteins undergo unusually rapid evolution, which makes the preparation of vaccines difficult.

Trypanosomes, schistosomes, and malaria parasites still represent major health problems. Malaria kills two to four million persons a year and endangers almost a third of the world's population. It has been impossible to produce suitable vaccines for any of these parasites. However, the cloning of genes for individual parasite proteins has given hope that effective vaccines can be devised. One problem is the lack of interest in financing the effort.

Many other protozoan parasites and bacteria invade cells and take up residence in macrophages. These include species of *Salmonella, Legionella,* and *Mycobacterium.* Bacteria often employ structural mimicry to gain access to a cell, *e.g.,* by mimicking the type III secretion system. Some bacteria have developed defenses against reactive oxygen species, allowing them to evade the action of phagocytes. *Borrelia burgdorferi,* the Lyme disease spirochete, synthesizes an unusual single-layer β-sheet outer surface protein, which becomes coated with complement protein H. This may protect the bacteria and allow them to live for a long time within cells.

Even the lowly *influenza virus* finds a way around our immunity so that it can strike us repeatedly. As this virus matures, it acquires a lipid membrane by budding from the host cell. Two virally encoded proteins are present in the membrane. One is a trimeric hemagglutinin, which forms small 7.6-nm spikes mat protrude from the virus surface. The hemagglutinin monomer is a 550-residue peptide containing four antigenic regions. The RNA genome of the virus undergoes rapid mutation. At least one amino acid substitution was found in each antigenic region, when hemagglutinins from influenza viruses causing epidemics in 1972 and 1975 were compared with the strain that caused a worldwide epidemic in 1968. Recently the type A influenza virus that caused the 1918-1919 pandemic, the greatest acute plague of the 20th century, has been "resurrected" and investigated using viral RNA from three victims. The globular part of the hemagglutinin appears to have come from a pig and the "stalk" from a human lineage. The virus takes advantage of the pool of virus in swine, humans, and birds to vary its structure and create new strains. The reason for the deadly nature of the 1918-1919 strain, which killed 20 – 40 million people, an unusually large number of whom were young, previously healthy adults, is not clear.

Viruses use a large variety of mechanisms to evade cellular defense mechanisms. Almost every aspect of the innate or adaptive immune systems provides some opportunity for evasion. The rapid mutation rate in a population of virus particles contributes greatly to this ability, allowing chronic infections such as those of hepatitis C or delayed catastrophic infections such as those of HIV.

3. Immunodeficiencies

More than 95 different problems of impaired immunity have been identified. They affect about 1 in 10,000 persons born. The defects may involve T cells, B cells, NK cells, or phagocytic cells. There may be problems in lymphocyte development. Complement proteins may be lacking or their regulation may be faulty. The immune system has specific "blind spots" and fails to recognize all dangerous foreign antigens. In the fatal *X-linked immuno-proliferative syndrome,* the immune system of susceptible males does not respond to the Epstein-Barr virus-induced mononucleosis by killing the persistently lymphoblastoid cells characteristic of that disease. Some individuals are born with *severe combined immunodeficiency* disease. This condition was made well known to the public by the plight of David, the "bubble boy," who lived 12 years in protective sterile rooms and a plastic bubble-like "space suit." The condition is often caused by a defect in the interleukin-2 receptor, but there are a variety of other causes. About 15% of cases arise from a defect in adenosine deaminase, an enzyme of the purine salvage pathway. Genetic therapy for this condition is being used. However, the most reliable treatment for these immunodeficiencies seems to be bone marrow transplantation. By 2000 more than 375 patients worldwide had received this treatment with up to 95% chance of survival. Virus-induced immunodeficiency is the prime characteristic of HIV infection and *AIDS*. Both the amounts of autoantibodies and of amyloid deposits increase with age, and immune complex disease is suspected of being a cause of aging.

AN INSECTICIDAL PROTEIN

During sporulation the bacterium *Bacillus thuringiensis* forms within its own cells large protein crystals, which are highly toxic to some insect larvae. The crystals account for 20-30% of the dry weight of the bacterial spores and contain more than one toxin and, curiously, a 20-kbp piece of DNA. Dusting of garden plants with dried spores from these bacteria has become a popular and effective way of combating cabbage worms and other insects. The toxic protein from one strain of bacteria is encoded by a 4222-bp gene. The corresponding 133-kDa 1176-residue polypeptide protoxin undergoes glycosylation and perhaps other modifications, presumably prior to crystallization. After ingestion by susceptible insect larvae (largely Lepidoptera) the protein is cleaved to form a smaller ~65-kDa protease-resistant core, which is the active toxin. Other strains of bacteria produce toxins specific for Diptera or Coleoptera.

X-ray crystallography of the 65-kDa form reveal a three-domain structure. The central domain varies among different strains and is probably involved in recognition and in binding to cell surface receptors. The toxin binds to a receptor, apparently an aminopeptidase N, after which the toxin is rapidly inserted into the membrane forming a 1- to 2-nm diameter pore. This leads to cell death.[j]

Because the toxins appear to be harmless to human beings and higher animals the toxin genes have been transferred into various other bacteria, which are symbiotic with plants and into plants themselves. Toxin genes in suitably modified form were first transferred into bacteria that live naturally in association with roots of *Zen mays* and into tobacco and tomato plants. The new host organisms expressed the toxin genes and protected the plants from damage by caterpillars.[k, l] Since then the toxin genes have been transferred into many crop plants, which are widely planted.

Two problems must be considered. Insects do develop resistance to the Bt toxin. This problem can be combated by protein engineering and by location of new sources of toxins. A second problem deals with the environmental impact. Will Bt toxin kill desirable insects? Will the gene be transferred in nature to other species and into the environment? The latter may seem unlikely, but as the toxins are applied to fight soil organisms such as nematodes, transfer into organisms of the largely unstudied soil ecosphere may pose problems.

4. Cancers of the Immune System

A major function of the immune system is thought to be destruction of cancer cells. In this case altered cell surface carbohydrates or proteins elicit an antibody response with destruction of the offending cells. That this process works imperfectly may explain why the incidence of cancer increases with age and also why the concentration of autoantibodies increases. The immune system is also susceptible to cancers, which include multiple myeloma, leukemias, and lymphomas. Some of these, such as Burkitt's lymphoma, involve rearrangement of chromosome segments that carry immunoglobulin genes. These may result from errors in the gene rearrangements involved in the development of lymphocytes.

G. DEFENSE MECHANISMS OF PLANTS

Plants make many compounds that repel or poison animals that eat them. Among such compounds are alkaloids, terpenes, calcium oxalate, fluoroacetate, cyanogenic glycosides, and phenolic compounds. The chewing of insects or other wounding of plant tissues releases phenolic glycosides and other reactive compounds from vesicles. Some of these compounds, which are often referred to as *phytoalexins*, are repellent to predators, have antimicrobial activity, and/or participate in chemical cross-linking and strengthening of the plant cell wall. Some are protease inhibitors that interfere with a predator's nutrition. Some released compounds attract insects that may assist in defense by feeding on predator eggs or by attracting wasps that deposit eggs in predator larvae. These can all be regarded as part of an innate defense system that in some respects resembles our own innate immune system. For example, plant defensins, most of which are directed against fungi, resemble those in our tissues.

A system of *receptor-mediated surveillance*, part of the innate system, triggers both immediate *local responses* and secondary immunity throughout the plant. Immediate responses include *programmed cell death* (called the *hypersensitive response*), tissue reinforcement, and production of antimicrobial metabolites. Secondary responses, known as *systemic acquired resistance*, develop immunity throughout the plant. The surveillance system of some plants consists of a series of receptors known as *resistance* (*R*) *proteins*, which recognize signaling molecules produced by pathogens. The R proteins are thought of as being paired with *avirulence* (*Avr*) *proteins* of the pathogen. If the resistance protein is missing, the plant will be susceptible to attack by the pathogen. The pathogen's *Avr* protein is thought to be part of the chemical attack on the plant, apparently assisting in the invasion. However, if the *Avr* gene has been lost or is mutated, the R protein won't detect the invasion, and the pathogen may have increased virulence. What are the characteristics of the *Avr* proteins? They are often small and may be crosslinked by S–S bridges. They may be taken up by plant cells via receptors that resemble the type III translocation system of bacteria. Similar small protein *elicitors* are released directly by wounding even in plants that do not have paired *R-Avr* genes.

The R proteins, which act as receptors for Avr, and other elicitor proteins, are usually leucine-rich-repeat proteins with a characteristic nucleotide binding site attached (NB-LRR proteins). Like other cell surface receptors they participate in signaling and utilize both ion channels and Ser/Thr protein kinases. The *Arabidopsis* genome contains ~150 sequences that may represent NB-LRR receptors.

What do these receptors do? Like other cell membrane receptors they may induce both rapid and slower responses. The rapid responses may result from transmembrane flow of ions, just as in neurotransmitter action. The first response observed is an oxidative burst, which within minutes generates reduced oxygen intermediates. These compounds may participate in crosslinking and lignification of cell walls. Together with nitric oxide (NO) and endogenous salicylic acid, they promote transcription of defense-related genes and participate in the hypersensitive response. A second pathway, utilized against some pathogens especially those that kill plants to obtain nutrients, involves production of jasmonic acid and ethylene.

Plants also have mechanisms for minimizing the damage from the over 500 known viruses. These don't often kill plants but can cause great damage. There are interferon-like responses and gene-silencing mechanisms. The latter often involve synthesis of dsRNA, cleavage by the enzyme Dicer, and interference with transcription. This defensive reaction can spread between cells and throughout a plant, apparently by transport of RNA through plasmadesmata and the phloem.

CHAPTER

9 Isolation of Neurons

As initiators of immune responses, dendritic cells (DCs) play a critical role in the defense against infectious agents. They are uniquely able to activate naive T cells, which are required for the initiation of adaptive immune responses to invading organisms. Optimal DC function requires appropriate development from bone marrow precursors, migration to peripheral lymphoid and nonlymphoid surveillance locations, maturation (including antigen processing/presentation, costimulatory molecule expression, cytokine production) after encounter with pathogens and migration to lymph nodes and/or spleen to interact with T cells.

Langerhans cells (LCs), an epidermal DC resident population, are one well studiea example of a peripheral DC subset that participates in antigen transport to and T-cell activation in peripheral LN. In addition, some DCs are able to suppress or down-regulate immune responses and thus serve an immunoregulatory role. Ethanol exposure has the potential to interfere with many of these DC functions and thereby increase the organism's susceptibility to infection. The aim of this chapter is to briefly review what is known about the effect of EtOH exposure on DC function, and to outline methodologies that are applicable to studying DC function following chronic EtOH exposure in vivo, in murine models.

Approaches to Studying DC Function After EtOH Exposure

It is impossible to precisely mimic in vitro the daily variability of EtOH exposure levels, local EtOH metabolite concentrations, or the microenvironment in which DCs normally reside in the intact animal. Thus, it is most relevant to study the effects of in vivo EtOH administration on DC function rather than exposing freshly isolated DC, or those generated from untreated animals, to EtOH in vitro. In some circumstances, no in vivo methods exist to adequately address the DC function under consideration. In these cases, in vitro experiments that measure events initiated in vivo or within a few hours of culture preparation (during which time DCs might be expected to maintain the deficits induced by EtOH in vivo) are reasonable surrogates.

Effect of Chronic EtOH Exposure on DCs

Very few studies have addressed the effect of chronic in vivo EtOH exposure on DC function. Some clues about the effect of chronic EtOH on DC function can be had from studies of unpurified "APC" function from the spleen. Macrophage phagocytosi is reduced by EtOH exposure. Ag presentation and immunostimulatory cytokine production of APC from EtOH mice is also reduced and defects are observed in the afferent phase of DTH in EtOH mice. EtOH mice also show decreased serum levels of inflammatory cytokines produced by DC and macrophages, (such as interleukin (IL-6), tumor necrosis factor-α, and IL-1.

Antigen-presenting cell (APC) functions can be ascribed to a number of different cell types, including monocytes, macrophages, B cells, and DCs. Only one study has specifically investigated the effects of chronic in vivo EtOH exposure on DC function and demonstrated decreased allostimulatory capacity and lower levels of costimulatory molecule (CD86) and major histocompatibility complex (MHC) class II expression on EtOH exposed DC, particularly after inflammatory (CpG) stimulation.

METHODS

Because of the advantages of an in vivo model of EtOH administration for determining pathophysiologically relevant effects on DC function, this chapter focuses on protocols that rely on this approach. All techniques have been performed after in vivo EtOH administration using the Meadows-Cook model described in the chapter by Coleman, et al. in this volume.

MATERIALS

Identification and Isolation of DCs and LCs

Purification of CD8α+ and CD8α– DCs from the Spleen

1. Dense BSA Solution: 186mL of phosphate-buffered saline (PBS), 29 mL of 1 N NaOH in 65 mL of water. Carefully layer 106 g of bovine serum albumin (BSA; Intergen, Purchase, NY) onto the surface of the solution, cover beaker with foil and place at 4°C overnight. Check density by measuring refractory index; correct BSA density of 1.08g/mL corresponds to an index of 1.384 – 1.385 at 25°C. To reduce the refractive index by 0.0005 add 5mL of PBS or to increase the index by 0.0005, add 5g of BSA. The pH of the solution should be 7.0 – 7.4. Filter-sterilize the solution using a self contained filtering unit. Dense BSA may be replaced by Nycoprep 1.077A (cat. no. 10122311).
2. Nycodenz buffer: 9g NaCl, 0.6055g Tris, 0.22368 g KC1, 0.11167g ethylene diamine tetraacetic acid (EDTA); bring to 1.12L with deionized (DI) water.
3. Nycodenz stock solution: add exactly 15.275g of Nycodenz powder to 50mL of DI water. Filter and keep in the dark at 4°C.
4. MACS buffer: PBS, 0.5% BSA, 500 mM EDTA.
5. HT-2 Media: Iscove's DMEM IX 439mL, 10% fetal calf serum (PCS) 50mL, L-glutamine (0.292mg/mL H_2O) 5mL, Na Pyruvate (0.11 mg/mL H_2O) 5mL, gentamycin sulfate (0.5 mg/mL) 0.5 mL, 2-ME (0.05 m*M*) 500μL.
6. Anti-CD8α + Microbeads (Miltenyi, Auburn, CA, cat. no. 120 – 000 – 298).
7. Anti-CD11c+ Microbeads (Miltenyi, cat. no. 130 – 052 – 001).
8. LS Columns (Miltenyi, cat. no. 130 – 042 – 401).
9. Collagenase.
9. Hanks' Balanced Salt Solution (HBSS, 10X) without sodium bicarbonate, calcium and magnesium.

11. Complete media: Iscove's DMEM IX, 439 mL, 10% FCS (final concentration) 50mL, L-glutamine (29.2mg/mL H_2O stock cone.) 5mL, Na pyruvate (100 m*M* stock cone.) 5mL, Pen/Strep (10μL/mL, 10μg/mL stock cone.) 5mL, nonessential amino acids (100X stock cone.) 5 mL, HEPES, 5 mL (1*M* stock concentration), 2-ME (89.3 m*M* stock cone.) 500 μL.

Isolation of LCs for Flow Cytometry from Mouse Ears

1. Trypsin (Amresco, Solon, OH; cat. no. 0458 – 25G). Prepare 0.5% solution in PBS fresh on the day of the assay.
2. Balanced salt solution (BSS): 10x BSS stock solution 1 (10g dextrose, 0.6g KH_2PO_4, 3.58g $Na_2HPO_4 \cdot 7H_2O$ [1.89g anhydrous], 0.1 g phenol red powder or 20mL of 0.5% phenol red solution), 10 x BSS stock solution 2 (1.86g $CaCl_2 \cdot 2H_2O$ [1.41 g anhydrous], 4g KC1, 80g NaCl, 2g $MgCl_2 \cdot 6H_2O$ [1.04g anhydrous], 2g $MgSO_4 \cdot 7H_2O$ [1.74 g anhydrous]). To make a working solution, combine 50mL of stock solution #1 with 50 mL of stock solution #2 and bring to 500 mL with deionized water.
3. Ficolite-LM (Atlanta Biologicals, Lawrenceville, GA; cat. no. 140650).
4. Staining buffer: 475 mL BSS, 25 mL bovine calf serum-defined, 0.5 g NaN_3
5. 60-mm Petri dish (Falcon; cat. no. 351007).
6. Fine forceps (VWR; cat. no. 25607–856 for straight, 25607–890 for curved).
7. 70-μm filter (Falcon; cat. no. 352350).
8. Glass microscope slides with frosted ends.
9. Dissecting microscope.

Purification of Pulmonary DC Subsets by FACS

1. Iscove's DMEM Ix.
2. Collagenase.
3. DNAse 1.
4. 0.5 M EDTA.
5. Dulbecco's modified Eagle's medium (DMEM).
6. 03 – 200/48 Nitex Mesh (Sefar, Kansas City, MO; cat. no. F03A0020048I043A02).
7. MACS buffer: PBS, 0.5% BSA, 500 mM EDTA.
8. Polypropylene round-bottom tubes (Falcon, Franklin Lakes, NJ; cat. no. 352063).
9. CDllc MACS Beads (Miltenyi Biotec, Auburn, CA; cat. no. 120 – 000 – 322).
9. FCS.
11. Cell Strainer (BD, Bedford, MA; cat. no. 352350).
12. LS Column (Miltenyi Biotec; cat. no. 130 – 042 – 401).

Migration of DCs and LCs

Immunofluorescent Staining of LCs in Epidermal Sheets

1. 0.5*M* ammonium thiocyanate: add 19 g of ammonium thiocyanate to 500mL of 0.1*M* phosphate buffer (pH 6.9) or PBS. Store at 4°C.

2. Blocking buffer: 1 mL of goat serum (Invitrogen, Carlsbad, CA; cat. no. PCN5000), 1 mL of 10X PBS, 8mL of deionized water. Mix, store at 4°C.
   (*a*) Alternate blocking buffer: 8mL of 100m*M* TBS, pH 7.3 (6.06 g of Tris base, 400 mL of deionized H_2O. Adjust pH to 7.3 with concentrated HC1. Adjust volume to 500 mL, then add 0.9 g of NaCl), 1 mL of rat antimouse FcγRII/III (clone 2.4G2), ImL of goat serum, 0.2g of BSA (Amresco, Solon, OH; cat. no. 0332-100G), 5μL of Tween 20.
3. Acetone.
4. Biotinylated or PE conjugated anti-mouse MHC class II.
5. Avidin-Cy5 (Jackson ImmunoResearch Laboratories, Inc., West Grove, PA; cat. no. 016–170–84) or Avidin-Cy3 (Southern Biotechnology Associates Inc. Birmingham, AL; cat. no. 7100–09L) if a biotinylated antibody is chosen in step 4.
6. Petri dishes (60 × 15 mm, BD Falcon, San Diego, CA; cat. no. 351007).
7. 24-well tissue culture plate (Corning Inc. Life Sciences, Acton, MA, Costar; cat. no. 3524) or 96 well plate (Costar cat. no. 3596).
8. Fine forceps.
9. Vecta Shield Slide Mounting Media (Vector Labs, Burlingame, CA cat. no. H-1000).
9. Microscope slides (Surgipath Medical Industries, Richmond, IL; cat. no. 00240).
11. Slide cover slips, 22 x 22 mm.
12. Clear fingernail polish.
13. Drugstore depilatory, *e.g.*, Nair.
14. Dissecting microscope.
15. Fluorescent microscope.

FITC Tracking of LCs and Dermal DCs to Lymph Nodes

1. BSS.
2. Liberase blendzyme 3 (Roche Diagnostics, Chicago, IL; cat. no. 1814176)/DNase (Roche Diagnostics, cat. no. 104 159) working solution.
   (*a*) Stock Liberase solution: dissolve Liberase in RPMI 1660 + L-glutarnine (Invitrogen, Carlsbad, CA; cat. no. 1 1875) at 1 mg/mL
   (*b*) Stock DNase solution: dissolve DNase in RPMI at 4000 U/mL
   (*c*) Working solution: 25 μL of Liberase stock solution, 10 μL of DNase stock solution, 20μL of bovine calf serum, 945 μL of RPMI. Make fresh on day of use, and keep on ice.
3. Fixative for mouse cells: 12.5 mL of 10 × PBS, 2.7 mL of 37% formaldehyde solution, 200 mg of NaN_3. Bring to a final volume of 100mL.
4. n-butyl phthalate.
5. Acetone.
6. FITC.
7. Cytofix/Cytoperm solution (Becton Dickinson, San Diego, CA; cat. no. 554722).
8. Perm/Wash Buffer 10 x (Becton Dickinson, cat. no. 554723). Dilute to Ix before use, according to manufacturer's instructions.

9. Polyclonal unconjugated rabbit antimouse langerin (Imgenex, San Diego, CA; cat. no. IMG-5102A).
9. Cy5-conjugated AffniPure goat antirabbit IgG (H + L) (Jackson Immuno-Research, West Grove, PA; cat. no. 111 – 175 – 144).
11. Anti-CD 11c (and other antibodies to cell surface antigens of interest).
12. 12-well tissue culture tray (Corning Inc. Life Sciences, Acton, MA; cat. no. Costar 3512).
13. 18- and 23-gage needles.
14. Glass vials.
15. 1-mL syringes.
16. Scalpels (1 per lymph node).

Assessment of rDC Migration from the Lungs to the Lymph Nodes/ by i.n. CFSE Labeling

1. 25 *mM* CFSE stock (Molecular Probes, Eugene, OR, cat. no. C1157): 25 mg dissolved in 1783.6μL of dimethyl sulfoxide (DMSO; stock solution).
2. Iscove's DMEM IX.
3. Anti-CD11c APC mAb (Caltag, Carlsbad, CA; cat. no. MCD11c05).
4. FACS buffer: 2% FCS, 0.02% sodium azide in PBS.

Antigen Presentation by DCs

DC Activation of Naive CD8 T Cells In Vitro

1. CD8α+ Microbeads (Miltenyi, Auburn, CA, cat. no. 120-000-298).
2. Anti-CD8β mAb biotinylated (Biolegend, San Diego, CA; cat. no. 118004).
3. Strepavidin PerCP (BD, San Jose, CA; cat. no. 554064).
4. CFSE (25*M* in DMSO; Molecular Probes, Eugene, OR; cat. no. C1157).
5. Bovine calf serum.
6. Complete medium: Iscove's DMEM IX, 439mL; 10% PCS (final concentration), 50mL; L-glutamine (29.2mg/mL H_2O stock cone.), 5mL; Na pyruvate (100m*M* stock cone.), 5mL; Pen/Strep (10μL/mL, 10μg/mL stock cone.), 5mL; nonessential amino acids (100X stock cone.), 5mL; HEPES, 5mL (1*M* stock cone.); 2-ME (89.3 m*M* stock cone.), 500μL.

Influenza Infection of DCs

1. Infect 1×10^6 purified DC/mL at the desired MOI (5, 10, etc.) by adding virus diluted in Iscove's media without serum to cultured cells.
2. Incubate cells and virus on ice for 30min to allow virus to attach to the cells and then continue the incubation at 37°C for 30min to allow infection.
3. Centrifuge cells at 460 g (1600rpm) for 10min at 4°C. Remove any remaining virus in the supernatent and resuspend cells in complete media containing serum.
4. Incubate for 18 h at 37°C.

CDs T Cell Activation,

1. Purify influenza specific HA Clone-4 CD8+ T-cells from the spleens of naive Clone-4 mice using CD8∞ microbeads.
   (*a*) Prepare single-cell suspension (spleen screen, slides, etc.) in Iscove's DMEM.
   (*b*) Centrifuge at 460 g (1600rpm) for 10min at 4°C and resuspend cells in MACS buffer at 90μL/10^7 cells.
   (*c*) Add 5μL CD8α+ beads per 10^7 cells and incubate at 4°C for 10 min (refrigerate not on ice). Add 2mL of Iscove's DMEM.
   (*d*) Centrifuge at 460 g (1600rpm) for 10 min at 4°C and resuspend cells in MACS buffer at 10 × 10^7 cells/mL.
   (*e*) Prewet column with 3 mL of MACS buffer solution.
   (*f*) Place cells over column (max of 5 lungs per column) and collect negative fraction.
   (*g*) Wash with 9 mL of MACS buffer.
   (*h*) Remove column from magnet and elute positive fraction with 6mL of media (2 mL of gravity, 2 mL of gentle plunge, 2 mL of forceful).
2. Centrifuge cells at 460 g (1600rpm) for 10min at 4°C and resuspend at 10 × 10^7 T-cells/mL in Iscove's DMEM.
3. Label purified T cells with 1.5 μM CFSE. Incubate 10 min at 37°C. At the conclusion of the incubation, quench CFSE by adding an equal volume of calf serum.
4. Purify desired DC populations by MACS or FACS sorting and then infect as described previously.
5. Coincubate purified T cells and DCs at a 5:1 ratio (*i.e.* 100,000 T celsl:20,000 DCs) in 200μL of complete media at 5% CO_2 at 37°C for 48-72 h.
6. Harvest cells, stain for CD8 β with anti-CD8β biotinylated (clone:CD8b) and strepavidin PerCP and measure CFSE dilution by flow cytometry.
7. Method can be combined with staining for T cells activation markers such as CD25, CD69, etc.

DC Activation of Antigen-Specific T-Cell Proliferation in Vivo

1. OTT or OTII mice as T-cell source (Jackson Laboratory, Bar Harbor, ME cat. no. 003831 or 004194)
2. C57B1/6 mice as DC source (donor) and as recipient of cell transfers.
3. 5mM stock solution of CFSE (5-(and -6) carboxyfluorescein diacetate succinimidyl ester, Molecular Probes, Eugene, OR; cat. no. c1157) in DMSO (store at –20°C, stable for 6mo.).
4. 10% PCS in PBS.
5. Super McCoy's medium: 85 mL of McCoy's 5A medium modified Ix with L-glutamine (Invitrogen, Carlsbad, CA; cat. no. 16600–082); 2mL of FCS; 2mL of essential amino acids 50x (Invitrogen, cat. no. 111–30–051); 1 mL of nonessential amino acids 100x (Invitrogen, cat. no. 11140); 1 mL of sodium pyruvate 100x (Invitrogen, cat. no. 11360); 1 mL of MEM vitamins 100x; 2mL of penicillin/streptomycin/glutamine 50 × (Invitrogen, cat. no. 5140 – 122); 50μL 0.1 *M* 2-mercaptoethanol.

6. Liberase/DNAse working solution.
7. OVA 258-264 peptide (for OTI assays) or OVA 323-339 peptide (for OTII assays) stock solution, 2mg/mL in sterile distilled water (Bio-Synthesis, Lewiston, TX; cat. no. 51023 –10).
8. 0.5% BSA (Amresco, Solon, OH; cat. no. 0332 – 100G) in PBS.
9. Anti-CD11c microbeads (Miltenyi Biotec, Auburn, CA; cat. no. 130 – 052 – 001).
9. Anti-mouse CD4 and CD8 antibodies, conjugated with a fluorochrome other than FITC.
11. 60-mm Petri dish (Falcon; cat. no. 351007).
12. Glass microscope slides with frosted ends.
13. ImL of insulin or tuberculin syringes.
14. 1-mL syringe with 23-gage needle.
15. AutoMACS separator

Allostimulatory Assay of DC Function

1. BALB/c mouse as source of responder T cells.
2. C57B1/6 mouse as source of DCs in first stimulation.
3. BALB/c, C57B1/6 and C3H mice as sources of DCs in restimulation.
4. Super McCoy's medium.
5. 0.5% BSA (Amresco, Solon, OH; cat. no. 0332-100G) in PBS.
6. Liberase/DNase working solution.
7. Fico-Lite LM (Atlanta Biologicals, Lawrenceville, GA; cat. no. 140650).
8. Anti-CD 1 Ic microbeads (Miltenyi Biotec, Auburn, CA; cat. no. 130–052–001).
9. Anti-biotin microbeads (Miltenyi Biotec, cat. no. 130–090–485).
9. Biotinylated anti-B220, DX5, GD11c, GD11b, and Ter1 19 antibodies.
11. ^{3}H thymidine (GE Healthcare, Piscataway, NJ; cat. no. 13-TRA120).
12. 96 well plate (Corning Inc. Life Sciences, Acton, MA; Costar, cat. no. 3596).
13. T25 culture flasks (Costar, cat. no. 430639).
14. 60-mm Petri dish (Falcon cat. no. 351007).
15. Glass microscope slides with frosted ends.
16. 1-mL syringe with 23-g needle.
17. AutoMACS separator
18. Scintillation counter.
19. Cell harvester (Cambridge Technologies, Cambridge, MA) and associated cell harvesting supplies.

METHODS

Identification and Isolation of DCs and LCs

Purification of Mature Bulk Splenic DCs

1. Prepare collagenase solutions in HBSS with Ca^{2+} and Mg^{2+}. (Dilute to 100U/mL. Also make 5 mL of 400U/mL dilution).

2. Collect 10 spleens (on Petri dish). Inject 1-2 mL of 100U/mL collagenase solution into each spleen.
3. Place the injected spleens into Petri dish. Collect the cells that were released in a 50-mL tube containing 10mL of ice-cold HBSS and keep on ice.
4. Add 6-8 mL of collagenase solution (100 U/mL) into the dish. Tease the spleen into very small pieces. Transfer released cells to collection tube and rinse twice more with 10mL fresh collagenase (100U/mL). Pipet up and down vigorously.
5. Add the high concentration collagenase solution (400U/mL) and incubate at 37°C for 30min.
6. Pipet up and down vigorously and mash dissociated tissue through the strainer.
7. Pool all the cells, filter through cell strainer and centrifuge at 460 g (1600rpm).
8. Resuspend cell pellet in dense BSA (no foaming!) Transfer 5mL of BSA containing cells from spleens into a 15-mL tube.
9. Overlay with 2mL of Ca^{2+} Mg^{2+}-free HBSS. Spin for 15min at 2192 g (3300 rpm) without braking.
9. Collect low-density fraction (all of overlaid medium and upper 1 mL of BSA) with Pasteur pipet. Transfer into a 15-mL tube and fill with Ca^{2+} Mg^{2+}-free HBSS, mix well, and centrifuge at 460 g.
11. Wash cells again in culture media and centrifuge at 300 g.
12. Plate low-density cells into 60-mm tissue culture dish in complete culture medium for 90min (37°C).
13. Rinse off nonadherent cells using a 10-mL pipet with prewarmed media (HT-2).
14. This wash must be done thoroughly. Pay attention to rims of the dishes. Aspirate the remaining media.
15. Repeat wash.
16. Culture adherent cells overnight (37°C).
17. Next day, collect nonadherent cells (>80% DCs, depends on thoroughness of wash steps 12–13).

Purification of Mature CD8α and CD8α⁺ DCs from Splenocytes by AutoMACS

1. Follow purification protocol for purification of bulk DCs through differential adherence.
2. Harvest bulk DCs off of the differential adherence plates (wash quadrants 3 times/2 washes pre-warmed media). Spin at 460 g (1600 rpm) 10 min 4°C.
3. Remove supernatant, and resuspend cells in 30μL of anti-CD8α (Ly-2)-Microbeads. Bring to 3 mL with MACS buffer.
4. Incubate at 4°C, for 15min (refrigerate, not ice). Spin 460 g (1600rpm) 10min 4°C.
5. Resuspend cells in 3mL of MACS buffer, and run on autoMACS under "posselds" program.
6. Positive select the CD8α population with anti-CD11c (N418)-Microbeads by repeating steps 3-5, and keep the $GD11c^{+}$ fraction.

Manual MACS Purification of Mature CD8α and CD8a+ DCs from Splenocytes

1. Follow purification protocol for purification of bulk DCs through differential adherence.
2. Harvest bulk DCs off of the differential adherence plates (wash quadrants 3 times/2 washes with pre-warmed media). Spin 460 g (1600rpm) for 10min at 4°C.
3. Remove supernatant, and resuspend cells in 30µL of anti-CD8α (Ly-2)-Microbeads. Bring to 3 mL with MACS buffer.
4. Incubate at 4°C (refrigerator, no ice), for 15min. Spin at 460 g (1600rpm) for 10 min at 4°C.
5. Prewet LS+ column with 3 mL of MACS buffer solution during the last 3 min of spin.
6. Resuspend cells in 3 mL of MACS buffer, and pass cells through LS column, wash with 9 mL of MACS buffer.
7. Remove column from magnet and elute CD8α+ cells with 6mL of MACS buffer (2mL gravity, 2m L gentle plunge, 2mL forceful).
8. Repeat steps 3–6 on CD8α⁻ or cells.
9. Pool CD8α⁺ cells, and then positive select the CD8α population with anti-CD11c Microbeads. Repeat steps 3 through 6, and keep the CD11c⁺ fraction.

Isolation of LCs for Flow Cytometry from Mouse Ears

1. Obtain ears from the mouse. With forceps, split each ear into dorsal and ventral halves.
2. Lay the ears in a 60-mm Petri dish so that the epidermal sides are facing up.
3. Pour 0.5% trypsin in PBS over the ears to submerge the ears in the Petri dish and place them in a 37°C water bath on a platform for 60 min.
4. Rinse the ears in BSS. Under the dissecting microscope, with fine forceps carefully peel off the epidermis taking care to preserve it as a sheet.
5. Prepare a single-cell suspension of the epidermis by gently rubbing it between the frosted ends of two microscope slides. Rinse the slides with BSS.
6. Pass the primary suspension through a 70-µm filter.
7. Transfer the suspension to 15-mL conical centrifuge tube.
8. Fill the tube with BSS. Centrifuge at 400 g at 4°C for 7min.
9. Resuspend the pellet by adding 3 mL of staining buffer, underlayer with 3 mL of Ficolite.
9. Centrifuge at 1700 g at room temperature for 20min.
11. Harvest cells from interface, and place them in a new 15-mL centrifuge tube.
12. Fill the tube with BSS. Centrifuge at 400 g, at 4°C for 7 min.
13. Resuspend the pellet. These cells are ready for study by flow cytometry.

Purification of Pulmonary DC Subsets by FACS

All steps and media to be used in a sterile manner.

1. Remove lungs from naive or infected mice.
2. Split the lungs into groups of two lungs/tube in Iscove's media.

3. Digest lungs with 1 mg/mL of collagenase and 0.02mg/mL of DNase 1 in 7.5 mL for 15 min at 25°C in rotator at 18 g (300rpm).
4. Prepare single cell suspension (spleen screens, slides, etc.). Note: remove debris by passing through Nitex mesh.
5. Continue incubation for another 10min at 25°C in rotator.
6. Add 50µL of 0.5 M EDTA, mix by inverting tubes, and incubate at room temperature for 5 min.
7. Centrifuge at 460 g (1600rpm) for 10min at 4°C.
8. Remove a small amount of cells to count.
9. Enrich for bulk DC by MACS.
   (*a*) Add 50 µL of anti-CD11c+ microbeads per 10^8 cells and incubate at 4°C for 10 min. (Do not incubate on ice as this will cap Ab-beads on cells leading to contaminating cell populations.)
   (*b*) Centrifuge at 460 g (1600 rpm) for 10 min at 4°C.
   (*c*) Resuspend cells in MACS buffer at 10×10^7 cells/mL.
   (*d*) Prewet LS+ column with 3 mL MACS buffer solution.
   (*e*) Place cells over column (max of five lungs per column) and collect negative fraction.
   (*f*) Wash with 9 mL of MACS buffer.
   (*g*) Remove column from magnet and elute positive cells with 6 mL media (2 mL gravity, 2mL gentle plunge, 2mL forceful.
9. Resuspend in Sterile Iscove's + 10% PCS at 1×10^7 cell/mL.
11. Stain enriched DC with anti-CD11b FITC (1µL/1×10^6 cells), anti-CD11c APC 1µL/2×10^6 cells), anti-I-A/I-E PE (1µL/2×10^6 cells) mAbs.
12. Incubate cells for 30 min on ice.
13. Centrifuge at 460 g (1600 rpm) for 10 min at 4°C.
14. Resuspend in Sterile Iscove's + 10% BCS at 1×10^7 cell/mL.
15. Run all cells through a cell strainer into new FACS tubes before sorting.
16. Sort into collection tubes of 1 mL of DMEM + 10% FCS.
17. Sort on CD11c + cells, then gate the aDC (CD11c+, I-A/I-E+, CD11b-), iDC (CD11c+, I-A/I-E+, CD11b+), and avelolar Macș (CD11c+, I-A/I-E inter, CD11b moderate (autofluorescent)).

Migration of DCs and LCs

Preparation and Fixation of Epidermal Sheets

1. Obtain ears from the mouse. With forceps, split each ear into dorsal and ventral halves. Lay the ears in a Petri dish so that the epidermal sides are facing up.
2. Apply depilatory (*e.g.*, Nair) to the ears for 20 – 30 s. Wash with PBS.
3. Submerge the ears in 0.5 M ammonium thiocyanate (in the Petri dish) and place them in the 37°C water bath on a platform for 15-20 min.

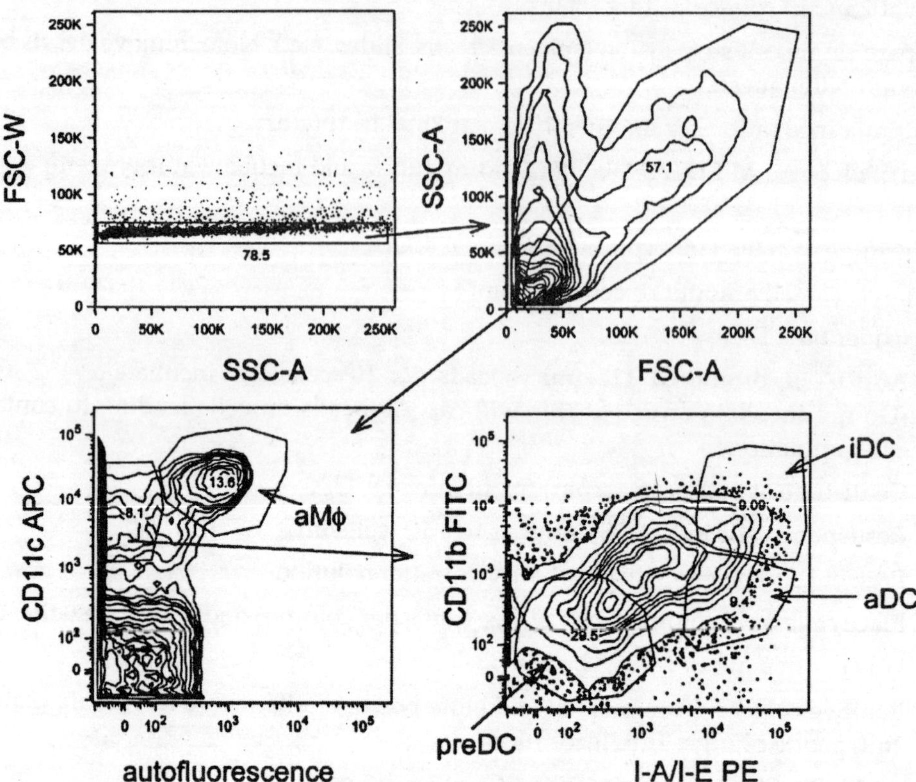

Fig. 9.1. Gating strategy for FACS purification of pulmonary dendritic cells. Cells are first gated upon 311 a singlet gate (upper left) and then on a large live cell gate (upper right). Next, autofluorescent⁺ CD11c cells are sorted (alveolar macrophages) away from the autofluorescent^{neg} CD11c cells (lower left). The autofluorescent^{neg} CD11c^{+} cell population contains interstitial DC (CD11b^{+}I-A/I-E^{+}), airway and alveolar DC (CD11 b^{-} 1 A/I-E^{+}), and precursor DC (CD11b^{-}I-A/I-E^{-}) cells (lower right).

4. Rinse the ears with PBS. Under the dissecting microscope, with fine forceps, carefully peel the epidermis off as a sheet.
5. Fix the epidermis in ice-cold acetone for 10-20min. Transfer to PBS for rehydration.

Staining the Epidermis for LCs

1. Incubate the tissue for 1 h at room temperature in 500 αL of blocking buffer in a well of a 24-well tissue culture plate or in 200mL of blocking buffer in a 96-well ELISA plate.
2. Aspirate the blocking buffer.
3. Incubate with 500 μL of biotinylated anti-MHC class II or MHC class II PE direct conjugate at appropriate titer, overnight at 4°C protected from light.
4. Wash three times in PBS by filling and emptying the well.
5. Incubate with 500 μL of Avidin-Cy5 or Avidin-Cy3 for 1 h at room temperature.
6. Wash three times in PBS.

7. Mount on a slide with assistance of fine forceps and the dissecting microscope. Do not use a bright light source during this step.
8. Cover completely with Vecta Shield (about 10-20μL) and a cover slip. Protect from light.
9. Let dry for 1 h and then seal the four sides of the cover slip to the glass slide with fingernail polish.
9. Store at 4°C until microscopic examination for up to 3 d. For longer storage, freeze slides at –20°C.
11. Examine the slides using a fluorescent microscope. PE fluoresces in the Cy3 channel.

FITC Tracking of LCs and Dermal DCs to Lymph Nodes

All cell washes in this procedure are performed at 4°C at 400 g for 7 min. 0.5% FITC solution preparation :

1. Weigh out 0.001 g of FITC into a glass vial.
2. Add 100 μL of acetone and 100 μL of n-butyl phthalate. Vortex intermittently for 5 min.

Animal Application

1. At desired time, (*e.g.*, 12 h before sacrifice) apply 25 μL of the FITC solution to one side of the mouse's abdomen. Apply 25 μL of vehicle alone to the opposite side as a control.
2. Isolate each inguinal LN. Place each LN individually in 200 – 400 μL of liberase/DNase working solution. Keep on ice. A 12-well tissue culture tray works well and facilitates identification of individual LNs.
3. Using a scalpel, individually mince each LN in the liberase/DNase working solution.
4. Incubate at 37°C for 30 min.
5. After digestion, prepare a single cell suspension from the LN by passaging the fragments repeatedly in and out of a syringe, initially with an 18-gauge needle followed by a 23-gauge needle.
6. Wash the cell suspension in BSS. Resuspend the cells in Perm/Wash buffer as desired (300–500 μL is recommended for each LN) and perform a cell count.

Intracellular Staining

1. Dilute only langerin and goat anti-rabbit secondary antibodies to the appropriate concentration for staining in Perm/Wash buffer.
2. Perform surface staining for at least CD11c (to identify DC) and other antigens of interest as usual. Remember to leave the Cy5 channel open for langerin.
3. At final wash, remove as much buffer as possible and resuspend the stained cells in 50 – 200 μL of Cytoperm/Cytofix solution (it is used undiluted from the manufacturer). Use 50μL for 0.6 × 10^6 cells, and use 200 μL for 3 × 10^6 cells. Vortex to mix and incubate at room temperature for 5 min.
4. Wash cells twice in Perm/Wash buffer. Add the diluted anti-langerin antibody (from step 1) and incubate at room temperature for 20 min.
5. Wash cells twice in Perm/Wash buffer. Add goat anti-rabbit IgG and incubate cells at room temperature for 20 min.
6. Wash cells twice in Perm/Wash buffer and transfer cells into mouse fix.

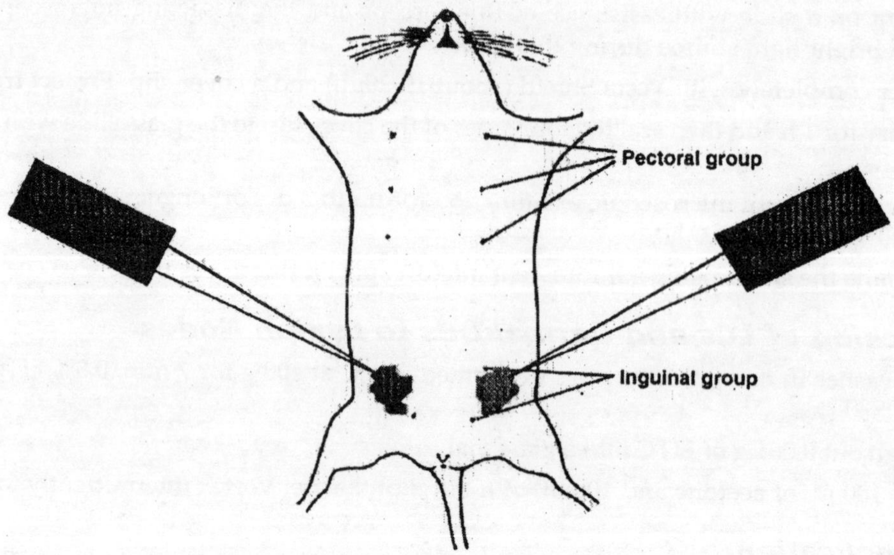

Fig. 9.2. Location of FITC/vehicle control application.

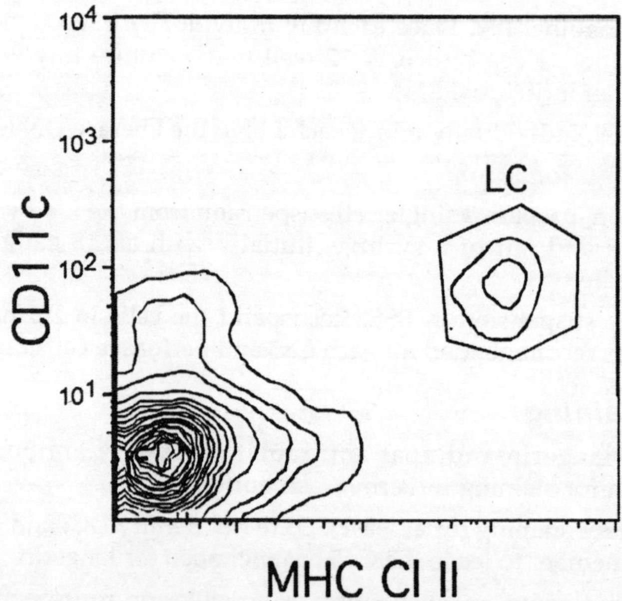

Fig. 9.3. Results of flow cytometric identification of LC. The gated population of CDl lc+ MHC Class ll + cells identifies LC as approximately 3% of the cell suspension.

Alternative Shortened Staining Procedure

This shortened procedure can only be used if the surface antigens are not sensitive to the fixation/permeabilization step.

1. Dilute all antibodies in Perm/Wash buffer.

2. After counting the cells, pellet them and remove the buffer. Depending upon how many cells are in the pellet, resuspend the cells in Cytoperm/Cytofix buffer (for 3 × 10^6 cells, use 200 μL). Incubate at room temperature for 5 min.
3. Wash the cells twice with Perm/Wash buffer. Aliquot cells into staining tubes (if not already done) and add antibodies to surface antigens and langerin. Vortex to mix. Incubate cells for 20 min at room temperature.
4. Wash the cells twice with Perm/Wash buffer and add fluorochrome-conjugated avidin (if biotinylated antibodies have been used) and the goat anti-rabbit IgG. Incubate at room temperature for 20 min.
5. Wash cells twice with Perm/Wash buffer and transfer cells into mouse fix.

Assessment of rDC Migration from the Lungs to the Lymph Nodes by i.n. CFSE Labeling

1. Prepare working CFSE solution from stock by diluting to an 8 *mM* solution with Iscove's DMEM Media (16 μL of CFSE + 34 μL media /mouse.
2. Anesthetize 1-4 mice.
3. Hold mouse vertical to 45°. Administer 50μL of CFSE i.n. alternating back and forth between nostrils.
4. Allow mice to recover.
5. Six hours later, i.n. administer DC maturation and/or migration stimulus (virus, inflammatory agent, etc.).
6. At various times after stimulus introduction (*i.e.,* 6, 12, 18, 24 h), remove the lung draining peribronchial and mediastinal lymph nodes. Prepare a single-cell suspension and stain cells for CD11c+ expression with 1μL/well of anti-CD 1Ic APC mAb (Caltag/Invitrogen, cat. no. MCD11c05) per 1 × 10^6cells/100 μL FACS buffer.
7. Analyze cells on flow cytometet. Monitor and calculate the number and percentage of recruited DC (*i.e.,* CFSE + CD11C + cells) in the lymph nodes.

Antigen Presentation by DCs

DC Activation of Antigen-Specific T-Cell Proliferation in Vivo

All steps must be performed with sterile technique. All wash steps are performed at 4°C for 7 min at 400 g.

Day 1

1. Prepare a single-cell suspension of splenocytes from OTI or OTII mice, wash in PBS, count, and resuspend at 5 × 10^7/mL in PBS.
2. Mix 5m*M* CFSE stock solution: cell suspension 1:10 (to achieve a final concentration of 5μ*M* CFSE), and incubate at 37°C for 10 min.
3. Wash three times in cold (4°C) 10% FCS in PBS.
4. Inject 5 × 10^7 CFSE-labeled splenocytes into each recipient C57B1/6 mouse intravenously using an insulin syringe.

Day 2

1. Remove spleen from donor C57B1/6 mice (EtOH or H_2O treated), being careful not to disrupt the capsule. Place into a 60mm Petri dish. Inject 1 mL of Liberase/DNase working solution into the spleen, and incubate 20min in a 37°C water bath, supported by a platform.
2. Prepare a single-cell suspension of splenocytes in the Petri dish by rubbing the spleen between the frosted ends of two microscope slides, and aspirate the suspension through a 1 mL syringe with a 23-g needle to break up remaining cell clusters.
3. Collect the cells in a 15-mL centrifuge tube. Wash with 0.5% BSA in PBS, and count the cells.
4. Label with GDI 1c+ microbeads as follows:
   (*a*) Resuspend the cells in 400 μL of 0.5% BSA in PBS/10^8 cells.
   (*b*) Add 100 μL of anti-CD lie microbeads per 10^8 cells, mix well, and incubate at 4°C (in the refrigerator, not on ice) for 15 min.
   (*c*) Wash with 0.5% BSA in PBS.
   (*d*) Resuspend cells in 500 μL of 0.5% BSA in PBS/10^8 cells.
5. Purify the DC on the autoMACS following the manufacturer's instructions (using "POSSELD" selection mode).
6. Wash the DCs once in 0.5% BSA in PBS and resuspend in Super McCoy's medium at 5×10^6/mL.
7. Pulse the purified DC with OVA 258-264 (for OTI) or OVA 323-339 (for OTII) stock solution to achieve a final concentration of 100n*M*.
8. Incubate at 37°C for 2 h, wash in PBS, and resuspend at 5×10^7/mL in PBS.
9. Transfer 5×10^6 peptide-pulsed DC intravenously into each recipient using insulin syringe.

Day 3

1. Prepare a single cell suspension of recipient splenocytes as in step 1 of day 1.
2. Stain with anti-CD8 (OTI) or anti-CD4 (OTII) using a fluorochrome conjugate other than FITC (this channel is needed to collect the CFSE fluorescence).
3. Run samples on flow cytometer.

Allostimulatory Assay of DC Function

All steps in this procedure must be performed under sterile conditions. All washes are performed in 0.5% BSA in PBS, 400 g at 4°C for 7min unless otherwise noted.

Preparation of T Cells

1. Prepare a single-cell suspension of splenocytes from BALB/c mouse. Wash and resuspend in 3 mL of 0.5% BSA in PBS:
2. Underlayer the cell suspension with 3 mL of Fico-Lite LM. Centrifuge the cells at 1700g for 20min at room temperature. Remove the cells from the interface. Wash and resuspend the cell pellet in 1 mL of 0.5% BSA in PBS.

3. Incubate splenocytes with appropriate concentrations of biotinylated anti-B220, DX5, CD11c, CD11b, and Ter1 19 antibodies for 20min on ice.
4. Wash three times. Resuspend cell pellet in 80 μL of 0.5% BSA in PBS per 10^7 cells. Add 20μL of anti-biotin microbeads per 10^7 cells. Mix well and incubate for 15 min at 4°C in the refrigerator (not on ice).
5. Deplete the antibody-coated cells on the autoMACS according to the manufacturer's instructions using the "DEPLETE" mode.
6. Wash the cells, resuspend in Super McCoy's medium, and count.

Preparation of DC for Initial Stimulation

1. Purify DCs from C57B1/6 mouse spleen by using anti-CD11c microbeads on the autoMACS according to steps 1-5.
2. Wash the cells. Resuspend the purified DCs in Super McCoy's medium, and count the cells. Put DCs and T cells into a T25 culture flask at a 1:10 ratio of DC to T cells (the density is generally around 1.1 × 10^6cells/mL, 10mL per flask). Incubate in 5% CO_2 at 37° C.

Restimulation of T Cells

1. On day 3 of culture, harvest the cells. Wash 3 times and resuspend in 400mL of 0.5% BSA in PBS per 10^8 total cells.
2. Deplete the DCs from the cell suspension. Use anti-CD11c microbeads to label the DCs as in step 1 of "Preparation of DCs for Initial Stimulation" in this protocol, but run the autoMACS in "DEPLETE" mode to remove rather than purify the DCs from the cell suspension.
3. Resuspend the re-purified T cells in Super McCoy's medium and place into a new T25 culture flask. Rest these cells for 2 d by incubating in 5% CO_2 at 37°C.
4. On day 3, purify DCs from the spleens of C57BL/6, C3H or BALB/c mice as in step 1 of "Preparation of DCs for Initial Stimulation" in this protocol.
5. Resuspend the cells in Super McCoy's medium, count, and place into culture in a 96-well plate at graded doses (*e.g.*, 1000, 5000, and 10,000 cells) with 1 x 10^5 T cells/well.
6. Incubate in 5% CO_2 at 37°C for 3 d.
7. Measure T-cell proliferation by adding 1μCi of ^{3}H-thymidine/well and incubating for an additional 18 h.
8. Harvest the cells using a cell harvester and measure the radioactivity incorporated into the divided cells with a scintillation counter using standard methods.

NOTES

1. Liberase and DNase stock solutions are stable at 4°C for 1 wk or at –20°C for at least 2 mo.
2. DC purification can also be done using the LS Column (Miltenyi Biotech, Auburn, CA; cat. no. 130–042–401) as described in Subheading 3.1.5
3. CD11c+, CD8α-fraction can be further enriched for CD11c + CD4 + DC by using the Miltenyi Biotec Multisort Release Kit to remove anti-CD11c Microbeads, stain with anti-CD4 Microbeads and positively select CD4+ cells.
4. Generally approx 1-3% of the cell suspension is LC, as identified by GD11c+ MHC Class 11 + cells by flow cytometry.

5. Negative fraction can be used for unstained and comps (be sure to lyse RBC using HBSS).
6. Anti-CD11c (clone N418) PE or biotin conjugates have also been used in our hands to label LCs in the ear. In general the biotinylated antibodies give brighter staining with less background than the direct conjugates. PE has a tendency to photobleach readily.
7. Avoid excessive handling of the epidermal sheets as they are quite friable.
8. Avidin-Cy3 is brighter than avidin-Cy5 in our hands. The latter is useful primarily if multicolor staining is performed.
9. Antigens that have been tested for sensitivity and found to be unaffected are epitopes recognized by the following antibodies: 4B12. (CCR7), C34–3448 (CCR5), G1-1 (CD86), 1C10 (CD40), and N418 (CDllc).
9. Performing the permeabilization in bulk will aid in the cell recovery if that is a concern.
11. LCs are identified in LN as FITC+langerin+, and dermal DC as FITC+CD1 1c+ langerin.

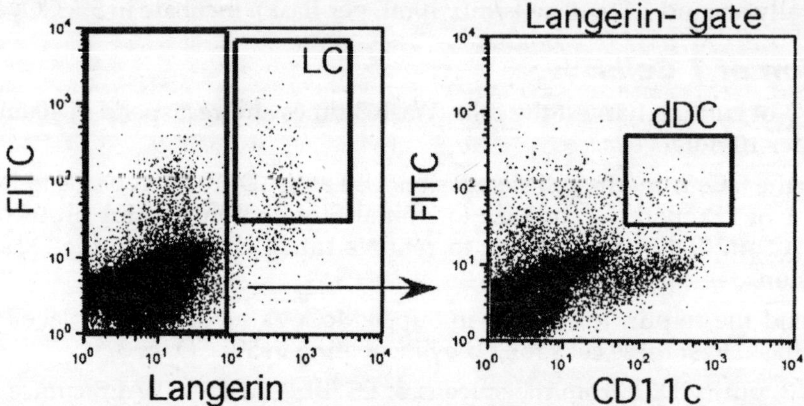

Fig. 9.4. Gating strategy for LC and dDC in draining LN following FITC skin painting. Recently immigrated LC are identified as langerin + FITC +, and dDC are langerin-CD11c + FITC +.

12. CFSE solution will be cloudy with some precipitate. However, it will still label cells.
13. Mouse must be anaesthetized enough to allow inhalation of CFSE without coughing or swallowing it.
14. Be careful not to get onto hands, skin, or other areas of mouse or those DCs will label and you will observe CFSE + DC in the appropriate non-lung draining lymph nodes.
15. There is substantial bleed-over of CFSE into the FL2 channel. Careful compensation is required.
16. To measure T cell cytokine production, you can choose to skip the CFSE labeling steps.
17. Supernatants can be used for ELISA.
18. Intravenous injections are easily accomplished by accessing the retro-orbital plexus immediately behind the medial aspect of the eye. Insulin (or tuberculin) syringes work best because there is no void volume left in the syringe (all the cells are transferred).
19. Restimulation with C57B1/6 DC measures the allogeneic response of the T cells, restimulation with BALB/c DC measures the syngeneic response, and restimulation with C3H DC measures the third party (primary) response.

CHAPTER

10 Growth and Development

One of the most fascinating of all biological phenomena is the development of an animal from a fertilized egg. From the early embryonic cells, which appear to be much alike, there arise during the course of a very few cell divisions differentiated organs and tissues such as liver, brain, kidney, muscle, skin, and red blood cells. The biochemical properties of differentiated cells are often highly specialized. Red blood cells make hemoglobin, while muscle cells make large amounts of myosin and actin. The endocrine cells of the pancreas make insulin or glucagon, while the exocrine cells form the digestive enzymes that are secreted into the intestinal tract.

Looking more broadly at the biological world we see many additional specialized features at every level of observation. Developmental patterns differ for every organism. Specialized organs abound. The structures of proteins, lipids, carbohydrates, and nucleic acids all vary, and every species has its own metabolic peculiarities. Even among bacteria we find extreme variation. Furthermore, many unicellular eukaryotic organisms undergo complex development within a single cell. The topics of this chapter are too complex for any detailed discussion. We will examine some aspects of growth and development for a few organisms and will ask whether there are simplifying generalizations. Comprehensive textbooks are available.

Enough has been learned about development to make it clear that the DNA contains genetically coded *developmental programs,* which are followed by cells. However, both transcription and translation are controlled by many chemical signals, which influence the execution of the genetic program. Such signals may arise from within a single cell, from the external environment, and from adjacent cells. The tight control is reflected by the fact that in most human tissues at any stage of development no more than about 10% of the total genes are transcribed at any one time. Chemical analysis makes it clear that most specialized cells contain a full complement of DNA but 90% of the genes are turned off.

A. BASIC CONCEPTS AND MOLECULAR ESSENTIALS

Listed in Table 10.1 are some essential aspects of growth and differentiation. Some of them are obvious. *Cohesion* between molecules provides the basis of specificity. *Receptors,* whether they be enzymes, hormone receptors, or receptors for chemotaxis, are essential. They are usually activated by a *conformational change* that accompanies binding. A cell must have receptors that can respond to a variety of *signals,* which may come from within the cell, from the external medium, or from neighboring cells. The receptor-signal pairs are essential to *local control,* which provides the basis for all of development.

To have any kind of spatial differentiation a cell must develop *directionality* (polarity). This permits *asymmetric cell division* and development of poles in a developing cell or embryo. *Adhesion* molecules hold cells together, allowing a cell to have a fixed position relative to other cells. Development of directionality and of developmental *patterns* are dependent upon *gradients* of concentration, of foods, heat, light, gravitation, etc., that can be detected by receptors. Gradients of compounds called *morphogens* help to provide a *positional identity* to cells. *Movement* of molecules of organelles and of intact cells is also essential. In multicellular organisms cells often migrate to new locations by following chemotactic signals.

Growth of individual cells enlarges them and often leads to *cell division.* The *cell cycle* describes this process with emphasis on replication of DNA. *Homeostasis* encompasses adaptation to altered nutrient or other environmental variables and to all processes which influence a cell to change. It provides for defensive reactions to many types of stress.

Table 10.1. Some Essentials for Growth and Differentiation

Cohesion of molecules, utilizing specific, hydrogen-bonding, and complementary surface shapes

Recognition, and conformational changes

- *Receptors* and *signals*
- *Local controls*

Polarity (directionality)

- Asymmetric cell division
- Poles

Gradients that can be sensed by receptors

- Food, physical qualities, morphogens

Adhesion between cells is required to hold an organism together and also, together with morphogen gradients, to provide a *positional identity*

Movement of molecules, organelles, and cells

Growth to enlarge cell size and numbers of cells

- *A cell cycle* for *replication of DNA* and *cell division*

Homeostasis to permit adaptation to changes in nutrient concentrations, stress

A developmental program, which is encoded in the genome

- Implementing this program usually requires many preformed compounds
- *Alternative developmental programs* often provide flexibility to an organism

Stem cells of *totipotent, pluripotent,* or multipotent nature supply new germ cells and other cells for multicellular organisms when needed

Programmed cell death (*apoptosis*) is part of many developmental programs and provides for removal of unneeded cells without inflammation

1. DNA and Developmental Programs

The genetic developmental program of an organism is encoded in the DNA. The expression of the program is implemented initially at the transcriptional level by a host of transcription factors that act at appropriate times and in appropriate places. Their action usually requires the presence of many preformed compounds. *Alternative developmental programs* may be used by unicellular organisms to adapt to new environmental conditions or by multicellular organisms to differentiate tissues.

Programmed Alterations and Rearrangements of DNA

Part of the developmental program may involve a temporary or permanent change in the DNA. One of the simplest of these changes makes use of the *transposable recombinational switch* in which a small piece of DNA is present in either of the two possible orientations. An example is the variation in "phase" of the flagella produced by *Salmonella.* A somewhat different example is provided by the unicellular yeast *Saccharomyces cerevisiae,* which changes the mating type of its haploid forms in a highly regulated pattern. The *a* mating type is expressed constitutively, but this is frequently switched to the *a*-type, which produces a different mating pheromone and responds to a pheromone from *a*-type cells. The change occurs through the transposition of different "cassettes" of DNA from "silent sites" into an expression site. The cassettes contain several genes, which are copied into the expression site called MAT. This site always contains genes of either *a* or α type. However, both *a* and α genes are present in other storage locations. When the mating type is switched, a copy is made of one of the stored cassettes and is placed into the MAT locus replacing the cassette already present. The MAT α genes encode two regulatory proteins, α1 and α2. Protein α1 is a positive regulator of the α-cell-specific genes, while protein α2 is a represser of the a-cell-specific genes. A similar mechanism appears to be employed by trypanosomes in changing their variable cell surface proteins.

Inactivation of Genes and Imprinting

Under some circumstances a chromosome or part of a chromosome is permanently inactivated but remains within the cell as compactly folded heterochromatin. Heterochromatin often consists of reiterated sequences of unknown function, but it may also contain groups of inactivated genes. The most impressive case is the total inactivation of one of the two X chromosomes in cells of female mammals. The entire chromosome appears as heterochromatin. The inactivation occurs early in embryonic development and is random with respect to the two X chromosomes. In some cells the maternal X chromosome, in others the paternal X chromosome, becomes inactive. However, upon further cell divisions the same chromosome in each clone remains inactive. As a consequence of the random inactivation, the female body is a mosaic with respect to genes in the X chromosomes. Formation of heterochromatin in X chromosomes. It depends upon *epigenetic markers,* which consist, at least in part, of methylated cytosines in 5'-CpG-3' sequences. These CG pairs are palindromic and can be methylated on both strands.

Selective inactivation of genes by methylation accounts for *genomic imprinting,* which occurs in mammals and which marks a gene of either maternal or paternal origin for silencing. About 50 imprinted genes have been identified in mice and in humans. Imprinting, known as *epigenetic inheritance,* also appears to arise in part by methylation of cytosines in CpG sequences. Four m^5C-methyltransferase genes have been identified in the human genome. One of these presumably provides the initial methylation of the cytosine in one strand of the DNA. A second methyltransferase (apparently encoded by gene *Dnmt*1) methylates the second strand and serves as the *maintenance methylase* that preserves the methylation pattern upon DNA replication. The faithful maintenance of imprinting has been demonstrated experimentally for genes of mice formed by nuclear transplantation, *i.e.,* mice grown from somatic cell nuclei transplanted into ova. Imprinted genes usually occur in clusters, which also contain imprinted genes for noncoding RNA molecules. The transcription of those genes often is correlated with repression *(silencing)* of protein-coding genes on the same chromosome. This observation is consistent with evidence that genes may be silenced by the binding of small RNA molecules to complementary sequences in the DNA or of mRNAs. The mechanism may be limited to *imprint control regions* or silencer units in the DNA, while the genes to be controlled are set off by *insulator elements*. The significance of genomic imprinting is not clear. However, it is very important to nuclear transplantation because the methylation state of imprinted genes is normally reset (*epigenetic reprogramming*) before embryonic development begins. Methylation of DNA is not essential to the life of filamentous fungi. Nevertheless, *Neurospora* does have a DNA methyltransferase, which methylates about 1.5% of the cytosines present.

Loss of DNA

While some genes are selectively inactivated, others may be irreversibly lost during development. The extreme case is that of the human red blood cell from which the entire nucleus is expelled. Loss of DNA may result from recombinational events. For example, crossing-over between sister chromatids during mitosis has been demonstrated for some cells and chromosomes. This does not alter the genetics of the progeny cells if equal amounts of genetic material are exchanged. However, if two or more similar base sequences occur in tandem in a DNA molecule, unequal crossing-over can occur with the loss of genetic material from a chromosome of one of the progeny cells. This may be a deliberately programmed route of differentiation for some cells.

Alternatively, loss of DNA from a chromosome may occur through a *looping-out excision* mechanism. Like the excision of a prophage from the chromosome of *E. coli* this loss of genes occurs at specific sites in the DNA. The best known case is the joining of gene segments during the differentiation of lymphocytes. The extent to which similar changes occur during terminal differentiation of other tissues or in nonmammalian species is uncertain.

Amplification of DNA of Chromosomes

During formation of oocytes parts of the DNA are "amplified" by repeated replication. This provides a way for the ovum to accumulate ribosomal RNA and various proteins in large amounts.

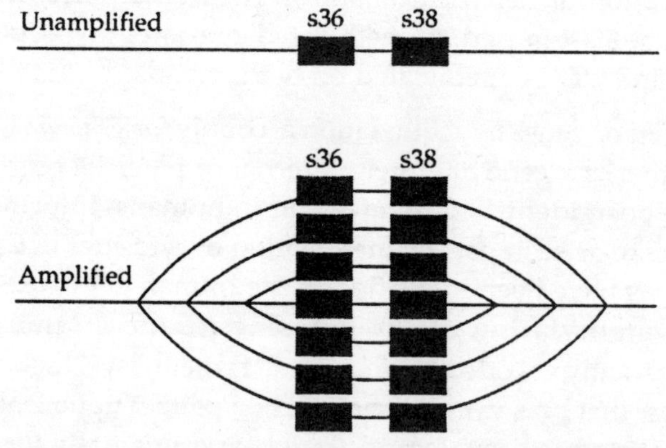

Similarly, genes for two abundant proteins of the egg shell or *chorion* of insects are amplified. Bidirectional replication initiated at discrete positions yields an "onion skin" structure containing many copies of an ~90-kb sequence containing the two genes. The polyploidy observed in some highly specialized cells such as the Purkinje cells of the cerebellum and of many cells of Diptera larva represent another way of amplifying genes. Polyploid cells of animals generally represent a terminal stage of differentiation and do not divide. They tend to contain their full complement of genes in each copy of their DNA, but most genes are not expressed.

2. Receptors and Signals

Receptors and their ligands are numerous, varied, and essential to all forms of life. Cell-surface receptors on bacteria detect feeding attractants as well as dangerous molecules. From bacteria to humans seven-helix receptors function to detect light, odors, hormones, and other molecules. The numbers of different receptors are impressive. For example, the tiny nematode C. *elegans* has 650 seven-helix transmembrane receptors and 411 protein kinases, many of which may be associated with receptors. Our bodies have thousands.

Every ligand that binds to a receptor is a signal of some kind. How many signaling molecules are there, and what are their structures? The number of proteins and small peptides affecting growth and development of cells may be enormous. In addition, the catecholamines, serotonin, histamine, and even bicarbonate ions may act as local hormones. For example, HCO_3^- determines whether a thin-walled sporangium or a thick-walled heat-resistant sporangium will appear in the phycomycete water mold *Blastocladiella.* We know that a very large number of different proteins and small peptides are secreted from cells, many in very small amounts. Some protein morphogens act at concentrations so low, *e.g.*, 50 pM, that it has been hard to detect them. T-cell receptors respond to extremely small amounts of short peptides that stream outward from many cells bound to surface MHC molecules. Do some of these same peptides act as signals between other cells during development?

Small RNA molecules are now being found to function in many ways within cells. They may also act as messengers within cells and between cells. The small 22 nt RNAs transcribed from genes *lin*-4 and *let*-7 of the nematode C. *elegans* control key developmental decisions. In green plants small RNAs move between cells and throughout the plants to trigger defensive responses. Much of the RNA transcribed from genes of any eukaryotic cell lacks any known function. Since evolution tends to act on all of the molecular constituents of a cell, this vast amount of RNA may have acquired vital roles in the control of metabolism and growth. A poorly understood intracellular structure, the *ribonucleoprotein vault* also remains a biochemical mystery. Differentiated cells may contain 10^4 and embryonic cells over 10^7 of these ~42 × 75 nm hollow objects. Vaults have internal cavities large enough to hold two intact ribosomes. Some vaults appear to be empty, but others contain materials, a fact that suggests some role in transport or storage.

Homeostasis

A large fraction of the receptors and signaling system of cells is devoted to the maintenance of a constant internal environment. This homeostasis is essential if a cell is to respond to changes in external conditions without disastrous consequences. Within mammalian cells the hypothalamus, pituitary, and adrenals have a primary responsibility for maintaining homeostasis.

Transcriptional Control in Differentiation

Development of an organism depends upon an orderly sequence of transcription of genes. Some genes are transcribed in germ cells, others within cells of an early embryo, and others later. As the embryo develops sequential waves of synthesis of the needed protein are observed. These are controlled by the actions of numerous transcription factors that act at a variety of *cis-regulatory modules* (CRMs) associated with promoter regions and also by controls on translation of mRNAs, and by negative feedback loops. The latter may involve hormones, morphogenic proteins, coregulatory proteins of various types, small RNAs, etc. More than 2000 transcription factors are encoded in the human genome. Most are positive-acting, *i.e.,* they *promote* transcription. Several families are specifically involved in regulation of development. These include the Zn^{2+}-containing GATA-1, which regulates globin synthesis, embryonic factors of *Drosophila (e.g.,* bicoid), vertebrate homeotic genes (Hox clusters), Pitl, the muscle-specific helix-loop-helix proteins MyoD and Myf5, and several forkhead (winged helix) proteins. A single CRM can bind many different regulatory molecules and single regulators can bind to a variety of CRMs. During development of the simple embryo of the sea urchin *Strongylocentratus purpuratus* a network of 40 highly regulated genes is needed to coordinate growth and development with the production of the proteins needed at each step.

Part of the control of differentiation lies in the interaction of proteins that regulate transcription with metabolites and hormones. For example, substrate depletion not only decreases growth rate of bacteria but also alters gene transcription. This occurs in *E. coli* as a result of a rise in internal cAMP concentration. The presence of an alternative energy source such as lactose induces changes in gene transcription. Such *physiological modulation* of a developmental pattern can also be seen in higher organisms.

More striking is the fact that environmental signals can trigger a cell to switch to an alternative developmental program by which enough new genes are activated to rebuild the cell into a new form. An example is spore formation, a process that occurs with some bacteria when external conditions become unfavorable for vegetative growth. Alternative developmental programs are also evident in eukaryotic organisms that undergo metamorphosis, and they may be important to development. Perhaps persistent states of repression of groups of genes can be passed through several generations of cells until a specific chemical signal triggers the unwrapping of the appropriate nucleosomes and transcription of formerly inactive genes.

3. Adhesion, Cell-Cell Recognition, and Cell Migration

Development of multicellular organisms depends upon both adhesion and on recognition of a correct interaction. Like enzyme-substrate, receptor-hormone, and antibody-antigen binding these interactions of macromolecules on cell surfaces often show a high degree of specificity. They may be accompanied by conformational changes and may trigger signaling cascades. We have already discussed some of these interactions, for example, the binding of a molecule of IgG attached to a surface antigen to protein C1q of the complement system and the binding of an MHC-antigen complex to a T-cell receptor.

There are many other cell-surface *adhesins,* among them are proteins that contain immunoglobulin-like domains and numerous *glycoproteins.* An example of the latter is the binding of a type of pili found in pathogenic strains of *E. coli* to epithelial cells of the urinary tract. The pilin subunits; like lectins, bind specifically to the disaccharide group Galpα1 $\rightarrow$ 4Gal. A lectinlike protein specific for *N*-acetylglucosamine rings is involved in invasion of erythrocytes by the malaria parasite *Plasmodium.* The unicellular alga *Chlamydomonas* produces sexual gametes of two mating types. When mixed together, gametes of opposite mating types, prior to fusion, adhere to each other via *agglutinins* present on their flagella. The agglutinins are glycoproteins rich in hydroxyproline, serine, glycine, arabinose, and galactose. Colored cells of different strains of the marine sponge *Microciona prolifera* find others of the same strain using highly specific proteoglycan-like aggregation factors. These compounds are highly polymorphic, and it has been suggested that they are part of a primitive immune system. The aggregation reaction requires calcium ions. In our own bodies Ca^{2+}-dependent lectins, the *selectins* (p. 187, 188), bind leukocytes and help to guide them to their sites of action.

Other adhesins include the *integrins,* cellular adhesion molecules *(CAMs), cadherins* and *fibronectin.* The CAMs, which are members of the immunoglobulin-like protein family, are glycoproteins bearing large 2, 8-linked sialic acid polymers. They promote Ca^{2+}-dependent aggregation. However, the effect of NCAM, which is widely distributed in a developmentally regulated fashion, can be antiadhesive if long chains of sialic acid are present. NCAM appears to play a role in remodeling and repair of tissues. Adhesion of molecules within cell membranes and the binding of substances to membrane surfaces provides another driving force in development. Within membranes molecules spontaneously sort themselves into lipid *microdomains,* often called *lipid rafts.* Related to lipid rafts are caveolae. These little craters arise in cholesterol-rich microdomains. They often contain the protein *caveolin* as well as glycosphingolipids and GPI-tailed proteins. ATP-dependent linking reactions may also occur to provide more permanent bonding. Membrane-associated molecules, in turn, become centers for attachment of cytoskeletal proteins and other protein complexes. As with the cytosol and extracellular fluids homeostatic mechanisms act to provide a relatively constant membrane-lipid environment.

Several types of cell junctions are associated with adhesion and participate in intercellular communication, The cadherins are transmembrane proteins with large extracellular domains. They are prominent components of adherens junctions in which they join the exterior surfaces of pairs of cells in a zipper-like manner. Another protein, *β-catenin,* links the short C-terminal tails of cadherin through a-catenin subunits to the actin cytoskeleton. In desmosomal junctions other specialized proteins including desmoglein have functions similar to that of cadherins. Tight junctions, from zebrafish to humans, depend upon a complex of several proteins including those of the claudin family. Significantly, the cohesive powers of some adhesins, *e.g.,* of cadherin, are altered during development. Cadherin E is nonadhesive in a four-cell mouse embryo but becomes adhesive after the eight-cell stage. It is obvious that many other changes in intercellular adhesion must also occur during growth and development.

The integrins comprise a large family of adhesive receptors that are found in animals from sponges to humans. They have both adhesive and signaling functions. Both subunits of their αβ heterodimeric structures have single transmembrane helices and short C-terminal cytoplasmic tails. The β1 subunit tails interact with cytoplasmic proteins. The distribution of integrins varies among cell types. Human leukocytes contain alpha subunits of types αd, αl, can, and αx with molecular masses of 150 – 180 kDa. Two ~95-kDa beta subunits (β1 and (β2) are present. However, T lymphocytes express β1, β2, and β7 integrins. Other patterns are observed for other leukocytes, in skin, and in other tissues. Integrin molecules tend to aggregate into clusters, which are found together with other proteins, at the ends of actin stress fibers. The largest of these clusters are known as focal adhesions. Signals may be sent through integrins in either of the two directions. The extracellular domains of integrins interact with a variety of proteins of the extracellular matrix. These include fibronectin, fibrinogen, vitronectin, collagen, and entactin. Other large cell surface adhesins include laminin and osteopontin, thrombospondin, von Willebrand factor, and related proteins. These adhesins appear to depend upon the sequence Arg-Gly-Asp (RGD), which binds noncovalently to integrins, which act as cell-surface receptors.

The functioning of the complex network of integrins, adhesins, and other components of the extracellular matrix is not understood in detail. One fundamental question is how the strength of adhesion can vary with time and stage of development. Roseman postulated an association of an oligosaccharide chain of a glycoprotein attached to one cell with a specific glycosyltransferase of another cell. The specific interaction would hold cells together, but addition of another glycosyl unit to the oligosaccharide by the transferase would alter the surface properties of the cell carrying the glycoprotein. This, in turn, could cause disaggregation of the cells. Glycosyltransferases can be found on the outer surfaces of cell membranes, and Roseman's proposal may correctly describe one aspect of cell adhesion.

Other molecules that are abundant on cell surfaces include heparan sulfate proteoglycans. Although they have often been regarded as providing a nonspecific "extracellular fly paper," recent evidence from studies of development in *Drosophila* suggest specific and important functions in signaling and in developmental patterning. Both hyaluronan and chitin also have been proposed to play an important role in vertebrate development. Proteoglycans of plant cell surfaces, as well as the hydroxyproline-rich proteins of cell walls, may function in plant development.

Movement of cells from one location to another is essential to embryonic development as well as to wound repair and to the immune response. Many brain structures are composed exclusively of immigrant cells. These *cell migrations* depend upon the cytoskeletal actin filaments, integrins, and focal adhesions. Chemotactic signals are also required. A great complexity of underlying chemistry is being elucidated.

4. Polarity, Asymmetric Cell Division, and Morphogens

Cells of *E. coli* usually divide exactly in the center to form two seemingly identical cells. However, under the right conditions some bacteria, *e.g., Caulobacter crescentus* and *Bacillus subtilis,* undergo asymmetric division to form two different types of cells. There is clearly an axial polarity. This polarity is evident even in *E. coli,* which has flagella streaming out at one end and its chemoreceptor-bearing "nose" at the other. Axial polarity is also obvious in other bacteria.

Polarity is evident in eukaryotic cells from protists to higher organisms. Cells of the yeast *S. cerevisiae,* whether haploid or diploid, divide in an asymmetric way by budding. Among body cells of higher animals those of the epithelium are among the most polarized. Polarity is always present in ova of eukaryotes, but the ova may initially be radially symmetric. The *anterior-posterior* axis, which is formed first, establishes a head-to-tail direction. In bacteria this major axis is determined as perpendicular to the division plane. In the tiny worm C. *elegans* the anterior-posterior axis of an ovum is determined by the position of entrance of the sperm. This marks the posterior end. In higher animals the axis, which is also known as the *animal-vegetal axis,* is established by uneven distribution of materials that include mRNAs and proteins in the unfertilized ovum. During embryonic development of bilaterial species two other axes, the *dorsal-*

ventral and *right-left* axes, are also developed and help to establish the body plan. Thoughout development polarized movements allow cells to intercalate be-tween one another to help shape the body.

Early studies of simple organisms such as Hydra and planaria showed that distinct chemical differences can be detected along the anterior-posterior axis. These organisms can be cut into pieces, many of which can regenerate complete individuals. Regions near the head regenerate most readily. These and other observations led to the concept of gradients of diffusible *morphogens,* or form-giving molecules. In *Drosophila* eggs an mRNA specified by the gene *bicoid* is localized at the anterior pole. The translation product, the bicoid protein, diffuses through the embryo, which in *Drosophila* lacks cell walls at this stage. Bicoid is a transcription factor and also one of a number of established morphogens. Many other morphogens are members of the *TGF-β* (transforming growth factor beta) family. Among them are proteins that establish the dorsal-ventral axis and also bone morphogenic proteins. Retinoids also appear to act as morphogens.

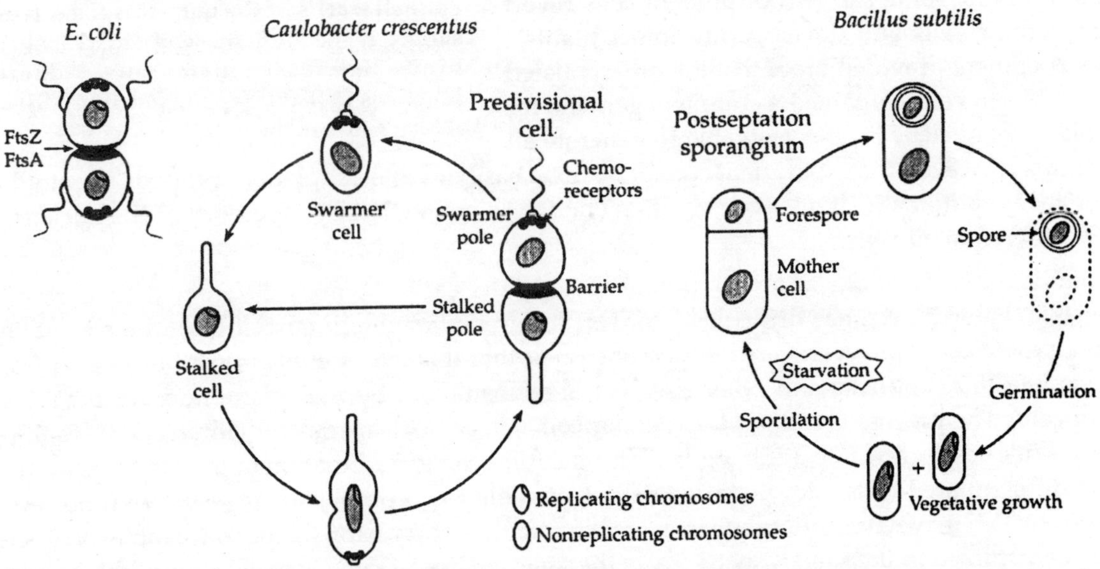

Fig. 10.1. Comparison of cell division in three species of bacteria. *Escherchia coli* divides symmetrically after forming a septum in a plane marked by a ring of FtsZ (tubulin-like) and other cell division proteins. *Caulobacter crescentus* divides asymmetrically to give one flagellated swarmer cell and one stalked cell. *Bacillus subtilis,* under starvation conditions, divide to form a mother cell and a forespore. The latter is engulfed by the mother cell, which promotes its conversion to a resistant spore.

5. Totipotency and Stem Cells

The cambium layer of plant stems differentiates continuously to form phloem on the outside of the cambium and xylem on the inside. At the same time, cambium cells are retained. Thus, at each cell division one daughter cell becomes a differentiated cell, while another remains the less differentiated cambium. This pattern of continuous differentiation from a line of *stem cells* with constant properties is found in animals as well as in plants. In the differentiation of cambium it appears that chemical signals obtained'from the surrounding cells on either the inside or the outside of the cambium layer determine whether the differentiated cell becomes phloem or xylem. Sucrose, auxin, and cytokinins are all involved.

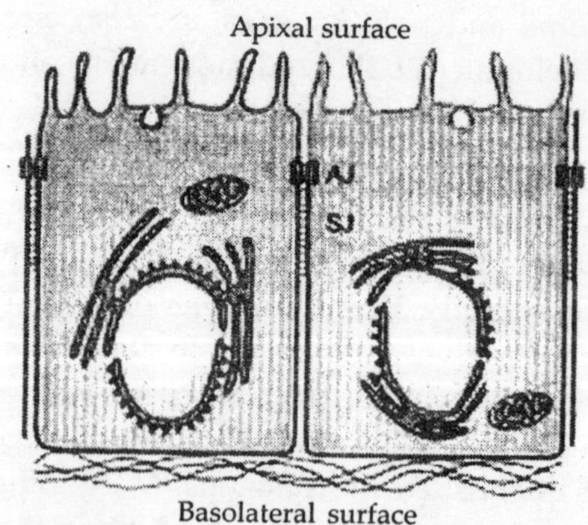

Fig. 10.2. A pair of epithelial cells of *Drosophila*. The apical surface (top), *e.g.*, of epithelial cells of the gut, faces the external surface, while the basolateral surface (bottom) binds to a basal membrane. Adheren junctions (AJ) and septate junctions (SJ) are shown between the cells.

Cloning

Asexual propagation (cloning) of plants ordinarily occurs by virture of the ability of embryonic meristematic tissue to differentiate into roots and shoots. If isolated phloem cells or other more differentiated cells are cultured, the result is often the formation of a *callus,* a dedifferentiated mass of cells somewhat reminiscent of embryonic cells. Under proper conditions, *e.g.*, in a coconut milk culture and in the presence of the correct auxin-to-cytokinin ratio, some carrot root phloem cells revert to embyronic cells and develop into intact plants." This experiment provided proof that the differentiated carrot phloem cells contained a complete genome for the plant. Its nucleus is *totipotent,* able to generate all cell types. However, the experiment cannot be done easily with most plants, and dedifferentiation is not always automatic. It does occurs often enough to establish the totipotency of the nucleus of many differentiated cells.

Similar considerations apply to animal cells. In the earliest stages differentiation is readily reversible. Later it becomes difficult to convert a differentiated animal cell into one resembling an embryonic cell. However, Gurdon demonstrated that this is sometimes possible. Nuclei from cells of intestinal epithelia and other tissues were substituted, by transplantation, for the nuclei of egg cells. The process is called *nuclear transplantation* or nuclear transfer cloning. The resulting eggs in some cases grew into adult toads. Thus, the full genetic information of the toad was present in the differentiated cells. However, it was not possible to accomplish this result with nuclei of neurons, which may have undergone irreversible differentiation. More recently mammalian nuclei have been utilized in the same way to create the famous sheep Dolly as well as mice, calves, pigs, and kittens. These animals are commonly said to have been *cloned,* a term that has long been used to denote asexual propagation, *e.g.*, in a colony of dividing cells or in propagation of plants by grafting.

In nuclear transplantation it is the DNA that is hoped to be the same in every individual in a clone. However, the ovum used for the transplantation contains mitochondria. Some mitochondria may also accompany the nucleus during the transfer. If the donors of the ovum and of the nucleus are different individuals the offspring will be mitochondrial hybrids. In addition, there are questions about the methylation state of DNA in the donated nucleus and about the age and health of the donated mitochondria. That these questions are significant is emphasized by a bit of 3000-year-old knowledge from mule breeders: a mare crossed with a donkey yields a mule but a stallion crossed with a donkey yields a hinny, which has shorter ears, a thicker mane and tail, and stronger legs than does a mule. There are worries because Dolly and many other animals produced by nuclear transplantation have not been completely healthy. Is something missing from the transplanted DNA or does it carry something extra, such as methyl groups? Recently it has been recognized that incorrect epigenetic marking of cytosine in CpG pairs that control maternally imprinted genes, especially those on chromosomes 11 or 15, may cause death of embryos or devastating human diseases. An important related question for those wishing to clone an animal by nuclear transplantation is "Should the cell that donates the nucleus be in the G_1 state of the cell cycle or the G_0 or paused state that precedes G_1?"

Stem Cells

For many years it has been appreciated that both erythrocytes and other blood cells arise throughout life from self-renewing stem cells in the bone marrow. Stem cells are also needed for renewal of bones, muscle, skin, neurons, etc. Stem cells appear to be present only in small numbers and in well-protected special *niches* in the body. They are able to live throughout an individual's lifetime, dividing quite rarely and always producing one or more highly differentiated cells as well as a new stem cell. A fertilized egg (zygote) is totipotent, able to generate all the cells of an animal including those of the placenta and other tissues that are not part of the embryo. However, the most capable stem cells are *pluripotent,* able to form more than one type of specialized cell. Mammalian pluripotent stem cells include tumor cells, *embryonic stem cells,* derived from preimplantation embryos, and *embryonic germ cells,* derived from the primordial germ cells of the postimplantation embryo. These germ cells are not only totipotent but, with good luck, may be immortal.

Recent results indicate that adultderived somatic cell nuclei may still retain full pluripotency. Some confusion has arisen because of the discovery that stem cells may sometimes fuse with differentiated cells. It is only recently that it has been possible to locate and to cultivate human stem cells. These cells, which may be recovered from both embryonic and mature tissues, include the blood-cell-forming *hematopoietic* stem cells, fetal *neuronal* stem cells, *melanocyte* stem cells, and *mesenchymal* stem cells (or marrow stromal cells). The last give rise to muscle, bone, cartilage, and tendons. Most stem cells may arise late in development and function principally in tissue renewal. Among the most abundant are those of epthelial tissues, whose cells provide 60% of differentiated tissue types in the mammalian body. Epidermal stem cells must provide for regular replacement of the outer skin surface but must also provide cells for rapid repair of wounds. The exact locations of epidermal skin cells have been difficult to find. The cells appear to be well-protected in areas deep in the skin. Some are located in hair follicles. Stem cells of plants are present in specialized structures called *meristems* A seedling typically has two meristems, at the tips of the shoot and root, respectively.

Cloning of human stem cells is of great medical interest because of the possibility of replacing defective cells or tissues. Tissue engineering may supply urgently needed differentiated cells for replacement purposes and may eventually lead to replacement organs. These efforts must be pursued with caution, but most researchers see a bright future for cloning of tissue cells. At the same time there is nearly universal agreement that nuclear transplantation cloning of human beings should not be attempted. One key objection is the near certainty that many seriously defective human embryos would be created.

How can we explain a pattern in which one daughter cell resulting from division of a stem cell undergoes differentiation while the other remains a stem cell? It depends upon methylation of CpG sequences in DNA. A methyltransferase, E_1, would modify a site in the DNA that controls differentiation by methylating a base in one of the two strands of a palindromic sequence such as a CpG pair. The maintenance methylase E_2 would further methylate the corresponding cytosine in the palindrome in the second strand. During tissue growth replication would produce a hemimethylated CpG in each daughter cell. These would also be methylated by E_2 to complete the replication process. In contrast, a stem cell would need to mark one DNA strand, *e.g.*, by action of a third methyltransferase (E_3) or some other type of DNA-modifying enzyme. E_3 would add a methyl group to DNA on some location outside of the palindromic region. Replication in this case would yield one hemimethylated palindome, which would allow one daughter cell to follow the normal growth and replication cycle. However, the other daughter cell would carry the mark designating it as a stem cell. The presence of both E_1 and E_2 in cells would lead to the continuous differentiation of the modified cells from unmodified ones, the situation found in stem line cells. A *"maintenance methylase"* with the properties of E_2 has been identified.

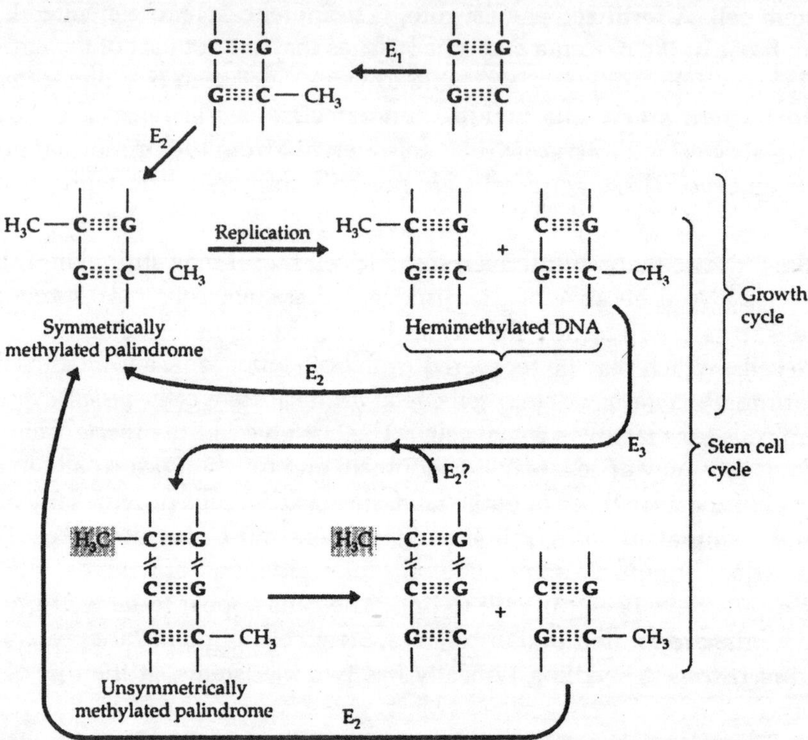

Fig. 10.3. Hypothetical way of controlling stem cell replication by methylation or other marking system. Methyltransferases E_1 and E_2 methylate the cytosine in a 5′-CpG-3′ or other palindromic sequence. In freely replicating cells these two enzymes keep the CG sequences methylated on both DNA strands. In stem cells another enzyme, perhaps a third methylase (E_3), marks a location outside the palindromic DNA on one strand (❑). Replication leaves the mark in the duplex, which is retained in a stem cell. The other strand will yield a hemimethylated duplex allowing the cell to follow the normal growth and replication pattern.

One question that has been asked is whether the 200 cell types of the vertebrate body all arise as a result of chemical interactions between cells and hormones and other external signals?

Alternatively, does a *developmental clock* count the number of cell divisions and at the appropriate time turn off one set of genes and turn on another? Methylases as well as other enzymes might modify DNA at specific times during development. For example, a hydrolase might deaminate the adenine in an AT pair to inosine. Upon replication and cell division one daughter cell would receive an unaltered DNA molecule, but the other would contain in place of the AT base pair an 1C pair. Following a second replication, a GC pair would be formed resulting in AT to GC mutation at a specific site in the DNA of some of the daughter cells. Such a simple change, occurring in response to an enzyme formed at a certain stage of development, could alter the expression of genes in a cell. Schemes involving palindromic sites and modification enzymes that could turn off specific genes after a given number of cell replications have been suggested.[136] In fact, there seems to be little support for such mechanisms.

Transcriptional controls and the sensing of a cell's location may provide adequate control.

If methylation and possibly other covalent modifications of DNA occur, how can one explain the totipotency observed for some nuclei of differentiated cells? During development of the ovum and the sperm there appears to be a "resetting" of the developmental clocks that led to differentiation. At this time all of the methyl groups on the CpG pairs of imprinted genes are removed. The mechanism is uncertain. Perhaps marking of newly replicated DNA stops. For example, in the case of the methylated DNA of Fig. 10.3 if E_1 and E_2 were absent in the cytoplasm of the ovum, no further methylation would occur during subsequent cell divisions. However, there may be active enzyme-catalyzed demethylation. Other mechanisms of gene silencing are also known.

6. Apoptosis or How the Tadpole Eats its Tail

Observers have long been fascinated by the rapid resorption of a tadpole's tail as it turns into a frog or toad. The process, designated by the Greek word apoptosis (whose "pop" is pronounced), or as *programmed cell death,* plays a major role in many aspects of development in nearly all organisms. For example, during human development about one-half of all the neurons generated die. Unneeded lymphocytes, some of which produce antibodies or T-cell receptors directed against a person's self, are also killed. Cells may die accidentally from injuries. In many cases the resulting death occurs by *necrosis* rather than apoptosis. Necrotic death is accompanied by swelling and bursting of the cell and a subsequent inflammatory response. In contrast, cells dying by apoptosis shrink, break into fragments, and are rapidly eaten by surrounding cells. There is no inflammation. Because of this it has been difficult to determine the extent to which apoptosis contributes to normal development. Apoptosis is also distinguished from *autophagy,* which is intracellular turnover under starvation conditions. Cells may need to scavenge unneeded proteins and organelles, recycling them within the cell.

Cells damaged by disease, *e.g.,* dopaminergic neurons in Parkinson disease, may die by apoptosis. A second form of self-destruction occurs when an axon is cut. Failure of the elaborate network of mechanism for repair of DNA and maintenance of the genome normally leads to apoptosis. In cancer essential steps in the apoptosis pathway are often inactivated.

Our view of apoptosis changed with the tracing of the origins and fates of all of the ~1000 cells of the nematode *C. elegans.* During development of the adult worm just 131 specific cells die by apoptosis. Studies of mutant worms revealed mutations in several cell-death *(ced)* genes. Three proteins, encoded by genes *ced-*3, *ced-*4, and *egl-*1, are essential for apoptosis. Somewhat surprisingly worms with defective Ced proteins are apparently healthy, even though they have 131 extra cells. On the other hand, in the fruit fly *Drosophila* mutations in similar death genes are sometimes fatal.

The nucleotide sequence of the *ced*-3 gene revealed that the Ced3 protein is closely related to the *interleukin-converting enzyme ICE*. ICE is a member of the *caspase family* of thiol proteases. At least 14 different caspases are found in the human body. Some

Table 10.2. Some Components of Apoptotic Systems.

Apafl	Mammalian homolog of Ced4; component of apoptosome
Apoptosome	Cytosolic complex: Apafl•caspase-9•cytochrome *c*
Bcl2	Mammalian homolog of Ced9 protein of *C.elegans;* inhibitor of apoptosis
Bcl-2 family	Group of regulators of apoptosis, both inhibitory and stimulatory (Bad, Bax, Bik, etc.)
CARD	
CD95 (AP-1, Fas)	One of most studied death receptors
Ced3	*C. elegans* thiol protease, related to mammalian caspase9
Ced4	Activator of Ced3, Related to mammalian Apafl
Ced9	Inhibitor of Ced3 and related caspases
DD	Death domain of a death receptor
DED	Death effector domain
DISC	Death-inducing signaling complex, formed in plasma membrane
FADD	Fas-associated death domain, an adapter protein
ICE	Interlecukin-converting enzyme, structurally related to Ced3
TNF	Tumor necrosis factor (a family of cytokines secreted by macrophages)
TNFR	Receptors for a TNF family member

of them function in apoptosis and others in maturation of pro-inflammatory cytokines. Most exist as proenzymes, which must be activated by proteolysis. The mammalian homology of Ced3 is caspase 9. The Ced4 protein of C. *elegans* is an activator for Ced3. Its mammalian counterpart is called *Apafl* (apoptotic protease-activating factor 1). Protein *Ced9* is an inhibitor of apoptosis, which probably protects the worm from erroneous deaths. Its mammalian equivalents are proteins of the *Bcl*-2 gene family.

It is well established that caspases participate in the final stages of apoptosis, but what initiates the process? There appear to be many ways in which apoptosis can be triggered. If every cell has a proper location in the body, which is determined by signals from adjacent cells, what will happen if the cell becomes detached? There is evidence that such detachment with the loss of survival signals causes apoptosis. Cell damage is also a major trigger. In other cases the cell is "instructed" to die. An example is the death of unneeded lymphocytes, one of many cellular processes induced by cytokines of the tumor necrosis factor (TNF) family. To allow for this process cells have surface receptors of the TNF receptor (TNFR) superfamily. Some TNFRs are *death receptors*, which are called by many names. One of the best known is CD95 (also called Fas or Apol). CD95 is involved in death of mature T lymphocytes at the end of an immune response and also in the killing of virus-infected cells and cancer cells by cytotoxic T cells or NK cells.

Members of the TNF family that activate CD95 (CD95 ligands or CD95Ls) are trimers. They bind to the cysteine-rich external domains of the transmembrane CD95 molecules inducing them to aggregate (Fig. 10.4). The cytosolic portion of each of these death receptors contain a *death domain* (DD). The bundle of aggregated receptors also bind to an adapter protein such as the Fas-associated death domain (FADD). It is one of many proteins involved in apoptosis whose structures are known. The FADD molecule contains a *death effector domain* (DED), which associates with a similar domain in the proenzyme procaspase 8. A rather large membrane-associated molecular complex, the *death-inducing signaling complex* (DISC; Fig. 10.4), is assembled in this way. Oligomerization of the procaspase domain causes activation via self-cleavage to give active caspase 8, which initiates the apoptotic response. Other "upstream" caspases (caspases 2, 9, and 10) also participate in initiation of apoptosis. In contrast, caspases 3, 6, and 7, the "executioner caspases," are thought to participate directly in demolition of the cell. A caspase-activated DNase also participates by degrading DNA.

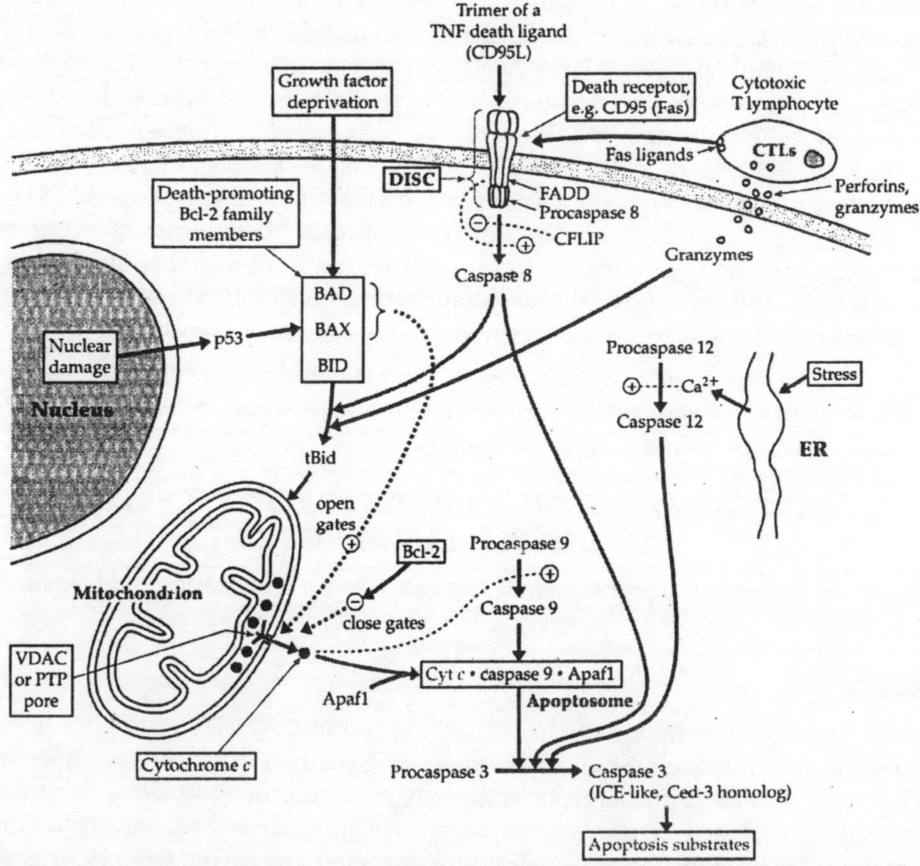

Fig. 10.4. Sketch illustrating only a few of the many aspects of apoptosis in a mammalian cell. Emphasis here is on the death receptor pathways and cytochrome c-activated apoptosis. A third pathway is initiated by stress in endoplasmic reticulum membranes. In addition, attack by cytolytic T cells sometimes causes apoptosis by action of a granzyme on protein Bid or via a death receptor. Objects in scheme are not drawn to a single scale.

A second major pathway of activation of apoptosis depends upon mitochondria. Various stresses such as lack of needed growth factors, exposure to ultraviolet light, or other apoptosis-inducing signals apparently open pores or gates in mitochondrial membranes allowing materials that would promote apoptosis to flow out into the cytoplasm and stimulate the effector caspases. This possibility was supported by the discovery that cytochrome *c* stimulates apoptosis. Cytochrome *c* is a small protein, which is present in inner mitochondrial membranes in a 1:1 ratio with other electron carriers. It is more mobile and less tightly bound than the other components. It carries electrons from complex III to cytochrome *c* oxidase within the intermembrane space. As a result of apoptotic stimuli cytochrome *c* rapidly flows out of the intermembrane space into the cytoplasm both interfering with respiration and triggering other changes in the cell. The outflow of cytochrome *c* may occur via the mitochondrial porin VDAC or under some circumstances via the mitochondrial permeability transition pore (PTP). Within the cytosol the escaped cytochrome *c,* together with caspase 9 and Apaf1, forms a large multimeric complex (cytc • caspase 9 • Apaf1) called an *apoptosome.* The apoptosome catalyzes activation of caspase 3, initiating the caspase cascade.

Control of the gates or pores by which cytochrome c escapes from mitochondria is poorly understood. Whereas in C. *elegans* a single protein Ced9 has been identified as an inhibitor of apoptosis, vertebrate animals have a large family of proteins that are related to the Ced9 homolog Bcl-2. Of these Bcl-2 and Bcl-x_L *inhibit* the flow of cytochrome c out of mitochondria, but several other members of the family, *e.g.,* Bad, Bid, Bik, and Bax, *promote* apoptosis. Bad carries a signal that indicates a lack of growth factor stimulation. Bid carries a death message from QD95R and other death receptors. Bax carries a signal from p53 indicating unacceptable DNA damage. However, a truncated form of Bax may prevent apoptosis of neurons. In every one of these pathways there are many complexities. In one of the best known pathways Bid is cleaved by caspase 8 to form a 15-kDa fragment t-Bid that becomes an integral membrane protein in the outer membrane of mitochondria. There it promotes the release of cytochrome *c.*

A quite different source of apoptotic signals are ER membranes, which respond to stress by releasing Ca^{2+}ions that activate caspase 12. Yet another type of apoptosis is sometimes induced by granzyme B, which is released from cytolytic T cells.

B. DIFFERENTIATION IN PROKARYOTIC CELLS AND IN SIMPLE EUKARYOTES

Every species undergoes developmental changes. Only a few of these will be considered here briefly.

1. Bacteria

Although they are usually regarded as unicellular, some bacteria develop more than one type of cell, and some even form colonies with filamentous growth or other distinct morphology. Many bacteria alter their development in response to changes in environment. For example, unfavorable conditions lead bacteria such as *Bacillus subtilis* to form compact endospores inside the vegetative cells. Many other bacteria including *E. coli* divide symmetrically. This fact also poses a question. How does a cell locate its center and divide? The answer is only partially understood. In all kinds of bacteria a protein known as *FtsZ* (filamentation temperature-sensitive protein Z), a GTP-binding protein homologous to eukaryotic tubulins, is essential. Prior to division FtsZ accumulates as a *septal ring* at the center of the *E. coli* cell. Contraction of the ring is thought to be an essential step in cell division. The FtsZ ring nucleates a growing complex of eight additional proteins known as FtsA, T, K, L, N, Q, W, and ZipA. While ZipA is not highly conserved among bacteria, in *E. coli* it is the first protein to add to the FtsZ ring. ZipA is somewhat related to eukaryotic actin. Another group of proteins is also needed for location of the midcell plane. These are known as MinC, MinD, and MinE. A MinC•MinD complex inhibits potential binding sites for FtsZ. MinD is an ATPase, which structurally resembles the Fe protein of nitrogenase (Fig. 10.2) and appears to propel the MinC•MinD complex in an oscillatory fashion from pole-to-pole. This behavior is not understood, but in some manner the 10-kDa MinE is able to overcome the inhibition and bind to FtsZ initiating division. Division in *E. coli* follows DNA replication by a constant time period (20 min at 37°C). The timing apparently depends upon diadenosine 5′-tetraphosphate (Ap_4A), which acts as a signal to couple division to replication.

Bacteria with Stalks

Caulobacter crescentis spreads to new areas while retaining a presence at home. As is illustrated in Fig. 10.1, asymmetric division produces two distinctly different cells. One, like the maternal cell, has a stalk, while the other has a rotary flagellum with which it travels. In time the motile "swarmer" cell sheds its flagellum and undergoes metamorphoses into a stalked cell. What controls this process? There is apparently a two-component system similar to the one that controls flagellar movement (Fig. 10.5). Two sensor histidine kinases phosphorylate a central response regulator CtrA. This represses initiation of replication in the swarmer cells as well as transcription of the cell division gene *ftsZ*. As mentioned on p. 1094 changes in DNA methylation may also occur. Flagellar biosynthesis and other steps of differentiation then occur in the swarmer cells but not in the stalked cells in which normal growth and replication take place. Action of proteases is also essential.

Sporulation

Bacteria of the genera *Bacillus* and *Clostridium*form metabolically inert spores when deprived of adequate nutrients (Fig. 10.1). Bacterial spores are remarkably resistant to heat and can survive boiling water for prolonged periods. Their metabolic rate is essentially zero, but they can be revived and can grow even after many years. For example, bacteria have been grown from a 118-year old can of meat. Some data suggest that spores can survive for no more than ~1000 years, but recent reports, not yet fully verified, say that spores have survived when embedded in salt crystals for 250 million years. At the onset of sporulation the synthesis of ribosomal RNA is turned off completely, and new classes of mRNA are made. More than 50 genetic loci are affected by mutations that cause spore formation. One or more specific forms of the 70-kDa δ subunit of RNA polymerase are produced and direct the initiation of the new mRNA molecules encoding new proteins. Prior to asymmetric cell division the first of the new δ factors, σF, is formed together with two regulatory proteins, Spo0A and Spo0B. Spo0B, a protein kinase, phosphorylates Spo0A, inactivating it. SpoAB also forms an inactive complex with δF. After asymmetric cell division σF remains inactive in the mother cell but is released in the prespore by action of Spo0E. This is a membrane-bound phosphatase, which dephosphorylates Spo0*A-P*, allowing it to form a complex with SpoOB with release of σF. Another protein, SpoIIIE, appears to direct one copy of the replicated DNA into the forespore. The δF factor then directs the transcription of genes in the forespore. In contrast, σE is produced only in the mother cell.

One of the most striking metabolic changes in metabolism during sporulation is the accumulation of large amounts of dipicolinic acid (Fig. 10.14). This requires the appearance of at least one new enzyme. In addition, as the spores develop the bacteria take up large amounts of Ca^{2+} and substantial concentrations of Mn^{2+} and other metal ions. In many bacteria 3-L-sulfolactic acid is also formed.

$$^{-}OOC-C(H)(OH)-CH_2SO_3^{-}$$

3-L-Sulfolactate

These components account for the following percentages of the total dry weight of spores of *B. subtilis:* dipicolinic acid, 10%; sulfolactic acid, 3-6%; Ca^{2+}, 3%; and Mn^{2+}, 0.3%. It is often suggested that the dipicolinic acid and other ions protect the proteins from denaturation. However, the heat resistance may arise from the maintenance of the core of the spore in a highly dehydrated state. When conditions become appropriate for growth again, the spore germinates, and the bacterium again follows the cell growth and division program.

More complex alternative developmental programs are followed by colonial forms of bacteria such as the myxobacteria. The life cycle involves aggregation of cells and formation of fruiting bodies as well as sporulation.

Signaling Among Bacteria

Even bacteria respond to signals from other bacteria. Individuals of a single species often react by secreting pheromones called *autoinducers* using a process called *quorum sensing.* Among the responses are swarming of cells, emission of light by luminous bacteria, synthesis of antibiotics, and formation of biofilms. As mention already that, autoinducers used by gram-negative bacteria are often N-acetylhomoserine lactones. A furanosyl borate diester may be a more nearly universal autoinducer. Programmed cell death can also be observed among bacterial populations.

2. Yeasts

The budding yeast *Saccharomyces cerevisiae* and the fission yeast *Schizosaccharomyces pombe* are the best known fungi. Although they usually grow as individual cells, they can grow with a filamentous form under some conditions. Other yeast, notably *Candida albicans,* are important pathogens and can also grow in either yeast or pseudohyphal filamentous forms. Like *E. coli, S. pombe* undergoes symmetric cell division. However, the strong $\beta(1 \rightarrow 3)$ linked glycans with their $\beta1 \rightarrow 6$ crosslinkages, mannose polysaccharides, and chitin provide a cell wall very different from those of bacteria. Cells of *S. pombe* grow mainly at their tips and begin early in mitosis to form a ring of actomyosin and other proteins at the center. This corresponds to the mammalian contractile ring. At the end of anaphase the ring contracts, and the septum that separates the two cells develops. In both types of yeast the septum is rich in chitin, which is secreted from the cell membrane or from vesicles known as chitosomes. After the septum is fully formed and thickened, a chitinase partially hydrolyzes the chitin releasing the cells. In *S. cerevisiae* the cell division is asymmetric. The position of the bud seems to be directed by the actin cytoskeleton.

All fungi form spores (gametes) during their haploid stage, which follows meiosis. The transcriptional program for *S. cerevisiae* involves at least four sets of genes, which are transcribed consecutively. During spore formation the mRNA levels of more than 1000 of the ~6200 protein-encoding genes are changed. About 50% are elevated and ~50% are depressed. Similar mating and sporulation pathways are observed for S. *pombe.*

3. The Cellular Slime Molds

The life cycle of *Dictyostelhun discoideum* is described briefly in Box 11-C. About 10^5 individual amebas aggregate to form a moving "slug" in response to the chemoattractant cAMP. Some other species of *Dictyostelium* are attracted to a folic acid derivative or to the ethyl ester of N-propionyl-γ-L-glutamyl-L-ornithine-δ-lactam.

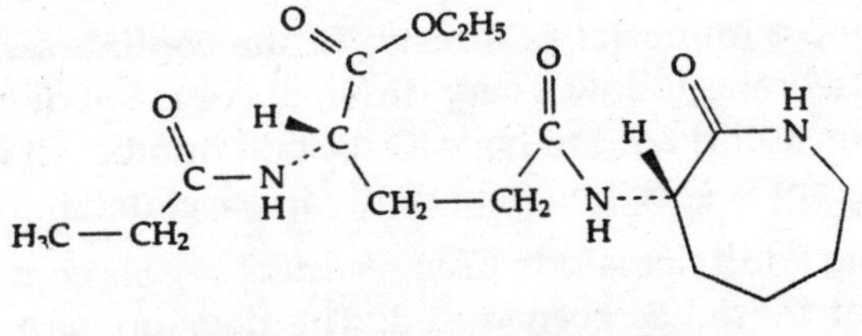

N-Propionyl-γ-L-glutamyl-L-ornithine-δ-lactam

In all cases the cells also utilize cAMP as an internal second messenger. For *D. discoideum* the components of the chemotactic-aggregation system include a 41-kDa cAMP receptor on the outside, adenylate cyclase, an extracellular diesterase that specifically hydrolyzes the cAMP to AMP, and a diesterase inhibitor protein. The inhibitor keeps the phosphodiesterase largely inactive initially, but when cAMP concentrations build up synthesis of the inhibitor is repressed and the cAMP is hydrolyzed, a necessary condition for retaining sensitivity of the receptors for the arriving pulses of cAMP.

The slug of aggregated amebas continues to move and to undergo differentiation into two cell types: about 80% of the cells become pre-spores and the remaining 20%, which are at the "head" of the slug, become pre-stalk cells. The front-to-back gradient of cAMP within the moving aggregate seems to be involved in differentiation. However, another "differentiation-inducing factor" as well as NH_3 may be involved in the formation of stalk. As the aggregate forms, the cells become cohesive, an 80-kDa surface glycoprotein being involved. Later other adhesive *discoidins,* 24- to 27-kDa RGD-containing galactose-binding lectins, also participate in holding the colony together. Some cells begin to produce cellulose. Trehalose is also formed and is stored in the spores. New enzymes have to be made to synthesize these materials. An alternative developmental pattern for some strains of *Dictyostelium* is formation of macrocysts between cells of two different mating types. A diffusible 12-kDa inducing factor appears to be re-leased by cells of one strain.

4. The Hydra

A well-fed hydra (Fig. 10.13) appears immortal. Its body cells are sloughed off and replaced at a steady rate so that within a month or so its body has been completely renewed. The hydra contains only ten cell types. These include two kinds of stem cells that give rise to the ectodermal and endodermal cells of the body wall as well as small *interstitial stem cells* (Fig. 10.13) that differentiate nerve cells, germ cells, and the nematocytes or stinging cells. Of the ~10^5 cells in a hydra about 3600 are interstitial stem cells. Each day they generate 400 nerve cells and 1800 nematocyte precursor cells as well as 3500 new interstitial cells. The nematocyte precursors move up the body of the hydra and take up residence in the tentacles. Their movement is thought to be guided by chemotaxis. The head activator peptide, which was identified following isolation from 3×10^6 hydras (3 kg), diffuses from the foot end of the animal forming a gradient. A foot activator may diffuse from the opposite end. The interstitial stem cells of hydra also give rise to clones that develop into the gametes. Female hydra always develop female gametes, but stem cells of male hydra give rise to both male and female gametes. This sex switching is reminiscent of the mating type variation of yeast.

5. Cell-constant Animals

While the hydra is almost immortal as a result of the continuous differentiation of its stem cell lines, other small invertebrates follow a very different course of development. Both the rotifers and the annelid worms (Fig. 10.14) tend to have a constant number of cells in the adult body. The entire developmental program is specified genetically in strict detail.

The one millimeter long adult nematode *Caenorhabditis elegans* contains only 959 somatic cells. The lineal descent of all of these has been traced. The development follows an almost exactly defined pathway with 113 programmed cell deaths during formation of the 558-cell newly hatched larva. In addition, each adult worm contains 302 neurons that make about 8000 synapses. This little nematode also has an alternative developmental pathway. The larvae shed their cuticles in four consecutive molts. If the food supply is inadequate, they enter a persistent nonfeeding state in which they may survive for months and are able to resume development when conditions are appropriate.

C. DEVELOPMENT OF ANIMAL EMBRYOS

The shapes and body plans of animals vary enormously. Consequently, the study of embryonic development of sea urchins, insects, frogs, chickens, mice, and humans might appear to lead to quite unrelated conclusions. However, there are many similarities as well as variety.

1. Germ Cells and Gametes

Throughout the animal kingdom from protozoa to human beings sexual reproduction predominates. It is true that there are about 1000 species that reproduce asexually. Among them are ~350 species of all-female rotifers and even a species of tiny mites, all of which are haploid females. Nevertheless, sex seems to have conferred some advantage on most species. There are two theories that may explain this: (1) Sex brings different combinations of genes together, allowing especially favorable combinations to survive, when changing conditions make life difficult. (2) Sex helps to remove deleterious mutations from a population. A large fraction of human fetuses (at least 10-25%) contain an "incorrect" number of chromosomes and as many as 20% of oocytes are defective. In contrast only 3-4% of sperm are chromosomally abnormal. Female meiosis I appears to be highly error-prone. Abnormal fertilized eggs or embryos are eliminated later in development.

Sex Determination

The sex of an individual is determined by the chromosomes. In humans and other mammals presence or absence of a Y chromosome determines the sex. However, in many organisms including *C. elegans* and *Drosophila* this is not true. Although *Drosophila* males like human males are XY, it is the ratio of the number of X chromosomes to the number of sets of autosome (A) that determines the sex. This is also true for *C. elegans,* which has no Y chromosomes. Apparently because of the differing ratios of X:A in the two sexes, organisms utilize a variety of *dosage compensation* methods. In cells of human females only one X chromosome is active. In *Drosophila* the rate of expression of genes from the X chromosome is roughly doubled in males. In C. *elegans* the expression from both X chromosomes of the (hermaphroditic) female is roughly halved. The biochemistry underlying these processes is quite complex.

The Mammalian Y Chromosome

The basic plan of the gonads prior to differentiation is female. However, if a Y chromosome is present (or if genes from a Y chromosome have been translocated to other locations) testes develop and begin to secrete androgen as early as the 60th day of gestation. A male-specific DNA sequence, *SRY* (sex determining region Y), constituting the gene for the *testes-determining factor,* is located in the small arm of the Y chromosome (Fig. 10.5). A small pseudoautosomal segment at the end of the short arm of the Y chromosome carries other genes and undergoes crossing-over during meiosis. The SRY gene lies between this and the centromere. The SRY protein is a member of the HMGA subgroup of *HMG* DNA-binding proteins. It binds tightly to the sequence AACAA (CT)(G/C) broadening the minor groove of the B-DNA and bending the DNA by more than 70°.

Both SRY and the related SOX proteins are critical developmental regulators. In early fetal life the mammalian embryo contains an indifferent gonad, able to differentiate into either a testis or ovary. Adjacent to the gonad are two simple ducts, the Mullerian (female) and Wolffian (male). In the male SRY acts in the developing gonads to induce differentiation into the Sertoli cells of the testis. In the mouse the *Sry* gene is active for only a brief period about ten days after fertilization. During that period cells of the genital ridge start to differentiate. In the absence of protein SRY they develop into the female follicle (granulosa) cells but in the presence of SRY into Sertoli cells. This is, in part, a result of production of the *Mullerian inhibitory substance* (MIS), which induces regression of the Mullerian duct, and later production of testosterone. MIS is a glycoprotein of the TGF-β family. Binding of SRY to a site in the Mis gene promoter appears to be involved in activation of the Mis gene. Recent evidence points to a role for both SRY and SOX proteins in pre-mRNA splicing. At least 25 other genes are also involved in spermatogenesis in the mouse. Many of these testis-specific genes have completely unmethylated CpG sequences. For example, a cAMP-responsive element present in a promoter sequence for a testis-specific subunit of pyruvate dehydrogenase must be demethylated for transcription to occur. The developing germ cells interact with the surrounding Sertoli cells at every stage both through direct cell-cell contacts and via secreted signals. Interstitial cells of the testis differentiate into Leydig cells, which secrete testosterone, promoting development of the Wolffian duct. A small population of germ-line stem cells provide for continuing spermatogenesis. In *Drosophila* their self-renewal depends upon signals from special hub cells. Other proteins needed for normal male development include the X-linked androgen receptor, whose absence causes testicular feminization, and dihydrotestosterone reductase.

The development of spermatozoa is unlike that of somatic cells. Extensive reorganization of chromatin occurs under the direction of cis-regulatory elements that are controlled by cells of the testis. Among specialized proteins that are synthesized is a testis-specific polyadenylate polymerase. During the remodeling histones are replaced by arginine-and cysteine-rich protamines. In mammals this occurs in two stages. Small intermediate proteins (TP1 and TP2) replace histones in the first stage and are displaced by protamines in the second. Sulfo-lipids, which are also present in myelin, are essential to spermatogenesis. Both the sphingolipid 3-sulfogalactosylceramide and *seminolipid,* a sulfate ester of monogalastosylalkylacylglycerol are present in large amounts. However, their functions are not clear.

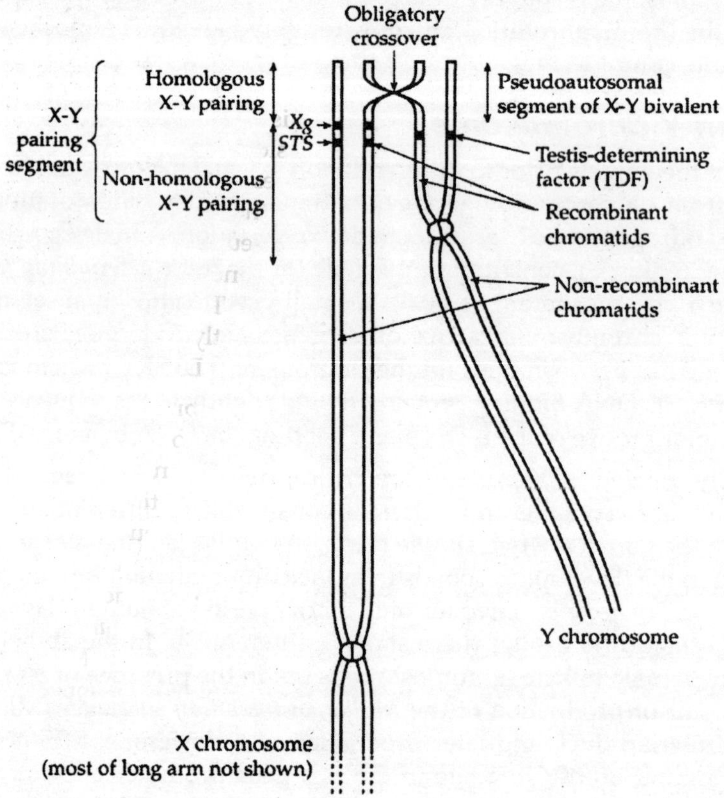

Fig. 10.5. Schematic diagram showing crossing-over between the human X and Y chromosome. The pseudoautosomal segment is that part of the X-Y bivalent where there can be X-Y exchange by crossing over. X-Y homology in this segment is maintained by, and may be necessary for, this crossing over. There is always one "obligatory" X-Y crossover, whose position varies. The length of the X-Y pairing segment varies with meiotic stage and can extend well beyond the pseudoautosomal segment into the Y long arm. Much (perhaps all) of the synaptonemal complex formed outside the pseudoautosomal segment represents non-homologous pairing.

Selenium plays a special role in development and protection of spermatozoa. The seleno-protein *phospholipid hydroperoxide glutathione peroxidase* (PHGPx) has a high activity in the testis and in spermatids. However, in manure spermatozoa it forms an enzymatically inactive oxidatively crosslinked capsular material around the midpiece of the cell perhaps providing mechanical stability. A similar 34-kDa selenoprotein is present in sperm nuclei and may be essential for condensation of DNA. Sperm tails contain specialized cytoskeletal proteins which form "outer dense fibers." In contrast to mammalian spermatozoa, nematode sperm move by ameboid motility that depends upon a specialized actin-like molecule. Sperm cells are unusually rich in polyamines, most of which are bound to RNA and DNA.

The X Chromosomes

The phenomenon of X chromosome inactivation in mammalian female cells is closely related to imprinting. The inactivation process is quite complex. It involves methylation of 5′-CpG-3′ sequences of DNA. It also depends upon an *inactivation center,* the *Xist* gene, which is expressed only from the inactivated X_i chromosome, whose *Xist* DNA is unmethylated. On the X_a chromosome this DNA is methylated, and the gene is silent. The *Xist* transcript is a long RNA that may bind to and coat much of the X_i DNA. The associated chromatin is enriched in a variant of histone H2A and is underacetylated on the tails of histones H2 A, H3, and H4 (Fig. 10.4). Also noteworthy is the fact that not all genes on X_i are inactivated. As many as 19% escape this control. Another gene *Tsix,* which is adjacent to *Xist* in the DNA, is also involved. *Tsit* encodes an RNA that is *antisense* to the *Xist* transcript but is transcribed from the active X chromosome X_a. One hypothesis is that the *Xist* transcript causes X inactivation and that the *Tsix* transcript acts in an opposite way to favor activation of the chromosome. A transcription factor known as *CTCF* has been identified as a possible regulator of the inactivation process. This is a trans-acting factor that is encoded on a chromosome other than X or Y. The process also depends upon methylation of histone H3. Methylation of H3 may also be a factor in gene silencing in other organisms. CTCF also regulates a number of other genes, *e.g.,* those of the globin gene cluster (Fig. 10.10). It binds to 60-bp sequences, perhaps in enhancer elements.

2. Development of the Ovum

In the early mammalian female embryo the absence of the Müllerian inhibitory substance MIS permits continuing development of the Müllerian duct, while the absence of testosterone permits the Wolffian duct to degenerate. However, positive developmental signals are also required. Among these is the protein *Wnt*-4, a member of a large family of locally acting signal molecules. Wnt-4 may be needed both for oocyte development and for further suppression of male development.

The earliest studies of oocyte development were done with sea urchins (often *S. purpuratus)* and with amphibians (often the South African clawed toad *Xenopus laeuis)* whose eggs are as much as 1000 times larger than those of mammals. However, despite the differences in size, modes of fertilization, and ovary development, oocytes of nematodes, sea urchins, frogs, insects, and mammals have much in common. Oocytes of C. *elegans* and of most other animals undergo a temporary arrest in development at the prophase stage of the first meiotic division. At this stage oocytes transcribe many genes. In some species chromosomes may develop a "lampbrush" appearance as a result of the transcriptional activity. Many mRNA molecules are stored in the expanding cytoplasm. Proteins are also synthesized and stored. Among these are specialized proteins of yolk granules and proteins used to construct an outer egg coat. Surrounding *follicle cells* also contribute nutrients to the oocyte. In insects, whose early embryonic development has some special characteristics, 16 surrounding *nurse cells* are connected to the oocyte by cytoplasmic bridges.

Oocytes may remain in arrest at the beginning of meiosis for prolonged periods before continuing through the *maturation* stage to form an ovum (egg). Women and other female mammals are born with thousands of oocytes, but only a few at a time develop into eggs. Maturation is often delayed until sexual maturity, when it is stimulated by hormones. In X. *laevis* progesterone stimulates maturation. In C. *elegans* and many other animals a signal from a sperm cell is needed to induce maturation. Maturation of the oocyte is often arrested again, this time at metaphase of the second meiotic division. Transcription is halted, and protein synthesis is slowed. Fertilization then induces rapid completion of meiosis. Penetration of the sperm leads to "activation" of the egg and completion of meiosis. In lower organisms activation can often be carried out by chemical or physical treatment in the absence of a sperm cell, with formation of parthenogenetic offspring.

3. Fertilization

Fertilization of the egg is a biochemically complex process. It involves recognition of sperm and egg, often in a species-specific manner. The jelly layer around sea urchin eggs contains peptides such as the *sperm-activating peptide* (speract), which stimulates increased respiration and motility of the sperm cells. Additional chemotactic peptides may also be released from the jelly layer of invertebrate eggs. Chemoattractants for vertebrate eggs are less well known, but a 21-kDa sperm attractant protein from X. *laevis* egg jelly has been characterized. Both in sea urchins and in mammals the jelly layer, which is called the *zona pellucida,* contains sperm cell receptors. These are glycoproteins that interact with proteins (spermadhesins) of the sperm cell membrane. One of these is the integrin-associated CD9, an integral membrane protein. Penetration of the sperm through the zona pellucida often involves a large specialized secretory vesicle, the *acrosome,* as well as the enzyme hyaluronidase. In some species the acrosome releases a large amount of monomeric G actin, which polymerizes suddenly into a tube of polymeric F actin, which in some way assists the penetration of sperm. In the horseshoe crab *Limuhis polyphemus* the acrosome in an unactivated sperm cell contains a twisted bundle of as many as 120 cross-linked actin filaments. When the sperm is activated by contact with the jelly coat of the egg, the acrosome straightens into a 50-μm-long crystalline bundle, which is driven into the egg coat. Of importance to all types of sperm cells are proteases and other materials that are also released from the acrosome and which help to etch a hole that allows the sperm to enter the ovum. The acrosome reaction also activates the egg, a process that may depend upon NO. In the sea urchin the acrosome contains a large amount of the protein *bindin,* which mediates a species-specific adhesion of the gametes and presumably fusion with the egg membrane.

Fusion of the sperm membrane with that of the ovum causes a rapid depolarization of the membrane of the ovum and an influx of calcium ions. This causes an immediate block to the entrance of any other sperm cells in most species. It also causes the fusion of the membranes of *cortical granules* (several thousand in a mouse ovum) with the cell membrane and release of their contents. The material released to the outside of the ovum includes various proteins and enzymes such as a peroxidase that catalyzes crosslinking of tyrosine side chains and hardens the material immediately around the ovum into a tough *fertilization membrane.* Within the ovum a respiratory burst resembling that of activated neutrophils provides $H_2 O_2$ to the peroxidase.

Fertilization also induces completion of meiosis and formation of a one-cell embryo containing a maternal pronucleus contributed by the egg and a paternal pronucleus derived from the sperm. Each pronucleus undergoes DNA replication and then enters the first mitosis, which yields a two-cell embryo containing one diploid zygotic nucleus per cell. Under the influence of the cyclin-dependent kinase cdc2 and a hyperphosphorylated form of the protein *nucleoplasmin,* the chroma tin of the compact sperm nucleus undergoes decondensation. The sperm basic proteins that coated the DNA are replaced by histones H_2 A and H_2 B.

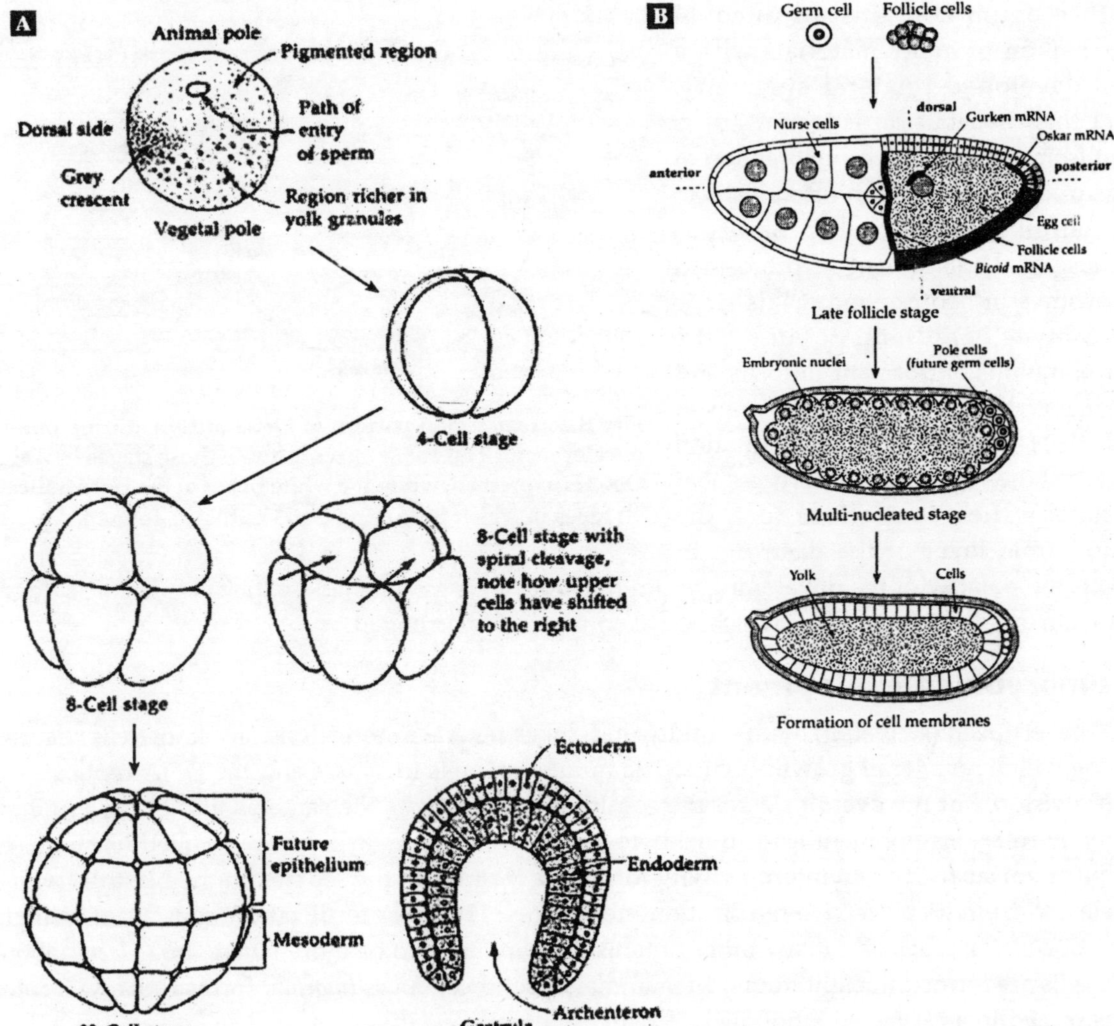

Fig. 10.6. Cleavage and gastrula formation. (A) Fertilization and cleavage of an amphibian egg is illustrated at the left. Spiral cleavage, usual among molluscs and other annelids, is also shown. The gastrula might be from an echinoderm or some lower invertebrate. (B) An egg of the fruitfly *Drosophila melanogaster*, showing early development. Anterior-posterior and dorsal-ventral axes are labeled.

Although an amphibian egg is nearly spherical, there is already a strong polarity. The nucleus lies nearer to the *animal pole* than to the *vegetal pole,* which in many eggs is rich in yolk granules. In eggs of amphibians the animal pole is highly pigmented, but the vegetal pole is less so. On one side above the equator, there is a gray crescent. In some animals this marking appears on the opposite side of the egg from the point of sperm cell entry. The gray crescent marks the future back (dorsal) side of the organism and the opposite part of the cell, the future ventral side. The point of sperm entry also marks the ventral side for the mouse, a fact that suggests that the plan of development of mammalian embryos may be basically the same as that of frogs. However, it is the internal components of the cell that actually determine the cell's axes. The cytoplasm of the mature ovum contains an unequal distribution of many materials with a well-developed bilateral symmetry. That this distribution is important is seen from the fact that centrifugation of eggs prior to fertilization often leads to formation of abnormal embryos because of displacement of preformed ribosomes and other materials. It is probably gradients in the concentrations of dormant mRNAs and other metabolites that lead to uneven growth of cells and to the indentation of cells at the vegetal pole, a process that initiates the formation of the endodermal layer of the gastrula. In insects the polarity of the developing ovum (the oocyte) is established by the cytoplasmic bridges from surrounding nurse cells, which are asymmetrically arranged.

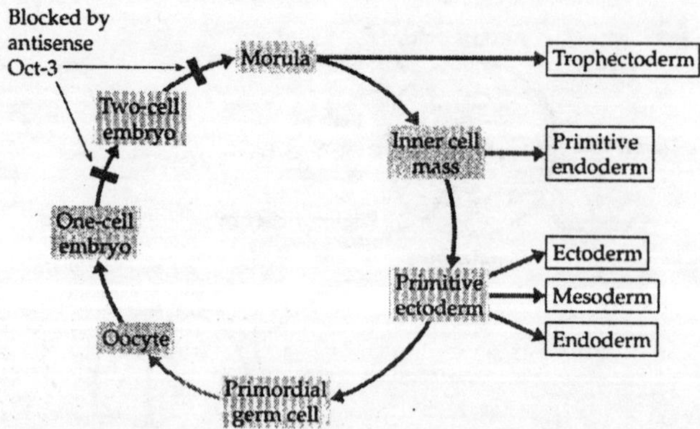

Fig. 10.7. Expression pattern of Oct-3 mRNA during mouse development. The green boxes indicate those stages in which Oct-3 is expressed, while the white boxes at the right indicate all types in which little or no Oct-3 mRNA is formed.

4. Embryonic Development

The fertilized (activated) ovum rapidly undergoes several mitotic divisions, known as *cleavage,* during which no overall growth occurs. The number of cells increases and the DNA replicates at each division, but the overall size of the resulting cell cluster is the same as that of the original ovum. Further development leads quickly to a stage in which a layer of cells (called blastomeres at this stage) surrounds an internal cavity forming a *blastula.* In the sea urchin the blastula, which is released from its protective fertilization membrane ~11 h after fertilization, consists of a single layer of cells. In frogs and many other organisms there are two or more layers. In X. *laevis* about 4000 cells are formed in eight hours. In mammals a solid cell mass *(morula)* forms first and is later transformed to a *blastocyst,* a hollow ball with an internal cavity.

Early mammalian development has been hard to study because of two facts: the ovum is very small, and a first priority is development of the placenta and of the layers of tissues that surround the embryo. This occurs in humans within the first week after fertilization. Both the trophectoderm and cells of the inner cell mass contribute to the extra embryonic tissues.

Development of a mouse beyond the one-cell stage is dependent upon a regulatory gene of the *Oct* family. The *Oct*-3/4 protein, which binds specifically to the DNA motif 5′-ATTTGGAT, consists of two domains, both of which are essential for tight binding to this sequence. One domain is a 75-residue *POU domain* that consists of a helix-turn-helix motif with an amino acid sequence that is highly conserved among mammalian *P*it and *O*ct regulatory proteins as well as some *Unc* proteins of C. *elegans.* The second domain is a 60-residue homeodomain. Oct-3/4 appears to be essential not only for cleavage of a one-cell egg but also for progression from a two-cell to a four-cell egg and also in other embryonic cells. Up to the two-cell stage very little transcription of zygotic genes is observed but further development requires zygotic genes and by the 8-cell stage protein synthesis from maternal mRNAs ends. Oct-3/4 continues to be synthesized in the embryo and is necessary for establishment of pluripotent stem cells in the embryo.

Development in *Drosophila* and other insects follows a somewhat different pathway. The egg, which is surrounded by follicle cells and 16 nurse cells, does not divide. However, its nucleus divides repeatedly, about once every nine minutes, to form ~6000 nuclei. Only then do separating membranes form to give individual cells. During the first two hours the nuclei form a *syncytium,* in which they are embedded in a common cytoplasm, that allows free diffusion of signaling compounds. At first the nuclei are in the center, but later most migrate to the periphery and form a single layer of cells comparable to the blastoderm of amphibian cells. A few nuclei remain in the central cavity to become yolk cells, and some at the posterior pole become separated into pole cells.

The next stage in embryo formation, which occurs universally, is the imagination of the blastula at the vegetal pole to form a *gastrula.* At this stage the embryo has distinct ectoderm and endoderm cell layers. The cavity, formed in the gastrulation process and connecting to the outside, is referred to as the *archenteron* and is the forerunner of the gastrointestinal tract or enteron. Gastrula formation is more complex in the frog embryo and still more so in the human embryo. In all but the most primitive of animals a third layer of *mesosomal* cells is formed between the endoderm and ectoderm. These three *germ layers* differentiate further as follows. The *ectoderm* yields the skin and nervous sytem; the *mesoderm* the skeleton, muscles, connective tissues, and circulatory system; and the *endoderm* the digestive tract, lungs, and other internal organs and germ cells.

Organ development occurs largely by infoldings of cells from the endoderm and ectoderm. These infoldings appear to be induced by chemical substances secreted by cells of an adjacent germ layer. Thus, ectodermal cells form the *neural plate,* the prospective brain, and spinal cord in response to induction by mesodermal cells lying beneath the neural plate area. The mammary glands also arise from interactions of mesodermal and ectodermal cells, while the formation of the pancreas, liver, and lungs depends upon interactions of groups of cells from endoderm and mesoderm. Because of their transparency zebrafish are especially useful for study of organ formation.

During the early stages of embryo growth, development seems to be directed largely by the polarity and gradients of the large amounts of mRNA, yolk and other constituents, which form a prepattern in the ovum. However, even at very early stages signaling from the nucleus of the ovum to surrounding cells is a necessary part of establishing the cell axes and developmental pattern.

The Anterior-posterior Axis

Only recently has it become possible to identify some of the specific mRNAs, the signaling molecules, and receptors that are involved in establishing the principal axes of the ovum. Even for this aspect of development the genetic program is very complex, there are many uncertainties, and the details go far beyond the scope of this book. Much of the most important work has been done with *Drosophila* for which numerous mutants have been identified and characterized. Many names of genes and of proteins are derived from a description of the phenotype of a mutant fly. These same names are often used for the corresponding genes or proteins in other organisms. However, a protein may be known by more than one name, depending upon the species.

A *Drosophila* mother deposits mRNA for ~80% of all of her genes in the egg, but not all of the encoded information is used. Because genetic experiments can be done so readily with *Drosophila,* it is possible to ask what *maternal effects* do come from this mRNA and what effects come from the genes of the zygote. For example, the *bicoid* mRNA that accumulates at the anterior end of the *Drosophila* egg is produced by nurse cells and is transported into the ovum. If the mother has defective *bicoid* genes the eggs die. A normal bi-coid gene present in the father does not prevent death. Another maternal effect protein is encoded by *nanos,* whose mRNA accumulates at the posterior pole of the ovum. The maternal *torso* gene, which acts on follicle cells, is also needed for the anterior-posterior axis formation. About 30 of these maternal effect genes are active in organizing the pattern of the embryo.

The *Drosophila bicoid* mRNA forms an anterior-posterior (A-P) concentration gradient, which controls early development along the axis. Another protein called *Staufen,* which forms a complex with the *bicoid* mRNA, assists in moving the RNA along on microtu-bules to form the gradient. The microtubules, together with associated transport proteins, are also essential to the prepatterning of the ovum. Bicoid protein is a transcription factor that is synthesized at the sites of its mRNA accumulation. It is absent from the eggs when they are laid but soon appears. Bicoid binds to the CRMs of target genes and activates them. Cooperative binding of multiple copies of the transcription factor may be necessary to provide the observed sensitivity to concentration. A transcription factor may bind at many sites on the DNA of the zygote or of nurse cells. For example, a gene called *even-skipped (eve)* is expressed in seven stripes in the blastoderm embryo. The enhancer that controls *eve's* second stripe contains at least five binding sites for Bicoid as well as others for the Hunchback, Giant, and Kruppel proteins.

A different gradient along the A-P axis is formed by the *Nanos* protein, whose mRNA localizes in the cytoplasm of the posterior part of the ovum. Like the *bicoid* mRNA, *nanos* mRNA forms a complex with the Staufen protein. Messenger RNA transcribed from a gene called *oskar* is also necessary for development of the posterior region of the ovum. An additional gene, called *gurken,* is also involved in establishing the A-P axis. The nucleus of the ovum secretes *gurken* mRNA, which is translated to the protein *Gurken,* a TGF-α-like protein that carries signals to follicle cells. They, in turn, help to organize the microtubules in the ovum. The nucleus of the ovum moves, sending the Gurken signal first to the posterior pole cells, then to a position that marks the dorsal side of the ovum. Thus, it participates in establishing both the A-P and dorsalventral axes.

The animal-vegetal axis of sea urchin eggs is established during oogenesis, but the mechanisms are still unclear. After fertilization a distinct pattern along the axis is established by the 60-cell stage. Signaling from the vegetal pole appears to play an important role. In the presence of lithium chloride, which is known to affect inositol triphosphate ($InsP_3$) metabolism, embryos develop an excessive fraction of endoderm and mesoderm (tissues of the vegetal half) at the expense of ectoderm (tissues of the animal half). Some data suggest that this effect implicates the *Wilt signaling pathway* and the protein *β-catenin.* This is the same protein that links C-terminal tails of cadherin to the actin cytoskeleton. Its cytoskeletal and signaling functions seem to be controlled separately.

In the frog *Xenopus* the ovum accumulates an RNA (Vg1 mRNA) that encodes a growth factor of the TGF-β family. Initially present throughout the ovum, it moves and forms a gradient of concentration that is highest at the vegetal pole.

The Dorsal-ventral (D-V) and Right-left Axes

Establishment of the D-V axis in *Drosophila* requires the participation of several genes. As mentioned above, *gurken* affects follicle cells. These cells cooperate with the ovum to set up a concentration gradient of the maternal gene *Dorsal,* the protein encoded by the *dorsal* gene. Dorsal is a transcription factor related to NF-κB. In the early embryo of the fly it forms a gradient in which it remains largely in the cytoplasm on the dorsal side but is mostly taken up into the nuclei on the ventral side. There it binds to a series of complex enhancers, each 300 – 1000 kbp in length, which act along the D-V axis. The enhancers interpret the concentration of Dorsal at five different threshold levels.

The establishment of a second axis in vertebrate embryos is a complex process, which depends upon prior formation of a mesoderm layer. Development of mesodermal cells (not shown in the more primitive gastrula illustrated in Fig. 10.6) is induced by diffusion of a growth factor from the vegetal pole. *Activin,* a member of the TGF-β family, has been proposed as the natural inducer of mesoderm. More recently *nodal,* a different TGF-β-like protein, has been suggested. The fact that induction can occur through thin (20μm) filters without any cell-cell contact indicates that specific chemical agents are responsible. Induction of the mesodermal layer in X. *laevis* appears to be an effect of an epidermal growth-factor-like protein. Additional factors are also needed to establish the D-V axis in vertebrates in which the dorsal side is homologous to the ventral region in *Drosophila.*

Expression of another set of genes establishes the *right-left* axis and characteristic asymmetries of the body. In the chick activin 2P, also a member of the TGF-β family, as well as Nodal and *Sonic hedge-hog* (Shh) participate in control. The gene *Pitx2* is a downstream transcription target for this signaling cascade in vertebrates.

Spemann's Organizer

In 1924, the German physiologist Hans Spemann with Hilde Mangold transplanted a small piece of tissue from the dorsal lip of a newt blastula to a site on the ventral side of an early blastula of a differently colored species. The embryo developed a small secondary embryo, most of the tissues of which came from the host, not from the transplanted piece. It was concluded from this famous experiment that the transplanted vegetal tissues had supplied a diffusible inducer. This morphogen caused the cells of the ventral surface of the blastula to secrete other morphogens. The signaling center in this part of the blastula surface is known as *Spemann's organizer,* or simply the organizer. It utilizes more than one inducer and a complex set of signaling interactions. The cascade that induces formation of the organizer involves the Wnt-β-catenin and TGF-β pathways and the transcription factor *Smad*4. The organizer secretes protein such as *noggin, folistatin,* and *chordin,* as well as nodal and other members of the TGF-β family. They establish the D-V axis and also direct the development of the head and the initial patterning of the central nervous system. Noggin seems to be a neural inducer.

Patterns, Signaling Pathways, and Homeotic Genes

While gradients in *Drosophila* eggs establish the anterior-posterior axis, products of other genes specify the developmental fates of cells of specific lineages and of cells found in particular spatial domains. Many *Drosophila* genes are needed to establish patterns, *e.g.,* spacings of similar elements such as hairs, components of compound eyes, and whorls of plants. Insects are organized into a pattern of consecutive segments from head to tail along the A-P axis. In *Drosophila* there are typically 17 of these segments, some carrying appendages such as antenna, legs, and wings. Each segment develops under a different set of influences from neighboring cells. Development is controlled by ~30 *segmentation genes,* which determine the number of segments and their internal organization. Of these genes one set of at least six *gap genes,* among them *hunchback* (Hb) and *Krüppel,* are expressed first. Both of these genes encode zinc-finger transcription factors. The Hunchback protein acts mainly on head parts and upper thorax, while Krüppel influences development of the thorax. Also among the segmentation genes are eight "pair-rule" genes and at least 16 segment polarity genes. The interactions of the products of these genes creates a prepattern that provides positional information and guides further development. Somewhat similar to the segmentation of the insect body is the development of skeletal muscle from a series of mammalian embryonic blocks known as somites. In *Drosophila* signaling pathways involving the secreted proteins EGF, Decapentaplegic (Dpp), Wingless (Wg), Hedgehog (Hh), and Notch are used repeatedly to provide positional information. The names of these proteins describe effects of mutations on the limbs of the insect, but the proteins have a much broader significance. They represent an evolutionarily conserved set of intercellular signaling pathways. Wingless is the first member of the previously mentioned Wnt family. Proteins of the Hh family, including the vertebrate Shh, control a large variety of processes that include development both of limbs and of the nervous system. The Hh and Shh proteins all carry a molecule of cholesterol covalently bound in ester linkage to the C termini of the biologically active N-terminal domains of these proteins. Defects in Shh signaling causes some human birth defects, and it is possible that drugs that inhibit cholesterol synthesis may have adverse effects on embryonic development.

The *Drosophila* Notch 1 is a 300-kDa integral membrane protein that contains 36 EGF-like repeats. Its activation by proteolysis produces a 200-kDa N-terminal portion and a 120-kDa C-terminal fragment, which contains a transmembrane domain. The small intracellular domain of this fragment is then released by protease action and travels to the nucleus where it activates several target genes. The Notch signaling pathway is conserved in all metazoans and influences many interactions that control cell fate during development. The proteolytic cleavages of the Notch protein parallel those of both the ErbB-4 growth factor receptor and the amyloid precursor protein APP. The same type of protease (γ-secretase or presenilin in the human brain) cleaves all three of these proteins. Actions of Notch are modulated by posttranslational alteration, *e.g.*, glycosylation by a fucosyltransferase encoded by gene *fringe*. A homolog of *fringe,* called *lunatic fringe,* encodes an essential component of somite formation in the mouse.

Functioning together with the signaling pathways are *selector genes* that determine which specific pathway is to be influenced. For example, eyes, antennae, legs, or wings of a fly may be selected. *Antennapedia (Antp)* is one of these genes. Several of the *Drosophila* selector genes as well as some pair-rule genes are also known as *homeotic genes*. Homeotic genes were first recognized by the fact that mutation causes conversion of one segment of an insect's body into the homologous tissues of another segment. For example, a mutation in *antennapedia (ante)* changes the antenna into a leg. Similar genes are also active in vertebrates, *e.g.*, in the development of the chick embryo limb bud, the very tip (the last 20 cell diameters of length) contains cells that differentiate into the various elements of the limb in a relatively autonomous manner. If this *progress zone* from one limb bud is grafted onto the end of another limb bud, the bones and cartilaginous elements of the limb are repeated. Both the number and morphology of fingers and toes are determined by homeotic genes as is the formation of sphincters in the developing gut. Homeotic genes (also known as Hox genes) contain a conserved sequence of 180 bp that specifies a 60-residue protein *homeodomain* (also known as a homeobox). The homeodomain folds into a helix-turn-helix motif characteristic of many transcriptional regulators. *Hox* genes are found among all forms of life. Hundreds have been described. They include genes for the previously mentioned transcription factors of the Oct and Pit families as well as for the yeast mating type proteins MATal and MATα2. The *Drosophila* genome contains eight *Hox* genes, while the human genome has ~40, which are organized into Hox clusters. Hox genes are also abundant in plants.

Despite intense interest the role of homeodomain proteins in development is not well understood. The highly conserved motif binds to DNA at many places in the genome. Current thinking is that homeodomain proteins interact with other regulatory proteins, and that various combinations of these proteins provide the information needed to direct development.

D. SPECIALIZED TISSUES AND ORGANS

Here are a few details about development of mammalian tisssues. Well begin with the blood and connective tissues, which arise from embryonic mesodermal cells.

1. Blood Cells and the Circulatory System

Every second of life a human must produce about 2.5 million red blood cells, about 2 million granulocytes, and many lymphocytes as well as other less numerous leukocytes. All of these arise from *multipotential stem cells* found in the bone marrow. Each of these stem cells divides to form one daughter stem cell and one *progenitor* cell. The progenitor cells are also stem cells but have differentiated into *myeloid, erythroid,* and *lymphoid* cells. Mature blood cells of most types have short lifespans and must be regenerated from stem cells continuously.

At all stages the differentiation process is regulated by the microenvironment which is rich in specific protein growth factors. Among the 20 interleukins, three stimulate growth of both multipotential progenitor cells and erythroid progenitor cells. The acetylated tetrapeptide Ac-Ser-Asp-Lys-Pro inhibits stem cell proliferation. Granulocyte-macrophage colony-stimulating factors stimulate proliferation of both granulocytes and macrophages. The kidney cytokine *erythropoietin,* a 30.4-kDa glycoprotein, is a primary regulator of red blood cell formation. Its action on a differentiated stem cell initiates massive hemoglobin synthesis and terminal differentiation of the erythrocyte. *Thrombopoietin* promotes formation of megakaryocytes and also their maturation and release of platelets to the blood. *Thymopoietin* promotes early T-cell differentiation. Activated macrophages secrete interleukin-1, which stimulates maturation and proliferation of B lymphocytes. Interleukin-2 T-cell growth factor; is produced by activated T lymphocytes and is needed by T lymphocytes for long-term helper and cytotoxic functions. Differentiation of the stem cells into the erythroid lineage requires transcription factor GATA-1, development into erythrocytes requires GATA-2, while development into T lymphocytes requires GATA-3.

Globin Genes

The genes that encode the human globins from which hemoglobin is formed are found in two clusters, the α-like genes on chromosome 16 and the β-like on chromosome 10. They are developmentally regulated, different genes in the clusters being active at different stages of development. Mammalian hemoglobins each contain two α chains or two related ζ chains and two other chains, β, γ, δ, or ε. Adult hemoglobin is mainly $\alpha_2 \beta_2$ but contains small amounts of α_2

δ_2. In early embryos the hemoglobin is $\zeta_2\varepsilon_2$ but during the second to sixth weeks of embryonic life the two fetal hemoglobin chains γ^G and γ^A replace the ε chains. The switch from fetal to adult hemoglobin begins a few weeks before birth and is complete by about ten weeks after birth. The β-like gene cluster contains five genes encoding globins ε, γ^G, γ^A, δ and β. Each gene consists of three exons separated by two introns and has rather similar control signals. These include CACCC at ~–100, CACA at ~–92, CCA AT at ~–75, and ATAAAA (TATA sequence) at ~–30, as well as AATAAA (cleavage and polyaderiylation). The CCAAT sequence appears twice in the γ^A promoter. A variety of transcription factors and chromatin modifiers influence the expression of these genes.

Only a small amount of fetal hemoglobin ($\alpha_2\gamma_2$) is produced after infancy. There are two genes, γ^G and γ^A for the β-like chains of fetal hemoglobin. A few adults make large amounts of fetal hemoglobin and this hereditary *persistence of fetal hemoglobin* has survival value for persons carrying thalassemia genes. This condition may result from a single base change in the CCAAT sequence found upstream (~75 bp) of the globin genes. Many other genes are also preceded by the same sequence, which in extended form is often TTGGPyCAAT. In one individual with persistence of fetal hemoglobin the first G in this sequence was replaced by A in one of the two CCAAT sequences present in the γ^A gene.

Thalassemias

In these important hemoglobin diseases the α or β chain either is absent or is present in far less than stoichiometric amounts. About 40 point mutations in the β globin gene have been described among patients with β thalassemia, in which β chains are missing (β^0 thalassemia) or are present in reduced amounts (β^+ thalassemia). These mutations sometimes occur in control regions at the 5′ end of the gene. For example, a change of the TATA sequence from ATAAAA to ATACAA causes decreased transcriptional efficiency and β^+ thalassemia. Other mutations result in abnormal splicing or in instability of the β globin. Deletions may result in the complete absence of the gene or in a frameshift that results in nonfunctional globin and β^0 thalassemia.

How do embryonic cells choose to transcribe only the embryonic globin genes? The decisions to switch from embryonic to fetal and from fetal to adult at appropriate times appear to be controlled by interactions with appropriately expressed transcription factors. Especially impressive is the total and permanent cessation of transcription of the embryonic globin ε gene at about the seventh week of gestation. Transcription of the β globin gene cluster is controlled by a powerful enhancer, the *locus control region* ~6 – 22 kbp upstream of the promoter. In addition, ~270 bp upstream of the ε gene promoter is a *silencer,* a DNA sequence to which inhibitory proteins may bind and, in cooperation with the enhancer, may completely silence the ε gene while allowing transcription of other genes in the cluster. The ζ gene, in the α globin cluster, is silenced by the binding of an NF-κB transcription factor to a 108-bp segment of DNA located 1.2 kbp to the 3′ end of the gene. The globin genes are shielded from action of nearby enhancers by *insulators,* DNA sequences that often contain CpG islands.

Blood Vessels: Vasculogenesis and Angiogenesis

Early in development of an embryo a network of blood vessels is formed from mesenchymal progenitor cells (vasculogenesis). Later, in either the embryo or the adult, new capillaries are formed (angiogenesis). As the organism develops these new vessels are "pruned," and the vascular

bed is remodeled to a tree-like form with vessels of both large and small diameter. A key activator of angiogenesis is the vascular endothelial growth factor *VEGF*. However, a number of other proteins including *endoglin*, a TGF-β binding protein, the clotting factor thrombin, and the ribonuclease *angiogenin* exert their influences. There are diseases that result from defective angiogenesis. However, a major interest in this process arises because of the essential role of angiogenesis in the growth of cancer and as a complication of inflammatory processes. Angiogenesis in the retina is a major cause of blindness resulting from diabetes mellitus or from macular degeneration. There are natural antagonists of angiogenesis, and efforts are being made to utilize them in therapy. A related project is development of completely tissue-engineered blood vessels for surgical use.

2. Cartilage, Tendons, Bone, Muscle, and Fat

Mesenchymal cells differentiate into cartilage, bone, muscle, adipose, and other connective tissues. *Chondrocytes* synthesize the variety of collagens that are needed for synthesis of cartilage and other connective tissues. The 32 or more genes encoding the polypeptides needed for synthesis of the 19 types of vertebrate collagens are developmentally regulated in a complex manner. Their promoters contain TATA and CCAAT sequences as well as other presumed regulatory codes. The first intron of several collagen genes has also been identified as a control region containing enhancers. The elastic fibers owe their properties in part to elastin. The control region of elastin genes lacks the TATA sequence but has SP1 binding sites. As with many other mammalian genes, a diversity of protein products, many in small amounts, are made by alternative splicing.

Collagen fibrils provide the scaffolding for formation of bone. Bone develops under the influence of *bone morphogenic proteins* BMP-2 to BMP-7. Most of these are cytokines of the TGF-β family. Noggin antagonizes the action of the BMPs. A characteristic of bone is rapid remodeling by which ten per cent of skeletal bone is replaced every year. The balance between action of the bone-forming osteoblasts and the bone-resorbing osteoblasts is regulated by surface proteins responding to *c-Fos* and to *interferon*-β.

Muscle, develops in response to four members of the myoD family. These include myoD, *myogenin, myf*5, and *MRF*4. All are muscle-specific transcription factors of the basic helix-loop-helix class. An unusual aspect of muscle development is formation of multinucleate *myotubes*. Apoptosis plays an important role in muscle development and can present significant complications in damaged cardiac muscle. Defects in several developmental control genes are responsible for congenital heart diseases.

3. Epithelia

Epithelial tissues, which line both internal and external surfaces, arise from all three cell layers of the blastula. The epidermis arises from ectoderm, while the lining of the digestive tract is formed by endodermal cells. Mesoderm provides the linings of blood vessels. About 60% of differentiated tissue types in the mammalian body are epithelia. Stem cells or progenitor cells are present and provide for renewal. While epidermal stem cells are located in deep layers of the skin, *keratinocytes* are readily cultured *in vitro* and can give rise to fully differentiated multilayered skin. Development appears to require transcription factors related to *Oct*-2 as well as *p*63, a homolog of the tumor suppressor p53. Mice deficient in the aspartyl protease *presenilin 1* (which is defective in some forms of Alzheimer disease) develop characteristic epidermal skin tumors. The β-catenin-Wnt signaling pathway seems to be involved. The gastrointestinal endoderm develops its highly convoluted surface under some control by the Notch signaling pathway. Endothelial progenitor cells *(angioblasts)* are responsive to many signaling molecules including thrombin. The *Drosophila* eye develops from the epithelium, again through signaling via Notch and other morphogens. Of outstanding importance to epithelial cells in general is their ability to form complex communicating junctions.

4. The Nervous System

Development of the vertebrate central nervous systems is initiated during gastrulation through an interaction between the dorsal ectoderm and an infolding of the dorsal mesoderm. Several different diffusible inducers are involved. These include noggin, folistatin, and other members of the TGF-β family as well as thyroid hormones, basic fibroblast growth factor (bFGF), and sonic hedgehog. The nervous system develops over a period of a few days with differentiation of precise numbers of neurons, astrocytes, and oligodendrites in successive waves. The order in which various cell types arise is determined by the order in which transcription factors such as Hunchback, Kruppel, and others are expressed. Multipotential neural stem cells provide the new cells that are required. Neural tissue from a region called the *neural plate* develops into a neural fold. The latter is closed to form the *neural tube,* within which the notochord, the precursor to the spine, as well as the neurons, glia, and other cells grow. The *neural crest* forms as an outgrowth from the dorsal surface of the neural tube under the influence of inducers of the Wnt and BMP families. Cells migrate from this crest to form the peripheral nervous system, melanocytes, and cranial cartilage. The pituitary, a central component of the neuroendocrine system, develops from tissues from the midline part of the anterior neural ridge.

The *floor plate,* which develops along the midline of the vental surface of the neural tube, is the source of Sonic hedgehog, *netrin*-1, and other secreted molecules. Some of these participate in axon guidance by which the growing tips *(growth cones)* of axons are able to connect to the correct "targets." For example, every visual receptor cell in the retina must send its signal to the correct locations in the visual cortex of the brain. How can this be accomplished? Over a century ago Ramon y Cajal proposed that chemoattraction, analogous to chemotaxis of bacteria, might be involved. A hundred years later Tessier-Lavigne and coworkers isolated the first of these attractants, *netrin*-1 and *netrin*-2, from 25,000 pulverized chick brains. Of these two closely related 75- and 78-kDa molecules netrin-1 is produced only in the floor-plate cells. Like the less well understood nerve growth factor, netrins induce outgrowths of neurites and also are chemoattractants for nerve growth cones. The netrin receptor is known as *DCC* (Deleted in Colorectal Cancer). A nematode protein UNC-6 is related to the netrins.

Growth cones are subject to both chemoattractant and chemorepellent effects of guidance molecules and also to attraction or repulsion resulting from cell-cell-contacts. To complicate the picture further, the netrins and also the brain-derived neurotrophic factor BDNF may first attract, and then after a period of adaptation or desensitization, repel a growth cone. Consecutive phases of desensitization and resensitization may result in a zigzag path of growth. The netrins were recognized first by observing growth of *commissural neurons.* These neurons originate within the spinal cord on one side or the other or the midline. They grow down toward the floor plate attracted by the netrin-1 or *BDNF* produced there. The neurons then cross the midline before turning and growing toward the brain. After crossing the midline the growth cones become insensitive to netrin but are repelled by a molecule (first recognized in *Drosophila)* called *slit*. Its receptor is appropriately named roundabout (*Robo*). A similar receptor in zebrafish (called *astray*) is required for retinal axon guidance. An important aspect of neuron guidance is apoptosis induced by misdirected growth. The Notch receptor apparently participates in this decision in the mammalian CNS. Positive signals for axon growth often involve the MAP kinase pathways, while inhibition may involve INS-2P-Ca^{2+} signaling.

Chemoattractants that function in development of the cerebral cortex include several *semaphorins*. A separate family of attractant and repellant compounds, the semaphorins have been identified in insects, chickens, and mammals. They play a role in regulation of communication between neurons. Because of the complexity of the brain, the study of growth of neurons with the brain is difficult. The $>10^{12}$ neurons each contact, on the average, 100 different cells. Some insight comes from mutant mice with names such as *reeler, scrambler, stargazer,* and *Yotari* (Japanese for tottering). The single defective gene in these mice can be identified and studied. For example, *reeler* mice are defective in *reelin,* a large glycoprotein of the extracellular matrix (ECM). The *reeler* phenotype can also result from mutation of the gene *disabled*-1, which encodes a cytosolic tyrosine kinase. Other mutations in mice implicate *VLDL* and *apoE* receptors in these developmental abnormalities.

The *stargazer* mutant mouse is ataxic and epileptic. It lacks functional *AMPA receptors,* which apparently are not delivered successfully to the synapses in the cerebellum in which they function. Mutation of a transmembrane protein *stargazin,* which may interact with the AMP receptor, causes the symptoms. *NMDA receptors* are involved in synapse formation in the brain. Filopodial extensions on dendrites, triggered by electrical activity, are essential for synapse formation, which occurs rapidly. Activation of NMDA receptors is apparently also necessary. Without this stimulation the excitatory glutamatergic neurons of the developing brain undergo apoptosis.

Why do neurons grow in the embryo but not in most parts of the adult CNS? Two proteins called *Nogo* were isolated from bovine brain. Their sequences were utilized in identifying the *Nogo* gene and three human isoforms of the protein. The large 250-kDa Nogo-A is present both in myelin and in the endoplasmic reticulum. Both Nogo-A and the diffusible 35-kDa Nogo-B are inhibitory of neurite outgrowth. This effect, as well as the crowding of regenerating neurons by the chondroitin and other matrix components, may provide obstacles to nerve regeneration.

E. Development of Green Plants

Green plants may have diverged from a common ancestor with animals ~1.6 billion (1.6×10^9) years ago. How do the genomes of presentday plants and animals compare? There are many similarities in basic metabolism. These arise from the intrinsic chemical properties and reactivities of cellular components and from the coevolution of plants and animals. Plants and animals also utilize similar structures and similar control of chromatin. However, in the control of development there are great differences. For example, the *Arabidopsis* genome contains no relative of the *Drosophila* Gurken, no receptor tyrosine kinases, no relatives of transcription factor NF-κB. However, there are similarities in parallel pathways utilized by plants and animals.

The alternating haploid (*n, gametophytic*) and diploid (2*n, sporophytic*) phases of the life cycle are diagrammed in Fig. 10.8A. Following flowering a diploid *mother cell* within the *ovule* undergoes meiosis to form four haploid *megaspores.* After mitosis a single egg cell is formed. Within pollen sacs in the anthers of each mother undergoes meiosis to yield four haploid *microspores.* Following mitosis these develop into pollen grains each of which contains two sperm cells as well as a vegetative nucleus. After falling upon the stigma surface and growth of the pollen tube, one of the sperm cells fuses with the egg to give the diploid zygote. The other sperm unites with the specialized diploid *central cell* in the ovule to form a triploid *(3n) endosperm nucleus,* which develops into the *endosperm,* the food storage tissue of the seed. Endosperm contains two tissues, a starchy inner layer and a protein- and oil-rich outer layer.

Variations of the life cycle occur. For example, a process called *apomixis* leads to asexual formation of seed. In many plants, including maize, separate flowers form the ovule and the pollen. This is one mechanism for avoiding inbreeding. In many plants systems of *self-incompatibility* have evolved. In some, *e.g., Arabidopsis* and other crucifers, pollen germination is disrupted unless it falls on a stigma possessing a different allele-specific receptor. In other cases development of the pollen tube is disrupted at a later stage. In maize and in more than 150 other species a mitochondrial and therefore maternally inherited trait prevents formation of a functional pollen. Malesterile plants, which carry this trait, are very useful in plant breeding. However, a near disaster occurred in the United States in 1970 when the fungal disease Southern corn leaf blight attacked the male-sterile maize that had been used for production of 85% of commercial hybrid maize. The mitochondrial defect in pollen formation also resulted in an increased sensitivity to the fungal toxin. The target of the toxin is a 115-residue pore-forming polypeptide in the inner mitochondrial membrane. The male sterility can be reversed if the plant carries two nuclear *restorer genes.* One of these encodes an aldehyde dehydrogenase, but its mechanism of action is unclear.

The plant embryo is a juvenile form, the seedling. In *Arabidopsis* the zygote, which is surrounded by maternal diploid tissue, divides asymmetrically. The resulting apical and basal cells differ in several ways. The small cytoplasm-rich apical cell is partitioned into eight proembryo cells by two rounds of vertical division and one horizontal division. The larger basal cell contains a vacuole and divides repeatedly horizontally to give 7 – 9 aligned cells. Only the uppermost of these becomes a part of the embryo. The others form an extra-embryonic suspensor. The apical part of the embryo develops the shoot meristem and the central part the radial pattern of tissue layers characteristic of plants. The root meristem develops from the basal portion of the embryo. Movements of proteins that provide positional cues are involved in the development of the embryo. Early embryonic and endosperm development is largely under maternal control. Most paternal genes may be initially silent.

Many angiosperms develop *fruit* from tissues of the ovary. The development and ripening of fruit is also complex and highly regulated.

Formation of seeds is a slow process. For example, in wheat the mature embryo, which consists of ~10^5 cells, develops over a seven week period. Seeds may live from a few years to 1000 years or more. Subsequent germination of the seed into a seedling requires only two days. The very dry embryo is converted into a highly hydrated plant whose further growth requires uptake of very large amounts of water. Many plants also synthesize large amounts of oxalic acid. This may arise from ascorbate or via oxidation of glycine. One of the earliest mRNAs to appear during seed germination encodes a 125-kDa glycoprotein called *germin.* This protein, which exists as multiple isoforms, is a copper-dependent oxalate oxidase (Eq. 10.1) which generates hydrogen peroxide. The latter is probably needed to crosslink cell-wall polymers. Germin may also be useful to plants in defense against oxalate-forming fungi.

$$C_2O_4^{2-} + 2H^+ + O_2 \rightarrow 2\,CO_2 + H_2O_2 \quad ...(10.1)$$

The rapid vegative growth, which includes development of shoots, leaves, and flowers, is controlled by a variety of transcription factors. Among these are homeodomain proteins that control differentiation of meristem cells. The induction of flowering is especially complex, involved day length, light quality, and effects of gibberellins. At the ends of their lives plant cells die slowly from *senescence.* In this process many materials are recycled for use by new cells. Other plant cells die via the *hypersensitive response,* a form of programmed cell death.

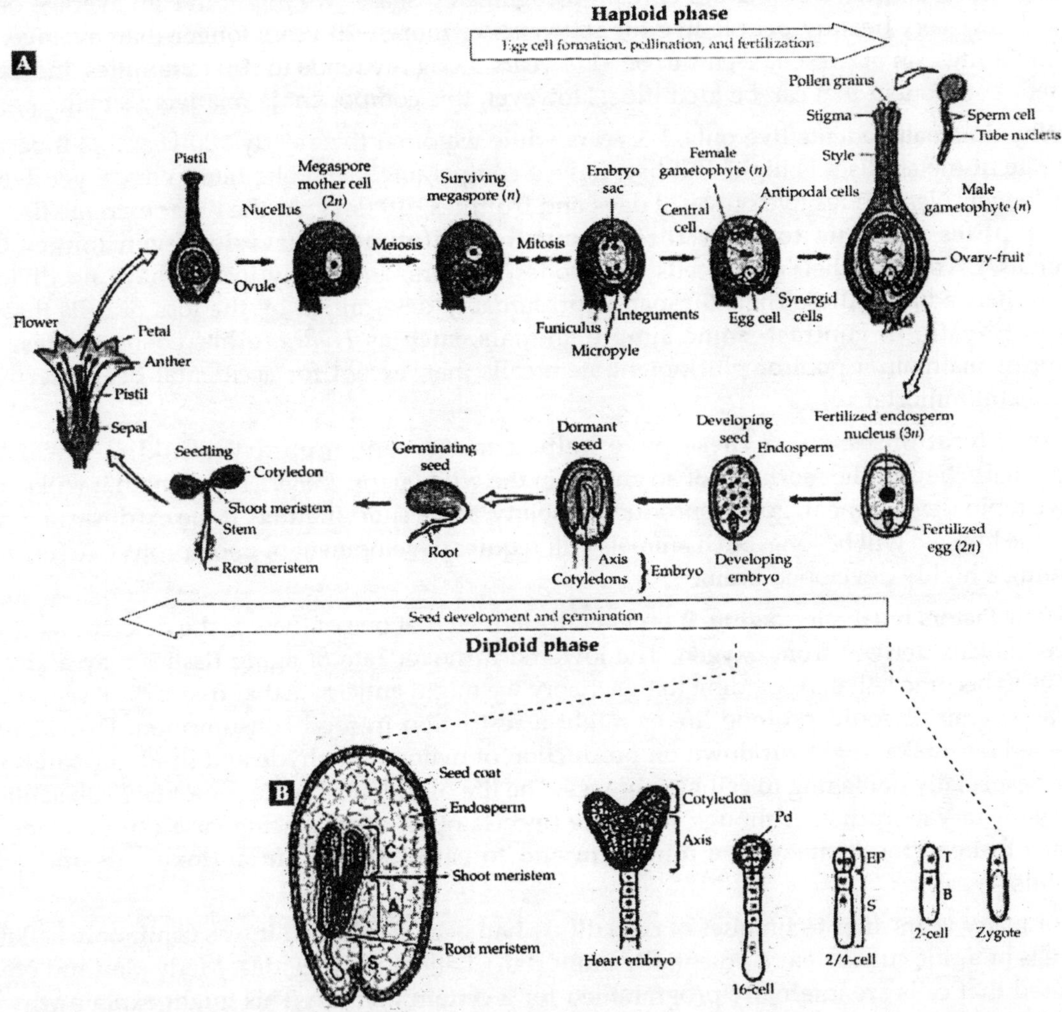

Fig. 10.8. (A) The life cycle of a flowering plant with emphasis on egg-cell formation and seed development. (B) Some further details of embryo development. T, terminal cell; B, basal cell; C, cotyledon; A, axis; SC, seed coat; En, endosperm; EP, embryo proper; S, suspensor; SM, shoot meristem; Pd, protoderm; RM, root meristem.

F. AGING

Why do we age? This question is often asked but the answers are not simple. Do our tissues deteriorate with age as a result of damage to proteins? From an accumulation of mutations in our DNA? From attacks by free radicals? From loss of hormone receptors? From misregulation of mitosis? From loss of telomeres on the ends of replicating DNA. From an internal genetic program that dictates our life span? All of these possibilities may be partially correct. The simple answer is that "we just wear out." However, different parts wear out at different rates and in different ways.

Perhaps we should be amazed that the human body can live for an average of 75-80 years. If we all avoided accidents and could cure all recognized diseases we might live an average of ~90 years. A few very healthy people live for 100 years or more, ~20 years longer than average. The maximum lifespan at present seems to be ~114 years. Long life tends to run in families, indicating a genetic component that can be identified. However, this component is relatively small.

Why do small rodents live only 2-3 years while we often live nearly 100 years? Is it because their rate of metabolism is high? But bats have a comparable metabolic rate to mice, yet live ten times longer. Nematodes live only ~20 days and fruitflies ~10 days. At the other extreme fish and some reptiles continue to grow throughout their lifetimes, surviving even longer than mammals. Except for their germ cells nematodes, rotifers, and many insects have no dividing cells in their adult bodies. Their lifespan is presumably determined by the loss of cells through injury or death. In contrast, some simple animals, such as *Hydra,* other coelenterates, and flatworms maintain a pool of pluripotent stem cells that, except for accidental death, seems to make them immortal.

Considerations such as these have helped move contemporary thinking toward an evolutionary view. If the mortality of an animal in the wild (*extrinsic mortality*) is high it will evolve to have rapid development, good reproductive ability, and a short lifetime. If the extrinsic mortality is low the lifetime will be long. Such animals will require development of good protective functions including a highly developed brain.

Many factors must affect aging. It is generally agreed that one of these is the deleterious effects of free radicals derived from oxygen. The lowered turnover rate of aging tissues may allow the damage to become lethal. According to this theory we might anticipate that free radical scavengers such as vitamin E could prolong life as might a restriction in food consumption. For example, decreased fat intake might cut down on production of malondialdehyde and lipid peroxides that may be especially damaging to cell membranes. The life of rodents can be prolonged substantially by a semi-starvation diet. Although there are uncertainties, a convincing case can be made for humans to keep food intake to a minimum and to eat foods rich in antioxidants and other nutrients.

For many years after techniques of cell culture had been developed it was commonly believed that cells in tissue culture were potentially immortal. Challenging this idea, Medvedev and others proposed that cells are internally programmed for a certain lifetime. This might explain why we have short-lived mammals as well as long-lived mammals. Support for the idea was supplied by Hayflick, who observed that animal cells in culture have a limited potential for doubling. For example, normal human diploid embryonic fibroblasts grow in culture and double their number aproximately 50 ± 10 times. Regardless of cultural conditions, the cells die after this number of doublings. Cells taken from older humans undergo a smaller number of doublings before dying, as do those taken from shorter lived animals such as the mouse (14-28 doublings). These experiments suggested that there is an internal program by which cells are scheduled to die from *replicative senescence*. Malignant transformation overrides this program and transformed cells appear "immortal."However, unlike fibroblasts some glial stem cells have been identified as possibly having unlimited proliferative capacity. These include cultured rat oligodendrocyte and Schwann cells. These results suggest that replicative senescence may not be inevitable.

As mentioned already that, "erosion" of the telomere ends on chromosomes is thought to be a major cause of cell senescence. Old cells have little or no telomerase. However, most human cancer cells, as well as those from immortalized cell cultures, do synthesize telomerase and maintain telomeres of adequate length. Inhibition of telomerase activity in immortalized cells causes telomere shortening and cell death. A second pathway for telomere maintenance is based upon homologous recombination. Experimental elongation of telomeres extends the lifespan of cells in culture. Furthermore, apparently healthy calves have been produced by nuclear transfer cloning from senescent fibroblast cells for which four or fewer cell doublings were expected to be possible. The cells of the cloned animals had the capacity for 90 or more cell doublings.

What information about aging can we obtain by study of the "model" organisms *S. cerevisiae, C. elegans, D. melanogaster,* and the mouse? In every case a variety of mutations may shorten or lengthen the lifespan. In every case dietary energy restriction can lengthen life. Yeast cells grown on 0.5% glucose instead of 2% glucose may undergo ~25% more cell doublings before a culture becomes senescent. However, mutant strains with a defect in the *SIR*2 gene have a shortened lifespan, which is not increased by caloric restriction. *SIR*2 encodes Sir2p, an NAD^+-dependent histone deacetylase. It is likely that caloric restriction causes the yeast to switch from anaerobic fermentation to oxidative metabolism. The resulting increase in the $[NAD^+]/[NADH]$ ratio activates Sir2p, thereby altering chromatin and silencing a group of genes. Mechanisms by which this shift in metabolism decreases deleterious mutations, even though respiration is increased, are probably complex.

Nematodes *(C. elegans)* usually live about three weeks but the simultaneous presence of life-extending mutations in two different groups of genes lengthen the lifespan as much as fivefold. One of the genes is the maternal-effect Clock gene *Clk*1. It has been found to code for a hydroxylase involved in the final step of synthesis of ubiquinone-9. The lifetime of wild-type nematodes is lengthened by ~60% by elimination of ubiquinone from the diet. The second group of genes that affect longevity regulate an insulin-like signaling system. In every case metabolism is slowed, an effect which may reduce the rate of harmful mutations. However, mutant animals may not be healthy. Some of these mutations induce formation of long-lived "dauer" larvae, providing a means for the larvae to survive for up to two months during periods of starvation. Others affect sensory cilia. Mutation of a cytosolic catalase gene *reduces* the lifespan. The nematode's lifetime is also affected negatively by its own germ cells, perhaps via a steroid hormone. The heat shock proteins, by chaperoning newly synthesized proteins and preventing aggregation, also increase life span both in *C. elegans* and *Drosophila.*

Some mutants of *D. melanogaster* with extended lifespans have a defective insulin/IGF signaling pathway. The *methuselah* mutant, whose lifespan is 35% greater than average, appears to involve a G-protein-coupled transmembrane receptor. Mutation of an insulin receptor homolog extends lifespan, apparently by causing a juvenile hormone deficiency. *Drosophila* lifespan is also lengthened by mutation of a transmembrane dicarboxylate transporter or by overexpression of a protein repair carboxyl methyltransferase.

Some mutant mice have extended lifespans. The Ames dwarf mouse has a mutation in $p66^{shc}$, a cell-surface protein that contains both Src-homology and collagen-homology domains. It lives almost one-third longer than do wild-type mice. Mice deficient in methionine sulfoxide reductase have a reduced life-span but fruit flies with overexpressed activity of the enzyme are more resistant than wild-type flies to oxidative damage.

In humans 100 or more years in age some mitochondrial mutations are associated with good health and longevity. Dietary factors doubtless play a role. For example, supplementation of rats' diet with lipoic acid improved mitochondrial function and increased the metabolic rate of old animals.

A number of genetic *progeroid diseases* result in premature aging. Several of these arise from deficiencies in repair of DNA. Among them are some types of cancer, and *Werner syndrome,* which arises from a defect in a 1432-residue protein with a central domain homologous to the RecQ family of DNA helicases. Defects in other RecQ homologs cause *Hutchinson-Gilford progeria* as well as the Bloom syndrome. Yet another DNA helicase, a subunit of transcription factor II (TFIIh), is defective in tri-chothiodystrophy (TTD). Another gene which helps to prevent aging is *KLOTHO.* First identified in mice, it encodes a transmembrane protein that has sequence similarities to β-glucosidases. Some mice with mutations in the tumor suppressor *p53* have enhanced resistance to rumors but age rapidly.

Aging seems to be inevitably linked to an increase in the incidence of cancer. This uncontrolled growth of cells appears to be allowed by the step wise accumulation of mutations that affect growth, differentiation, and survival. Several aspects of cancer are discussed in other chapters of this book. However, the topic is so complex and research so active that it is hard to give even a thumbnail sketch of more recent discoveries.

Much effort is being dedicated to identifying the many signaling pathways that control growth, the mechanisms that cells employ to recognize problems in the control of growth, and the means by which cells can correct the problems or undergo apoptosis and avoid cancer. Some of the complexity arises because of the large number of signaling pathways in which mutations may produce activated proto-onco-genes or faulty tumor suppressors. A large network of these suppressors is present in human cells. Among the relevant signaling pathways are the following:

RAS – RAF – ERK
p53
the PtdIns 3-kinase - PKB/Akt pathway,
which is opposed by PTEN
EFG receptor (EFGR) signaling
Wnt-Catenin signaling
E. Cadherin

The importance of oncogenes and tumor suppressors has been demonstrated by conversion of human cells in culture into tumor cells in vitro. Introduction of an activated *ras* gene, an SV40 viral protein that inhibits formation of both p53 and the Rb gene, and an active telomerase gene sufficed. However, there is some doubt about the relevance of this work to human cancer.

Most cancerous cells have extra chromosomes. The karyotype is rarely normal. This and other evidence suggest that *genomic instability* may be the major cause of cancer. In healthy cells stalled RNA polymerase is removed by transcription-coupled repair and lesions in DNA are either repaired or the cell undergoes apoptosis. Telomere dysfunction is also a factor. The two breast cancer susceptibility genes *BRCA*1 and *BRCA*2 are apparently responsible for about half of all hereditary breast and ovarian cancers. Protein BRCA1 is an 1863-residue nuclear protein, which is thought to function in transcription. However, recent evidence indicates that BRAC2 is directly involved in repair of double-strand breaks in DNA by homologousrecom-bination. Other data implicate the Neu-Ras pathway, proto-oncogenes *c-myc* and *Wnt*-1, and cyclin D1 in breast cancer.

Yet another aspect of cancer is the *aberrant glycosylation* observed for many proteins. The state of glycosylation of cell-surface proteins is one of many factors that affect metastasis, which is critical to growth of tumors. The recognition that causes of cancer are numerous has led to a new large-scale project to identify as many cancer-associated mutations as possible within the entire human genome. One early success from this effort is identification of mutations in the gene *BRAF*, one of the three human *RAF* genes. These mutations are present in 15% of human malignant melanomas.

INDEX

H

I

❑❑❑